A Concrete Introduction to Real Analysis

Second Edition

TEXTBOOKS in MATHEMATICS

PUBLISHED TITLES

ABSTRACT ALGEBRA: A GENTLE INTRODUCTION
Gary L. Mullen and James A. Sellers

ABSTRACT ALGEBRA: AN INTERACTIVE APPROACH, SECOND EDITION
William Paulsen

ABSTRACT ALGEBRA: AN INQUIRY-BASED APPROACH
Jonathan K. Hodge, Steven Schlicker, and Ted Sundstrom

ADVANCED LINEAR ALGEBRA
Hugo Woerdeman

ADVANCED LINEAR ALGEBRA
Nicholas Loehr

ADVANCED LINEAR ALGEBRA, SECOND EDITION
Bruce Cooperstein

APPLIED ABSTRACT ALGEBRA WITH MAPLE™ AND MATLAB®, THIRD EDITION
Richard Klima, Neil Sigmon, and Ernest Stitzinger

APPLIED DIFFERENTIAL EQUATIONS: THE PRIMARY COURSE
Vladimir Dobrushkin

APPLIED DIFFERENTIAL EQUATIONS WITH BOUNDARY VALUE PROBLEMS
Vladimir Dobrushkin

APPLIED FUNCTIONAL ANALYSIS, THIRD EDITION
J. Tinsley Oden and Leszek Demkowicz

A BRIDGE TO HIGHER MATHEMATICS
Valentin Deaconu and Donald C. Pfaff

COMPUTATIONAL MATHEMATICS: MODELS, METHODS, AND ANALYSIS WITH MATLAB® AND MPI, SECOND EDITION
Robert E. White

A COURSE IN DIFFERENTIAL EQUATIONS WITH BOUNDARY VALUE PROBLEMS, SECOND EDITION
Stephen A. Wirkus, Randall J. Swift, and Ryan Szypowski

A COURSE IN ORDINARY DIFFERENTIAL EQUATIONS, SECOND EDITION
Stephen A. Wirkus and Randall J. Swift

DIFFERENTIAL EQUATIONS: THEORY, TECHNIQUE, AND PRACTICE, SECOND EDITION
Steven G. Krantz

PUBLISHED TITLES CONTINUED

DIFFERENTIAL EQUATIONS: THEORY, TECHNIQUE, AND PRACTICE WITH BOUNDARY VALUE PROBLEMS
Steven G. Krantz

DIFFERENTIAL EQUATIONS WITH APPLICATIONS AND HISTORICAL NOTES, THIRD EDITION
George F. Simmons

DIFFERENTIAL EQUATIONS WITH MATLAB®: EXPLORATION, APPLICATIONS, AND THEORY
Mark A. McKibben and Micah D. Webster

DISCOVERING GROUP THEORY: A TRANSITION TO ADVANCED MATHEMATICS
Tony Barnard and Hugh Neill

DISCRETE MATHEMATICS, SECOND EDITION
Kevin Ferland

ELEMENTARY DIFFERENTIAL EQUATIONS
Kenneth Kuttler

ELEMENTARY NUMBER THEORY
James S. Kraft and Lawrence C. Washington

THE ELEMENTS OF ADVANCED MATHEMATICS: FOURTH EDITION
Steven G. Krantz

ESSENTIALS OF MATHEMATICAL THINKING
Steven G. Krantz

EXPLORING CALCULUS: LABS AND PROJECTS WITH MATHEMATICA®
Crista Arangala and Karen A. Yokley

EXPLORING GEOMETRY, SECOND EDITION
Michael Hvidsten

EXPLORING LINEAR ALGEBRA: LABS AND PROJECTS WITH MATHEMATICA®
Crista Arangala

EXPLORING THE INFINITE: AN INTRODUCTION TO PROOF AND ANALYSIS
Jennifer Brooks

GRAPHS & DIGRAPHS, SIXTH EDITION
Gary Chartrand, Linda Lesniak, and Ping Zhang

INTRODUCTION TO ABSTRACT ALGEBRA, SECOND EDITION
Jonathan D. H. Smith

INTRODUCTION TO ANALYSIS
Corey M. Dunn

INTRODUCTION TO MATHEMATICAL PROOFS: A TRANSITION TO ADVANCED MATHEMATICS, SECOND EDITION
Charles E. Roberts, Jr.

PUBLISHED TITLES CONTINUED

INTRODUCTION TO NUMBER THEORY, SECOND EDITION
Marty Erickson, Anthony Vazzana, and David Garth

INVITATION TO LINEAR ALGEBRA
David C. Mello

LINEAR ALGEBRA, GEOMETRY AND TRANSFORMATION
Bruce Solomon

MATHEMATICAL MODELLING WITH CASE STUDIES: USING MAPLE™ AND MATLAB®, THIRD EDITION
B. Barnes and G. R. Fulford

MATHEMATICS IN GAMES, SPORTS, AND GAMBLING–THE GAMES PEOPLE PLAY, SECOND EDITION
Ronald J. Gould

THE MATHEMATICS OF GAMES: AN INTRODUCTION TO PROBABILITY
David G. Taylor

A MATLAB® COMPANION TO COMPLEX VARIABLES
A. David Wunsch

MEASURE AND INTEGRAL: AN INTRODUCTION TO REAL ANALYSIS, SECOND EDITION
Richard L. Wheeden

MEASURE THEORY AND FINE PROPERTIES OF FUNCTIONS, REVISED EDITION
Lawrence C. Evans and Ronald F. Gariepy

NUMERICAL ANALYSIS FOR ENGINEERS: METHODS AND APPLICATIONS, SECOND EDITION
Bilal Ayyub and Richard H. McCuen

ORDINARY DIFFERENTIAL EQUATIONS: AN INTRODUCTION TO THE FUNDAMENTALS
Kenneth B. Howell

PRINCIPLES OF FOURIER ANALYSIS, SECOND EDITION
Kenneth B. Howell

REAL ANALYSIS AND FOUNDATIONS, FOURTH EDITION
Steven G. Krantz

RISK ANALYSIS IN ENGINEERING AND ECONOMICS, SECOND EDITION
Bilal M. Ayyub

SPORTS MATH: AN INTRODUCTORY COURSE IN THE MATHEMATICS OF SPORTS SCIENCE AND SPORTS ANALYTICS
Roland B. Minton

A TOUR THROUGH GRAPH THEORY
Karin R. Saoub

PUBLISHED TITLES CONTINUED

TRANSITION TO ANALYSIS WITH PROOF
Steven G. Krantz

TRANSFORMATIONAL PLANE GEOMETRY
Ronald N. Umble and Zhigang Han

TEXTBOOKS in MATHEMATICS

A Concrete Introduction to Real Analysis

Second Edition

Robert Carlson

CRC Press is an imprint of the
Taylor & Francis Group, an **informa** business
A CHAPMAN & HALL BOOK

CRC Press
Taylor & Francis Group
6000 Broken Sound Parkway NW, Suite 300
Boca Raton, FL 33487-2742

First issued in paperback 2022

Version Date: 20171023

ISBN 13: 978-1-03-247643-8 (pbk)
ISBN 13: 978-1-4987-7813-8 (hbk)

DOI: 10.1201/9781315152721

Publisher's Note
The publisher has gone to great lengths to ensure the quality of this reprint but points out that some imperfections in the original copies may be apparent.

Visit the Taylor & Francis Web site at
http://www.taylorandfrancis.com

and the CRC Press Web site at
http://www.crcpress.com

Contents

List of Figures

Preface

This book is an introduction to real analysis, which might be briefly defined as the part of mathematics dealing with the theory of calculus and its more or less immediate extensions. Some of these extensions include infinite series, differential equations, and numerical analysis. This brief description is accurate, but somewhat misleading, since analysis is a huge subject which has been developing for more than three hundred years, and has deep connections with many subjects beyond mathematics, including physics, chemistry, biology, engineering, computer science, and even business and some of the social sciences.

The development of analytic (or coordinate) geometry and then calculus in the seventeenth century launched a revolution in science and world view. Within one or two lifetimes scientists developed successful mathematical descriptions of motion, gravitation, and the reaction of objects to various forces. The orbits of planets and comets could be predicted, tides explained, artillery shell trajectories optimized. Subsequent developments built on this foundation include the quantitative descriptions of fluid motion and heat flow. The ability to give many new and interesting quantitatively accurate predictions seems to have altered the way people conceived the world. What could be predicted might well be controlled.

During this initial period of somewhat over one hundred years, the foundations of calculus were understood on a largely intuitive basis. This seemed adequate for handling the physical problems of the day, and the very successes of the theory provided a substantial justification for the procedures. The situation changed considerably in the beginning of the nineteenth century. Two landmark events were the systematic use of infinite series of sines and cosines by Fourier in his analysis of heat flow, and the use of complex numbers and complex valued functions of a complex variable. Despite their ability to make powerful and accurate predictions of physical phenomenon, these tools were difficult to understand intuitively. Particularly in the area of Fourier series, some nonsensical results resulted from reasonable operations. The resolution of these problems took decades of effort, and involved a careful reexamination of the foundations of calculus. The ancient Greek treatment of geometry, with its explicit axioms, careful definitions, and emphasis on proof as a reliable foundation for reasoning, was used successfully as a model for the development of analysis.

A modern course in analysis usually presents the material in an efficient but austere manner. The student is plunged into a new mathematical environment, replete with definitions, axioms, powerful abstractions, and an overriding emphasis on formal proof. Those students able to find their way in these new surroundings are rewarded with greatly increased sophistication, particularly in their ability to reason effectively about mathematics and its applications to such fields as physics, engineering and scientific computation. Unfortunately, the standard approach often produces large numbers of casualties, students with a solid aptitude for mathematics who are discouraged by the difficulties, or who emerge with only a vague impression of a theoretical treatment whose importance is accepted as a matter of faith.

This text is intended to remedy some of the drawbacks in the treatment of analysis, while providing the necessary transition from a view of mathematics focused on calculations to a view of mathematics where proofs have the central position. Our goal is to provide students with a basic understanding of analysis as they might need it to solve typical problems of science or engineering, or to explain calculus to a high school class. The treatment is designed to be rewarding for the many students who will never take another class in analysis, while also providing a solid foundation for those students who will continue in the "standard" analysis sequence.

1

Real numbers and mathematical proofs

Simple facts of arithmetic or geometry are often tested by common experience. If you throw six nuts into a basket, and then add ten more, you get the same total as if ten went in first, followed by six. The commutativity of addition is thus testable in a meaningful way.

The same cannot be said for many of the results of mathematics. What direct experience suggests that there are infinitely many prime numbers, or that the square root of two is not the quotient of two integers? Is the scarcity of solutions to the equation

$$x^n + y^n = z^n$$

convincing evidence that this equation has no positive integer solutions if n is an integer bigger than 2?

Experience can also mislead. We can easily see 6 nuts, but finding -5 nuts is more challenging; even professional mathematicians had trouble with negative numbers through the eighteenth century [5, p. 592-3]. Familiarity with integers, rational numbers, and a geometric understanding of real numbers as lengths makes it hard to believe that there is a number whose square is -1.

How then do we develop sound mathematical reasoning? Statements, ideas, or algorithms need not be valid simply because they are phrased in a precise way, have been tested in a few cases, or appeal to common intuition. In times past experts believed that every length could be represented as the ratio of two integers, and that squares of numbers are necessarily greater than or equal to 0. Brilliant minds believed, along with most calculus students, that except for isolated exceptional points, all functions have derivatives of all orders at every point. Even professional logicians seemed unwary of the traps in statements such as "there is a barber in town who shaves someone if and only if they do not shave themselves."

The development of a sophisticated system of reliable mathematical thinking was one of the greatest achievements of the ancient Greeks. The accomplishment had several components. One part was the development of logic, so that one has a way to construct valid arguments and to analyze arguments to assess their validity. In addition the Greeks were able to apply these ideas in the development of geometry, thereby creating a rich mathematical discipline with numerous applications that could serve as a model for subsequent mathematics. Comments on the development of these mathematical ideas may be found in [5, pp. 45, 50, 58–60, 171–172].

The modern view is that reliable mathematics is the result of a rigorous process emphasizing proofs. A small number of foundational axioms are accepted as true based on their effectiveness as building blocks, their simplicity, and on extensive experience. The foundational axioms for mathematical proofs may be judged by our experience and intuition, but in the end it must be admitted that their truth is assumed. The same is true for the logical procedures which allow us to generate new results based on the axioms. Based on the axioms and accepted rules of logic, proofs should consist of a careful and complete sequence of arguments; each step should either be previously established, or be a logical consequence of previously established results.

The modern concept of mathematical proof is quite mechanistic. In fact the ideal is to create a system whereby the validity of any proof can be checked by a computer, and in principle every proof can be generated by a computer. This development has gone quite a way beyond the original conception of the Greeks. It must also be admitted that in practice it is rare to see one of these ideal mathematical proofs. They tend to be long and tedious, with an astounding effort required to achieve results of mathematical significance. The mechanistic view is taken as a guide, but in practice proofs are provided in a more informal style.

Our treatment of analysis begins with a set of axioms for number systems, including the real numbers. Elementary arguments give us a first taste of the use of proofs to extend the axioms to a richer algebraic theory. To avoid getting hopelessly bogged down in technical details, a considerable amount of foundational material is assumed to be known in advance. Some of this material includes properties of sets, functions, the equality predicate $=$, and the elementary properties of the integers and rational numbers.

A more extensive discussion of proofs follows this introduction to the algebraic properties of the real numbers. The proof by induction method provides several illustrations of the power of proof techniques. Propositional logic allows us to assess the truth of complex statements based on the truth of simple constituent parts. This introduction to proof techniques concludes with a discussion of quantifiers and predicates.

The formal treatment of mathematics, with its emphasis on careful proofs, is a very cautious and sometimes difficult approach, but the resulting structure has a durability and reliability rarely matched in other subjects. The choice of the real numbers as the focus for axiomatic characterization is efficient, but it should be noted that there are alternative treatments which place the emphasis on the integers and rational numbers. Alternatives which define real numbers in terms of rational numbers are attractive because they minimize the assumptions at the foundations of mathematics. Such an alternative development may be found in [7, pp. 1–13] and in [9, pp. 35–45].

1.1 Real number axioms

1.1.1 Field axioms

The first axioms for the real numbers concern the arithmetic functions addition $+$ and multiplication $\cdot$. These properties of $+$ and $\cdot$ are shared with the rational numbers and the complex numbers, along with other algebraic structures. A set $\mathbb{F}$ such as the real numbers $\mathbb{R}$, with arithmetic functions $+$ and $\cdot$, is called a *field* if for any $a, b, c \in \mathbb{F}$ the following axioms F.1–F.10 hold.

There is an addition function $+$ taking pairs of numbers $a, b \in \mathbb{F}$ to the number $a + b \in \mathbb{F}$. Properties of addition are

$$a + b = b + a, \quad \text{(commutativity)} \tag{F.1}$$

$$(a + b) + c = a + (b + c). \quad \text{(associativity)} \tag{F.2}$$

There is a number 0 such that

$$a + 0 = a. \quad \text{(existence of an additive identity)} \tag{F.3}$$

For every number a there is a number b, written $(-a)$, such that

$$a + b = 0. \quad \text{(existence of an additive inverse)} \tag{F.4}$$

There is a multiplication function $\cdot$ taking pairs of numbers $a, b \in \mathbb{F}$ to the number $a \cdot b \in \mathbb{F}$. Properties of multiplication are

$$a \cdot b = b \cdot a, \quad \text{(commutativity)} \tag{F.5}$$

$$(a \cdot b) \cdot c = a \cdot (b \cdot c). \quad \text{(associativity)} \tag{F.6}$$

There is a number 1 such that

$$1 \neq 0, \tag{F.7}$$

$$a \cdot 1 = a. \quad \text{(existence of a multiplicative identity)} \tag{F.8}$$

For every number $a \neq 0$ there is a number b, written a^{-1} or $1/a$, such that

$$a \cdot b = 1. \quad \text{(existence of a multiplicative inverse)} \tag{F.9}$$

Finally, there is an axiom describing the interplay of multiplication and addition.

$$a \cdot (b + c) = a \cdot b + a \cdot c. \quad \text{(distributive law)} \tag{F.10}$$

Before turning to the order axioms, several consequences of the field axioms will be developed. These results will hold equally well for real numbers, rational numbers, complex numbers, and certain other algebraic structures (see Problem 1.4).

Proposition 1.1.1. *For every $a \in \mathbb{F}$,*

$$a \cdot 0 = 0.$$

Proof. Axioms *F*.3 and *F*.10 lead to

$$a \cdot 0 = a \cdot (0 + 0) = a \cdot 0 + a \cdot 0.$$

By F.3 and F.4, adding an additive inverse of $(a \cdot 0)$ to both sides gives

$$0 = a \cdot 0.$$

□

The expression $-a$ will be used to denote an additive inverse of $a \in \mathbb{F}$. A multiplicative inverse of $a \neq 0$ will be denoted a^{-1}. Both are unique, so the notation is not misleading. As a first step in answering questions about uniqueness, consider the uniqueness of the multiplicative identity.

Proposition 1.1.2. *Suppose that $a, b \in \mathbb{F}$, $a \neq 0$, and $a \cdot b = a$. Then $b = 1$.*

Proof. Multiply both sides of $a \cdot b = a$ by a^{-1}, which exists by *F*.9. Using the associativity axiom *F*.6 for multiplication,

$$(a^{-1} \cdot a) \cdot b = 1 \cdot b = b = a^{-1} \cdot a = 1.$$

□

Here is the uniqueness result for multiplicative inverses.

Proposition 1.1.3. *Suppose that $a, b_1, b_2 \in \mathbb{F}$, and $a \neq 0$. If $a \cdot b_1 = a$ and $a \cdot b_2 = a$, Then $b_1 = b_2$.*

Proof. Multiply both sides of $a \cdot b_1 = a \cdot b_2$ by b_1. Using the associativity axiom *F*.6 for multiplication,

$$(b_1 \cdot a) \cdot b_1 = b_1 \cdot a \cdot b_2.$$

Since $b_1 \cdot a = 1$, $b_1 = b_2$. □

Proposition 1.1.4. *Suppose that $a, b \in \mathbb{F}$ and $a \neq 0$. Then there is exactly one number $x \in \mathbb{F}$ such that $a \cdot x + b = 0$.*

Proof. For the equation

$$a \cdot x + b = 0$$

it is easy to find a formula for the solution. First add $-b$ to both sides, obtaining

$$a \cdot x = -b.$$

Then multiply both sides by a^{-1} to get

$$a^{-1} \cdot a \cdot x = (a^{-1} \cdot a) \cdot x = 1 \cdot x = x = a^{-1} \cdot (-b).$$

Thus there is at most one solution of the equation, and it is simple to check that $a^{-1}(-b)$ is a solution. □

The uniqueness of the additive inverse follows from Proposition 1.1.4 by taking $a = 1$. If instead $b = 0$, then Proposition 1.1.4 shows that if the product $x \cdot y$ of two elements of a field is zero, then at least one factor must be zero.

Proposition 1.1.5. *Suppose that $x_1, b \in \mathbb{F}$, and $x_1^2 = b$. If $x \in \mathbb{F}$ and $x^2 = b$, then either $x = x_1$ or $x = -x_1$ (or both if $x_1 = 0$).*

Proof. First notice that

$$x_1^2 + x_1 \cdot (-x_1) = x_1 \cdot (x_1 + (-x_1)) = x_1 \cdot 0 = 0,$$

by the distributive law, so $x_1 \cdot (-x_1) = -x_1^2$. If $x^2 = b$ then another application of the distributive law leads to

$$(x + x_1) \cdot (x + (-x_1)) = x^2 + (-x_1^2) = b - b = 0.$$

Either $x + x_1 = 0$ or $x + (-x_1) = 0$. □

As is typical in arithmetic, write b/a for $a^{-1}b$, and ab for $a \cdot b$. In any field the positive integers can be defined recursively, $2 = 1 + 1$, $3 = 2 + 1$, $4 = 3 + 1$, etc, but, unlike the rationals or reals (see Problem 1.4), some of these "integers" may not be different from 0. With this caveat in mind, here is the quadratic formula.

Proposition 1.1.6. (Quadratic formula): *Suppose that $b, c, x_1 \in \mathbb{F}$, and x_1 satisfies the equation*

$$x^2 + bx + c = 0. \tag{1.1}$$

Suppose in addition that $2 \neq 0$. Then there is a $d \in \mathbb{F}$ satisfying $d^2 = b^2 - 4c$ such that the numbers

$$x_1 = \frac{-b + d}{2}, \quad x_2 = \frac{-b - d}{2},$$

are solutions of (1.1)*, and every solution $x \in \mathbb{F}$ is one of these.*

Proof. Since $2 \neq 0$, the number 2 has a multiplicative inverse and $2^{-1} \cdot b = b/2$ is defined. For $x \in \mathbb{F}$,

$$(x + \frac{b}{2})^2 = x^2 + (2^{-1} + 2^{-1})bx + \frac{b^2}{2 \cdot 2} = x^2 + bx + \frac{b^2}{4},$$

and (1.1) is equivalent to

$$(x + \frac{b}{2})^2 = \frac{b^2 - 4c}{4}. \tag{1.2}$$

If

$$d = 2(x_1 + b/2),$$

then $d \in \mathbb{F}$,

$$(d/2)^2 = \frac{b^2 - 4c}{4},$$

and $x_1 = (-b + d)/2$ satisfies (1.2). Similarly, if $x_2 = (-b - d)/2$, then

$$(x_2 + \frac{b}{2})^2 = (-d/2)^2 = d^2/4 = (b^2 - 4c)/4,$$

so x_2 also satisfies (1.2).

Finally, if x is any solution of (1.1), then

$$(x + \frac{b}{2})^2 = \frac{b^2 - 4c}{4} = (d/2)^2.$$

By Proposition 1.1.5

$$x + \frac{b}{2} = \pm\frac{d}{2},$$

so x must have the form

$$x = -\frac{b \pm d}{2}.$$

□

1.1.2 Order axioms

The existence of an ordering relation $\le$ for the field of real numbers is one of the ways to distinguish it from the field of complex numbers, as well as many other fields. For instance, in establishing the quadratic formula we had to worry about the possibility that $2 = 1 + 1$ and 0 were the same number. This will not be the case if $\mathbb{F}$ satisfies the order axioms. By definition a field $\mathbb{F}$ is an *ordered field* if for any $a, b, c \in \mathbb{F}$ the following axioms O.1–O.6 hold.

There is a relation $\le$ satisfying

$$a \le a, \tag{O.1}$$

$$a \le b \quad \text{and} \quad b \le a \quad \text{implies} \quad a = b, \tag{O.2}$$

$$a \le b \quad \text{and} \quad b \le c \quad \text{implies} \quad a \le c, \tag{O.3}$$

$$\text{either} \quad a \le b \quad \text{or} \quad b \le a, \tag{O.4}$$

$$a \le b \quad \text{implies} \quad a + c \le b + c, \tag{O.5}$$

$$0 \le a \quad \text{and} \quad 0 \le b \quad \text{implies} \quad 0 \le a \cdot b. \tag{O.6}$$

As additional notation, write $a < b$ if $a \le b$ and $a \ne b$. Also write $a \ge b$ if $b \le a$ and $a > b$ if $b \le a$ and $b \ne a$.

Proposition 1.1.7. *If $\mathbb{F}$ is an ordered field and $a \le 0$, then $0 \le -a$.*

Proof. Add $-a$ and use O.5 to get

$$0 = a + (-a) \leq 0 + (-a) = -a.$$

□

Proposition 1.1.8. *If $\mathbb{F}$ is an ordered field, then $0 < 1$.*

Proof. By O.4 either $0 \leq 1$ or $1 \leq 0$. Let's show that $0 < 1$ in either case. First suppose $0 \leq 1$. Axiom F.7 says $0 \neq 1$, so $0 < 1$ in this case.

If instead $1 \leq 0$, then $0 \leq -1$ by Proposition 1.1.7. Now by Proposition 1.1.1

$$0 = (-1) \cdot (1 + (-1)) = -1 + (-1) \cdot (-1),$$

and so $(-1) \cdot (-1) = 1$. By O.6 it then follows that $0 \leq (-1) \cdot (-1) = 1$. Again, $0 \neq 1$ by F.7, so $0 < 1$.

□

Proposition 1.1.9. *Suppose $\mathbb{F}$ is an ordered field. If $a, b, c \in \mathbb{F}$, $a < b$, and $b < c$, then $a < c$.*

Proof. Axiom O.3 gives $a \leq c$, so the case $a = c$ must be ruled out. If $a = c$, then by O.5,

$$a - a = 0 < b - a = b - c.$$

This gives $c - b \leq 0$, and the hypotheses give $c - b \geq 0$, so O.2 leads to $b = c$, contradicting our assumptions. □

Proposition 1.1.10. *If $\mathbb{F}$ is an ordered field then all positive integers n satisfy $n - 1 < n$. In particular $n \neq 0$.*

Proof. Start with $0 < 1$, the conclusion of Proposition 1.1.8. Adding 1 to both sides $n - 1$ times gives $n - 1 < n$. The combination of $0 < 1 < \cdots < n - 1 < n$ and Proposition 1.1.9 gives $0 < n$. □

It is helpful to establish some facts about multiplication in ordered fields in addition to O.6.

Proposition 1.1.11. *In an ordered field, suppose that $0 \leq a \leq b$ and $0 \leq c \leq d$. Then $0 \leq ac \leq bd$.*

Proof. The inequality $a \leq b$ leads to $0 \leq b - a$. Since $0 \leq c$, it follows from O.6 that

$$0 \leq bc - ac,$$

or

$$ac \leq bc.$$

By the same reasoning,

$$bc \leq bd.$$

Axioms O.3 and O.6 then give

$$0 \le ac \le bd.$$

□

Proposition 1.1.12. *In the ordered field $\mathbb{F}$, suppose $b \le 0 \le a$. Then $a \cdot b \le 0$.*

Proof. By Proposition 1.1.7

$$0 \le -b,$$

so by O.6

$$0 \le a \cdot (-b) = -(a \cdot b).$$

and

$$a \cdot b \le 0.$$

□

Proposition 1.1.13. *In an ordered field $\mathbb{F}$, if $0 < a \le b$, then $0 < 1/b \le 1/a$.*

Proof. Proposition 1.1.12 implies that the product of a positive and negative number is negative. Since $a \cdot a^{-1} = 1$, $a^{-1} > 0$, and similarly for b^{-1}. Applying O.6 gives

$$b - a \ge 0,$$
$$a^{-1}b^{-1}(b - a) \ge 0,$$
$$a^{-1} - b^{-1} \ge 0$$

or

$$a^{-1} \ge b^{-1}.$$

□

It is also convenient to be able to compare numbers with integers. For this we need another order axiom, called the *Archimedean Property.*

$$\text{For every } a \in \mathbb{F} \text{ there is an integer } n \text{ such that } a \le n. \qquad (O.7)$$

Proposition 1.1.14. *Suppose $\mathbb{F}$ is an ordered field satisfying the Archimedean property. If $a > 0$ and $b > 0$, then there is an integer k such that $a \le k \cdot b$.*

Proof. Since $b \ne 0$, it has a positive multiplicative inverse. For some integer m then, $0 < b^{-1} \le m$. Proposition 1.1.11 yields $1 \le b \cdot m$. Similarly, if $a \le n$ then

$$a \cdot 1 \le n \cdot m \cdot b,$$

and we may take $k = n \cdot m$. □

Another consequence of O.7 is used quite often in analysis.

Lemma 1.1.15. *In an ordered field satisfying the Archimedean property, suppose $a \ge 0$ and $a < 1/n$ for every positive integer n. Then $a = 0$.*

Proof. The only choices are $a = 0$ or $a > 0$. If $a > 0$ there is an integer n such that $1/a \le n$, or $a \ge 1/n$. Since this possibility is ruled out by the hypotheses, $a = 0$. □

The axioms F.1–F.10 and O.1–O.7 discussed so far describe properties expected for the real numbers. However, these axioms are also satisfied by the rational numbers. Since our axioms do not distinguish between the rational and real numbers, but these number systems are distinct, there must be some properties of the real numbers that have not yet been identified.

1.2 Proofs

The idea of carefully reasoned mathematical proofs dates back to the ancient Greeks. In rough outline the plan is to present a sequence of statements in order to reach a correct conclusion. Each statement in the sequence should be true, and its truth can be established in two ways. First, a statement may be true because its truth was established before the current proof was started. Second, a statement may be true because it follows from true statements by a rule of inference.

Probably the simplest and most familiar examples of such arguments involve algebraic manipulations. As an illustration, consider the proof that the square of an odd integer is odd. If t is an odd integer, then

$$t = 2n + 1. \qquad s.1$$

Since both sides are equal, their squares are equal, so

$$t^2 = (2n + 1)(2n + 1). \qquad s.2$$

The distributive law says that for all real numbers x, y, z,

$$(x + y)z = xz + yz. \qquad s.3$$

Apply this property with $x = 2n$, $y = 1$, and $z = 2n + 1$ to find

$$t^2 = 2n(2n + 1) + (2n + 1) = 2[n(2n + 1) + n] + 1. \qquad s.4$$

If

$$m = n(2n + 1) + n, \qquad s.5$$

then m is an integer and

$$t^2 = 2m + 1, \qquad s.6$$

which is odd.

1.2.1 Proof by induction

There is an amusing story from the childhood of the famous mathematical scientist Carl Friedrich Gauss (1777–1855). His elementary school teacher, wanting to keep the class busy, assigns the problem of adding the numbers from 1 to 100. Gauss's hand goes up more or less instantly, and the correct answer

$$(100 \times 101)/2 = 10100/2 = 5050$$

is produced. Actually Gauss is supposed to have solved the more general problem of finding the sum

$$\sum_{k=1}^{n} k = 1 + 2 + 3 + \cdots + n = \frac{n(n+1)}{2}.$$

This problem has the following elementary solution.

$$2 \times (1 + 2 + 3 + \cdots + n)$$

$$= \begin{array}{cccccccccc} & (1 & + & 2 & + & \ldots & + & (n-1) & + & n) \\ + & (n & + & (n-1) & + & \ldots & + & 2 & + & 1) \end{array}$$

$$= n(n+1).$$

Each vertical sum is $n+1$, and there are n such sums.

What about higher powers? For instance you could be planning to build a pyramid of height n with cubes of stone. At the top you will use 1 cube of stone. Moving downward, each layer is a square of stones, with the k-th layer having $k \times k = k^2$ stones. In order to make labor and transportation plans it will be helpful to know the total number of stones if the height is n, which amounts to

$$\sum_{k=1}^{n} k^2.$$

There is a simple formula for this sum,

$$\sum_{k=1}^{n} k^2 = \frac{n(n+1)(2n+1)}{6}.$$

Along with many other problems, the verification of this formula for every positive integer n can be handled by the technique known as proof by induction. After introducing proofs by induction, several important applications of the method are given. The more general problem of finding formulas for the sums of powers

$$\sum_{k=1}^{n} k^m$$

is treated in a later chapter.

Proof by induction is one of the most fundamental methods of proof in mathematics, and it is particularly common in problems related to discrete mathematics and computer science. In many cases it is the method for establishing the validity of a formula, which may have been conjectured based on a pattern appearing when several examples are worked out. The following formulas provide a pair of classic illustrations:

$$\sum_{k=1}^{n} k = 1 + 2 + \cdots + n = \frac{n(n+1)}{2}, \tag{1.3}$$

$$\sum_{k=1}^{n} k^2 = 1 + 2^2 + \cdots + n^2 = \frac{n(n+1)(2n+1)}{6}. \tag{1.4}$$

The first formula (1.3) had an elementary noninductive proof, but the second (1.4) is more of a challenge. Let's test some cases. If $n = 1$ then the sum is 1 and the righthand side is $1 \cdot 2 \cdot 3/6 = 1$. If $n = 2$ then the sum is 5 and the right-hand side is $2 \cdot 3 \cdot 5/6 = 5$. The first two cases are fine, but how can the formula be checked for every positive integer n?

The brilliant induction idea establishes the truth of an infinite sequence of statements by verifying two things:

(i) that the first statement is true, and

(ii) whenever the n-th statement is true, so is statement $n + 1$.

Sometimes it is convenient to modify this idea, showing that whenever the first n statements are true, so is statement $n + 1$.

Let's try this out with the list of statements

$$S(n): \quad \sum_{k=1}^{n} k^2 = 1^2 + 2^2 + \cdots + n^2 = \frac{n(n+1)(2n+1)}{6}.$$

As is often the case, verification of the first statement is simple.

$$\sum_{k=1}^{1} k^2 = 1^2 = 1 = \frac{1(2)(3)}{6}.$$

Now suppose that n is one of the numbers for which the n-th statement $S(n)$ is true:

$$\sum_{k=1}^{n} k^2 = 1^2 + 2^2 + \cdots + n^2 = \frac{n(n+1)(2n+1)}{6}.$$

The next case is a statement about the sum

$$\sum_{k=1}^{n+1} k^2 = 1^2 + 2^2 + \cdots + n^2 + (n+1)^2.$$

Since the n-th statement is true, we may make use of it.

$$\sum_{k=1}^{n+1} k^2 = 1^2 + 2^2 + \cdots + n^2 + (n+1)^2$$

$$= [1^2 + 2^2 + \cdots + n^2] + (n+1)^2 = \frac{n(n+1)(2n+1)}{6} + (n+1)^2$$

$$= (n+1)[\frac{n(2n+1)}{6} + \frac{6(n+1)}{6}] = \frac{(n+1)(n+2)(2n+3)}{6},$$

since the identity $n(2n+1) + 6(n+1) = (n+2)(2n+3)$ is easily checked. This shows that if $S(n)$ is true, then so is the statement

$$S(n+1): \quad \sum_{k=1}^{n+1} k^2 = 1^2 + 2^2 + \cdots + n^2 + (n+1)^2 = \frac{(n+1)(n+2)(2n+3)}{6}.$$

If the formula (1.4) is true in case n, then it is true in case $n+1$. But since it is true in the case $n = 1$, it must be true for every positive integer n. Why does this follow? Suppose some statement $S(n)$ is not true. Then there must be a smallest positive integer n_0 such that the statement $S(n_0)$ is true, but $S(n_0 + 1)$ is false. This is impossible, since it has been shown that whenever $S(n)$ is true, so is $S(n+1)$. The false statement $S(n)$ can't exist!

The proof by induction method commonly arises in questions about concrete mathematical formulas which are clearly either true or false. When trying to establish the general procedure for proof by induction a bit of care is required. There are statements which do not have a well defined truth value (T or F). One example is *This statement is false.* If the statement is true, then it is false, and if it is false, then it is true! We shouldn't run across any of these self referential statements.

Here are some additional formulas that can be proved by induction.

$$\sum_{k=0}^{n-1} x^k = \frac{1-x^n}{1-x}, \quad x \neq 1.$$

Case $n = 1$ is

$$1 = \frac{1-x^1}{1-x}.$$

Assuming case n is true, it follows that

$$\sum_{k=0}^{(n+1)-1} x^k = \sum_{k=0}^{n-1} x^k + x^n$$

$$= \frac{1-x^n}{1-x} + x^n = \frac{1-x^n + x^n(1-x)}{1-x} = \frac{1-x^{n+1}}{1-x}, \quad x \neq 1.$$

It is a bit more challenging to establish the next result, the Binomial

Theorem. Recall that n factorial, written $n!$, is the product of the integers from 1 to n:

$$n! = 1 \cdot 2 \cdot 3 \cdots n.$$

It turns out to be extremely convenient to declare that $0! = 1$, even though it is hard to make sense of this by means of the original description of $n!$.

It will also be necessary to introduce the binomial coefficients which are sufficiently important to get a special notation. For any integer $n \geq 0$, and any integer k in the range $0 \leq k \leq n$, define the binomial coefficient

$$\binom{n}{k} = \frac{n!}{k!(n-k)!} = \frac{1 \cdot 2 \cdots n}{(1 \cdot 2 \cdots k)(1 \cdot 2 \cdots [n-k])}.$$

The symbol on the left is read n choose k, because this is the number of distinct ways of choosing k objects from a set of n objects.

Theorem 1.2.1. (Binomial Theorem) *For positive integers n, and any numbers a and b,*

$$(a+b)^n = \sum_{k=0}^{n} \binom{n}{k} a^k b^{n-k}.$$

Proof. The case $n = 1$ amounts to

$$(a+b)^1 = a + b,$$

which looks fine.

Assuming the truth of $S(n)$, the expression in $S(n+1)$ is

$$(a+b)^{n+1} = (a+b)(a+b)^n = (a+b) \sum_{k=0}^{n} \binom{n}{k} a^k b^{n-k}$$

$$= \sum_{k=0}^{n} \binom{n}{k} a^{k+1} b^{n-k} + \sum_{k=0}^{n} \binom{n}{k} a^k b^{(n+1)-k}.$$

To give the sums a common form, replace $k+1$ with j in the first of the last two sums to get

$$(a+b)^{n+1} = \sum_{j=1}^{n+1} \binom{n}{j-1} a^j b^{n-(j-1)} + \sum_{k=0}^{n} \binom{n}{k} a^k b^{(n+1)-k}.$$

Of course the variable of summation, like that of integration, is a "dummy," so

$$(a+b)^{n+1} = \sum_{k=1}^{n+1} \binom{n}{k-1} a^k b^{(n+1)-k} + \sum_{k=0}^{n} \binom{n}{k} a^k b^{(n+1)-k}$$

$$= a^{n+1} + b^{n+1} + \sum_{k=1}^{n} [\binom{n}{k-1} + \binom{n}{k}] a^k b^{(n+1)-k}.$$

All that remains is to show that for $1 \le k \le n$,

$$\binom{n}{k-1} + \binom{n}{k} = \binom{n+1}{k}.$$

A straightforward computation gives

$$\binom{n}{k-1} + \binom{n}{k} = \frac{n!}{(k-1)!(n+1-k)!} + \frac{n!}{k!(n-k)!}$$

$$= \frac{n!k}{k!(n+1-k)!} + \frac{n!(n+1-k)}{k!(n+1-k)!} = \frac{n!}{k!(n+1-k)!}(k+n+1-k)$$

$$= \binom{n+1}{k}.$$

□

1.2.2 Irrational real numbers

The first distinction between the real and rational numbers was discovered by the Greeks, probably in the fifth century B.C. [5, p. 32]. Recall that a number is *rational* if it can be written as the ratio of two integers, m/n. The Greeks were familiar with the length $\sqrt{2}$ from geometry, but for some time thought all lengths could be expressed as rational numbers. To establish that $\sqrt{2}$ is not rational requires a bit of number theory.

Recall that a positive integer p is said to be *prime* if $p > 1$ and whenever p is written as the product of two positive integers, $p = j \cdot k$, one of the factors is p. When considering products of integers, a single number n is taken as a product with one factor. The following result is basic in arithmetic.

Theorem 1.2.2. *(a) Every positive integer $n \ge 2$ can be written as the product of prime factors. (b) If the (possibly repeated) factors are written in nondecreasing order, this prime factorization is unique.*

Proof. (a): Let's prove the first part of the theorem by induction on n, with the first case being $n = 2$. In this first case 2 is prime, so there is single factor 2.

Now assume the result holds for all integers k with $2 \le k < n$. If n is prime then the factorization has a single prime factor. If n is not prime, then $n = p \cdot k$, where $2 \le p < n$ and $2 \le k < n$. By the induction hypothesis both k and p have prime factorizations, and so n has a prime factorization. □

It takes more work [4, p. 3] to show that the factorization is unique if the factors are listed in nondecreasing order.

An argument by contradiction is used to show there is no rational number whose square is 2. The idea is to start with the assumption that some rational number has square 2. By deducing a false statement this assumption is shown to be incorrect.

Theorem 1.2.3. *There is no rational number whose square is* 2.

Proof. Suppose there is a rational number

$$\sqrt{2} = \frac{m}{n}$$

whose square is two. There is no loss of generality if m and n are taken to be positive integers with no common prime factors. (If there are common prime factors, terms in the numerator and denominator can be cancelled until the desired form is obtained.) Multiplying by n and squaring both sides leads to

$$2n^2 = m^2.$$

Obviously m^2 is even.

Notice that if $m = 2l + 1$ is odd, then $m^2 = 4l^2 + 4l + 1$ is odd. Since m^2 is even, m must have a factor 2. It follows however, that $n^2 = m^2/2$ is also even. This means m and n have the common factor 2.

Starting with integers m and n with no common prime factors, we have deduced that they have a common prime factor. This contradiction shows that the assumption $\sqrt{2} = m/n$ must be incorrect. □

1.2.3 Propositional logic

This section treats the construction of new statements from old ones using the following collection of *propositional connectives*:

logical symbol	English equivalent
$\neg$	not
$\wedge$	and
$\vee$	or
$\Rightarrow$	implies
$\Leftrightarrow$	is equivalent to

(1.5)

Propositional logic formalizes aspects of deductive reasoning illustrated by the following story. Two children, George and Martha, are left at home with no companions. When their parents return, they discover that a bag of cookies has been consumed. If George, who always tells the truth, reports that he did not eat any cookies, then we may reasonably conclude that Martha is the culprit. To distill the logic from this tale, use the letter A to represent the statement "George ate cookies" and B to represent "Martha ate cookies". Based on George's testimony, our deduction is that since A or B is true, and A is false, the implication that B is true must be valid. Notice that if George was not questioned, but instead Martha confesses to eating some cookies, we're left in the dark about George's culpability.

The use of propositional connectives to construct statements is based on a starting collection of statements, represented by letters A, B, C, etc., whose internal structure is not of concern. Since such statements are indivisible and provide the basic building blocks for more complex expressions, they are called *atoms*, or *atomic statements*, or *atomic formulas.* With proper use of the propositional connectives, other *composite formulas* can be generated. Suppose R and S are formulas, which may be atomic or composite. By using the propositional connectives, the following new formulas may be generated:

$$\neg R, R\wedge S, R\vee S, R\Rightarrow S, R\Leftrightarrow S.$$

Formulas generated by these rules are said to be well formed, to distinguish them from nonsense strings of symbols like

$$R\Leftrightarrow\neg.$$

Starting from the atoms, repeated application of the rules allows us to generate complex formulas such as

$$[\neg(A\vee B)]\Leftrightarrow[(\neg A)\wedge(\neg B)].$$

Some of these constructions arise often enough to merit names. Thus $\neg A$ is the *negation* of A, $B\Rightarrow A$ is the *converse* of $A\Rightarrow B$, and $(\neg B)\Rightarrow(\neg A)$ is the *contrapositive* of $A\Rightarrow B$.

To minimize the need for parentheses in composite formulas, the propositional connectives are ranked in the order

$$\neg, \wedge, \vee, \Rightarrow, \Leftrightarrow.$$

To interpret a formula, the connectives are applied in left to right order ($\neg$'s first, etc) to well formed subformulas, and from left to right in a particular expression. For example the statement

$$A\vee B\Leftrightarrow\neg(\neg A\wedge\neg B)$$

should be parsed as

$$[A\vee B]\Leftrightarrow[\neg((\neg A)\wedge(\neg B))],$$

while

$$A\Rightarrow B\Rightarrow C$$

should be parsed as

$$(A\Rightarrow B)\Rightarrow C.$$

1.2.3.1 Truth tables

In propositional logic the atomic formulas A, B, C, ..., are assumed to be either true or false, but not both. The truth of (well formed) composite formulas may be determined from the truth of the constituent formulas by the use of truth tables. Here are the truth tables for the propositional connectives.

A	$\neg A$
T	F
F	T

A	B	$A \wedge B$	$A \vee B$	$A \Rightarrow B$	$A \Leftrightarrow B$
T	T	T	T	T	T
F	T	F	T	T	F
T	F	F	T	F	F
F	F	F	F	T	T

The truth tables of the propositional connectives are intended to have a close connection with natural language usage. In addition to *A implies B*, statements of the form *if A, then B*, or *B if A*, or *A only if B* are interpreted formally as $A \Rightarrow B$. The formula $A \Leftrightarrow B$ corresponds to the statement forms *A is equivalent to B* or *A if and only if B*.

In several cases the natural language usage is more complex and context dependent than indicated by the corresponding truth table definitions. First notice that $A \vee B$ is true if either A is true, or B is true, or if both A and B are true. This connective is sometime called the *inclusive or* to distinguish it from the *exclusive or* which is false if both A and B are true. In English, sentence meaning often helps determine whether the "inclusive or" or the "exclusive or" is intended, as the following examples illustrate:

John eats peas or carrots. (inclusive)

Mary attends Harvard or Yale. (exclusive)

The logically meaning of implication can also have some conflict with common interpretations. Thus the logical implication

$$\textit{two wrongs make a right} \Rightarrow \textit{all the world's a stage}$$

is true if the first statement is false, regardless of the truth or meaning of the second statement.

To illustrate a truth value analysis, consider the composite formula

$$[A \vee B] \Leftrightarrow [\neg((\neg A) \wedge (\neg B))]. \tag{1.6}$$

To determine how the truth value of this composite formula depends on the truth values of its constituents, a truth table analysis can be carried out. Introduce the abbreviation

$$C = ((\neg A) \wedge (\neg B)).$$

For example (1.6) the truth table is

A	B	$\neg A$	$\neg B$	C	$\neg C$	$[A\vee B]$	$[A\vee B]\Leftrightarrow[\neg C]$
T	T	F	F	F	T	T	T
F	T	T	F	F	T	T	T
T	F	F	T	F	T	T	T
F	F	T	T	T	F	F	T

(1.7)

Notice that the composite formula (1.6) is true for all truth values of its component propositions. Such a statement is called a *tautology.* The tautologies recorded in the next proposition are particularly important in mathematical arguments. The proofs are simple truth table exercises left to the reader.

Proposition 1.2.4. *The following are tautologies of propositional logic:*

$$A\vee\neg A \quad (\textit{Law of the excluded middle}) \tag{1.8}$$

$$(A\Rightarrow B)\Leftrightarrow(\neg B\Rightarrow\neg A). \quad (\textit{Contraposition}) \tag{1.9}$$

1.2.3.2 Valid consequences

Let's return to the story of George, Martha, and the missing cookies. If A represents the statement "George ate cookies" and B represents "Martha ate cookies," the logic of our analysis can be described as

$$(A\vee B)\wedge(\neg A)\Rightarrow B.$$

Using the abbreviation

$$C=(A\vee B)\wedge(\neg A),$$

the following truth table show that this formula is a tautology.

A	B	$\neg A$	$A\vee B$	C	$C\Rightarrow B$
T	T	F	T	F	T
F	T	T	T	T	T
T	F	F	T	F	T
F	F	T	F	F	T

(1.10)

Consider the stronger interpretation

$$(A\vee B)\wedge(\neg A)\Leftrightarrow B,$$

whose corresponding truth table is

A	B	$\neg A$	$A\vee B$	C	$C\Leftrightarrow B$
T	T	F	T	F	F
F	T	T	T	T	T
T	F	F	T	F	T
F	F	T	F	F	T

(1.11)

Since the truth value of (1.11) is sometimes false, this is not a tautology. However the formula is true whenever $\neg A$ is true. That is, if George can be trusted, the conclusion is sound. When a statement S is true whenever the list of propositions P is true, we say that S is a *valid consequence* of P. Thus $(A \lor B) \land (\neg A) \Leftrightarrow B$ is a valid consequence of $\neg A$.

The valid consequence concept is typically employed when considering the soundness of arguments presented in natural language. As another example, consider the following natural language discussion of taxation.

If state revenue does not increase, then either police services or educational services will decline. If taxes are raised, then state revenue will increase. Therefore, if taxes are raised, police services or educational services, or both, will not decline.

As taxpayers concerned about police and educational services, we have an interest in understanding whether the statement

if taxes are raised, police services or educational services, or both, will not decline

follows from the premises. To analyze the question, let's formalize the presentation. Use the letters A, B, C, D, to represent the statements

A: state revenue increases,

B: police services will decline,

C: educational services will decline,

D: taxes are raised.

For the purposes of logical analysis, a reasonable translation of the example into symbols is

$$\Big([\neg A \Rightarrow (B \lor C)] \land [D \Rightarrow A]\Big) \Rightarrow \Big(D \Rightarrow [\neg B \lor \neg C]\Big). \tag{1.12}$$

Let's consider a truth table analysis of the question of taxation in this example. The truth values of the basic propositions $A - D$ are not given. Rather, it is claimed that the composite formulas

$$\neg A \Rightarrow (B \lor C), \quad \textit{and} \quad D \Rightarrow A$$

are true. The question is whether the claim (1.12) is a valid consequence of these statements.

To show that the logic is faulty, it suffices to find truth values for $A, \dots, D$ for which these composite assertions are true, but (1.12) is false. Suppose that all the statements $A, \dots, D$ are true. Then the statement $\neg A \Rightarrow (B \lor C)$ is true since $\neg A$ is false, and $D \Rightarrow A$ is true since both D and A are true. Thus $[\neg A \Rightarrow (B \lor C)] \land [D \Rightarrow A]$ is true, while $D \Rightarrow [\neg B \lor \neg C]$ is false. Consequently, the implication (1.12) is false, and the logic of the argument is flawed.

In this case an exhaustive analysis of the truth table was not needed. Since each of the propositions $A, \dots, D$ could be independently true or false, a complete truth table would have $2^4 = 16$ rows. More generally, a composite formula with n atomic formulas would have a truth table with 2^n rows. The exponential growth of truth tables with the number of atomic formulas is a serious shortcoming.

1.2.4 Rules of inference

Returning again to the tale of George and Martha, we would like to conclude that Martha is the cookie thief. In this example the assumption that $A \vee B$ is true is reasonable since no one else was home. Our trust in George leads us to accept the truth of $\neg A$ as well. Then the definition of $\wedge$ allows us to conclude that $(A \vee B) \wedge (\neg A)$ is true. Finally, the formula

$$(A \vee B) \wedge (\neg A) \Rightarrow B$$

is true because it is a tautology.

To take advantage of this information to conclude that B is a true statement, a *rule of inference* is used. A rule of inference allows us to draw certain conclusions based on premises such as George's veracity. The rule of inference useful here is *modus ponens*, which allows deductions of the following form.

$$\begin{array}{c} R \\ R \Rightarrow S \\ \hline S \end{array}$$

Taking $R = (A \vee B) \wedge (\neg A)$ and $S = B$, the truth of R and $R \Rightarrow S$ allow us to assign guilt to Martha. That is, B is a true statement. Some additional rules of inference appear in the exercises.

1.2.5 Predicates and quantifiers

In the propositional logic discussed above, the propositional connectives $\neg$, $\wedge$, $\vee$, $\Rightarrow$, and $\Leftrightarrow$ were used to construct composite formulas from a collection of atomic formulas. There was no consideration of the internal structure of the basic statements; only their truth value was important. In many mathematical statements there are aspects of the internal structure that are quite important.

As examples of typical mathematical statements, consider the assertion

$$\textit{for every number } x, \ x^2 \ge 0, \tag{1.13}$$

or the statement of Fermat's last theorem:

$$\textit{there are no positive integers } x, y, z, n \textit{ with } n > 2 \tag{1.14}$$

$$\textit{such that } z^n = x^n + y^n.$$

There are three aspects of these statements to be considered: the domain of the variables, the predicate or relationship of the variables, and the quantifiers.

First, each statement has variables which are expected to come from some domain $\mathcal{D}$. In (1.13) the variable is x, and its domain has not been specified. The statement is true if the domain $\mathcal{D}$ is the set of real numbers, but it is false if $\mathcal{D}$ is the set of complex numbers. Fermat's last theorem (1.14) has a clear statement that x, y, z, n are all positive integers, which may be taken as the domain $\mathcal{D}$.

Second, each statement has a predicate. A *predicate* is a function of the variables whose value is true or false for each choice of values for the variables from the domain $\mathcal{D}$. In (1.13) the predicate is

$$P(x) : (x^2 \geq 0),$$

while in (1.14) the predicate is more complex,

$$Q(x, y, z, n) : (n > 2) \wedge (z^n = x^n + y^n).$$

The third ingredient is the quantification. Are the predicates expected to be true for all values of the variables, or for only some values of the variables? The symbols $\forall$ and $\exists$ represent our two quantifiers. The symbol $\forall$ is read "for all," and is called the *universal quantifier*. A statement of the form $\forall x P(x)$ is true for the domain $\mathcal{D}$ if $P(x)$ has the value T for all x in the domain $\mathcal{D}$, otherwise the statement is false. The symbol $\exists$ is read "there exists," and is called the *existential quantifier*. A statement of the form $\exists x P(x)$ is true for the domain $\mathcal{D}$ if there is some x in $\mathcal{D}$ for which $P(x)$ has the value T, otherwise the statement is false. The new symbols $\forall$ and $\exists$ are added to the previous set of propositional connectives to allow us to generate composite formulas. With the aid of these symbols we may formalize our mathematical statements as

$$\forall x P(x), \quad P(x) : (x^2 \geq 0), \tag{1.15}$$

and

$$\neg(\exists (x, y, z, n) Q(x, y, z, n)), \quad Q(x, y, z, n) : (n > 2) \wedge (z^n = x^n + y^n). \tag{1.16}$$

Just as with propositional logic, there is a collection of formulas that may be generated from variables, predicates, propositional connectives, and quantifiers. The atomic formulas are simply the predicates with the appropriate number of variables. For instance if P, Q, R are predicates with one, two, and three arguments respectively, then

$$P(x), Q(x, y), R(x, y, z)$$

are atomic formulas. Then, if S and T are formulas, so are

$$\neg S, S \wedge T, S \vee T, S \Rightarrow T, S \Leftrightarrow T,$$

as well as

$$\forall x S, \quad \exists x T,$$

where x is a variable.

When formulas involve quantifiers and predicates, there can be a question about the appropriate selection of variables. Consider the example

$$[\exists x P(x) \wedge \exists x Q(x)] \Rightarrow [\exists x (P(x) \wedge Q(x))].$$

This formula has the same meaning as

$$[\exists x P(x) \wedge \exists y Q(y)] \Rightarrow [\exists z (P(z) \wedge Q(z))],$$

since the introduction of the new variables does not change the truth value of the formulas. $\exists x Q(x)$ and $\exists y Q(y)$ have the same truth value in any domain. In contrast the formulas $\exists x(P(x) \wedge Q(x))$ and $\exists x(P(x) \wedge Q(y))$ are not equivalent; in the second case the quantification of the variable y has not been specified.

The introduction of predicates adds a great deal of complexity to our formulas. For instance, in propositional logic it was possible, at least in principle, to consider the truth value of a formula as a function of the truth values of its atoms. In that context we singled out certain formulas, the tautologies, which were true regardless of the truth values of the arguments. There is an analogous idea in the predicate calculus. Say that a formula S is *valid* if the truth value of S is true for every assignment of its variables to values in every domain $\mathcal{D}$. Since the domain $\mathcal{D}$ might be an infinite set such as the integers, it is not possible, even in principle, to construct and examine a complete truth table.

To show that a formula is not valid it is only necessary to find a single domain $\mathcal{D}$ and an assignment of the variables to elements of $\mathcal{D}$ such that the formula is false. But to establish the validity of a formula S we would have to argue, without an exhaustive table, that S is always true. This is not always difficult. For instance it is not hard to show that $P(x) \Leftrightarrow P(x)$. In general, however, establishing which formulas are valid will be more of a challenge.

Here are some valid formulas involving quantifiers and predicates. The proofs are omitted.

$$[\neg \exists x P(x)] \Leftrightarrow [\forall x \neg P(x)] \tag{1.17}$$

$$[\neg \forall x P(x)] \Leftrightarrow [\exists x \neg P(x)]$$

$$[\forall x P(x) \wedge \forall x Q(x)] \Leftrightarrow [\forall x (P(x) \wedge Q(x))]$$

$$[\exists x P(x) \vee \exists x Q(x)] \Leftrightarrow [\exists x (P(x) \vee Q(x))]$$

$$[\exists x (P(x) \wedge Q(x))] \Rightarrow [\exists x P(x) \wedge \exists x Q(x)]$$

$$[\forall x P(x) \vee \forall x Q(x)] \Rightarrow [\forall x (P(x) \vee Q(x))]$$

As a final topic in this discussion of predicate calculus, some brief remarks about *equality* are in order. Certainly one of the more common symbols in mathematics, equality is a two place predicate. To put it in the context of our previous discussion we might write $E(x, y)$ instead of $x = y$. As a predicate, $E(x, y)$ has a truth value when x and y represent elements of the domain $\mathcal{D}$; $E(x, y)$ is true if x and y are the same element, otherwise it is false.

Among the properties of equality are the following:

$$x = x, \quad \text{(reflexive)}, \tag{1.18}$$

$$(x = y) \Rightarrow (y = x), \quad \text{(symmetric)},$$
$$[(x = y) \wedge (y = z)] \Rightarrow [x = z], \quad \text{(transitive)}.$$

It is common in mathematics to encounter two place predicates sharing the reflexive, symmetric, and transitive properties of (1.18). Such predicates are called *equivalence relations.*

To construct an example of an equivalence relation $P(x, y)$ which is distinct from equality, suppose our domain $\mathcal{D}$ is the set of integers. Define the predicate $P(x, y)$ with the value T if $x - y$ is even, and let $P(x, y) = F$ if $x - y$ is odd (see Problem 1.29).

Another example of an equivalence relation appears when rational numbers are considered. Usually one would say that a rational number is the quotient m/n, where m and n are integers and $n \ne 0$. Of course a wide variety of such fractions are intended to represent the same number: $1/2 = 2/4 = 3/6 = \dots$. Let the domain $\mathcal{D}$ be the set of ordered integer 4-tuples (m_1, n_1, m_2, n_2) with $n_1, n_2 \ne 0$. Define a predicate

$$R(m_1, n_1, m_2, n_2) = \begin{Bmatrix} T, & m_1 n_2 = m_2 n_1, \\ F, & \text{otherwise.} \end{Bmatrix} \tag{1.19}$$

A more suggestive notation

$$(m_1, n_1) \sim (m_2, n_2)$$

for the idea that m_1/n_1 represents the same number as m_2/n_2 is probably easier to manage than (1.19). The equivalence properties (1.18) are easily verified.

1.3 Problems

1.1. *Suppose that a and b are elements of a field $\mathbb{F}$.*
a) Show that if $a \cdot b = 0$, then $a = 0$ or $b = 0$.
b) Show that $(-a) \cdot b = -(a \cdot b)$.
c) Show that $-(-a) = a$.
d) Show that every element a has a unique additive inverse.

1.2. *Suppose that $a \neq 0$ and $b \neq 0$ are elements of a field $\mathbb{F}$.*
a) Show that $a^{-1} \neq 0$ and $(a^{-1})^{-1} = a$.
b) Show that $ab \neq 0$ and $(ab)^{-1} = a^{-1}b^{-1}$.

1.3. *Suppose that $a \neq 0$ is an element of an ordered field. Show that $a \cdot a > 0$.*

1.4. *Let $\mathbb{Z}_p$ be the set of integers $\{0, 1, 2, \ldots, p-1\}$, and suppose that addition $x \oplus y$ and multiplication $x \otimes y$ are carried out modulo p. That is, if xy is normal integer multiplication and $xy = pn + r$, with $0 \leq r < p$, then $x \otimes y = r$. Addition modulo p is similar.*

a) Construct addition and multiplication tables for $\mathbb{Z}_2$ and $\mathbb{Z}_3$. For instance, here is the addition table for $\mathbb{Z}_2$:

$\oplus$	0	1
0	0	1
1	1	0

b) Show that $\mathbb{Z}_2$ and $\mathbb{Z}_3$ are fields.
c) Is $\mathbb{Z}_4$ a field?

1.5. *Show that if $2 = 1 + 1 \neq 0$ in a field $\mathbb{F}$, then $4 = 1 + 1 + 1 + 1 \neq 0$ in $\mathbb{F}$.*

1.6. *Suppose that p, q are elements of an ordered field $\mathbb{F}$. Show that $q \geq 1$ and $p \geq 0$ implies $pq \geq p$.*

1.7. *If p is a prime number, prove that $\sqrt{p}$ is not a rational number. (You may assume the uniqueness of prime factorization.)*

1.8. *Suppose $\mathbb{F}$ is an ordered field. Show that if $0 < a < b$, then $0 < a^2 < b^2$.*

1.9. *Suppose $\mathbb{F}$ is an ordered field satisfying the Archimedean Property O.7. Show that if $a, b \in \mathbb{F}$ and $a < b$, then there is a rational number q satisfying $a < q < b$. (Hint: Consider the numbers m/n where $1/n < b - a$.)*

1.10. *Consider the quadratic equation*

$$x^2 + bx + c = 0, \quad x \in \mathbb{R}.$$

Suppose that b and c are rational, and $b^2 - 4c$ is prime. Show the equation has no rational solutions.

1.11. *Use induction to show that*

$$\sum_{k=1}^{n} k = \frac{n(n+1)}{2}.$$

1.12. *Use induction to show that*

$$\sum_{k=0}^{n-1} \frac{1}{(k+1)(k+2)} = 1 - \frac{1}{n+1}.$$

1.13. *Use induction to show that*

$$\sum_{k=0}^{n-1} k2^{-k} = 2[1 - (n+1)2^{-n}].$$

1.14. *Show that*

$$2^n = \sum_{k=0}^{n} \binom{n}{k}.$$

1.15. *Find the flaw in the following logic.*

Let's prove by induction that if you have a collection of N horses, and at least 1 of them is white, then they are all white.

Clearly if the collection has only 1 horse, and at least 1 is white, then they are all white.

Suppose the statement is true for K horses. Assume then that you have a collection of $K+1$ horses, and at least 1 is white. Throw out one horse, which is not the chosen white one. Now you have a collection of K horses, and at least 1 is white, so all K are white. Now bring back the ejected horse, toss out another one, repeat the argument, and all $K+1$ horses are white.

Since there is a white horse somewhere in the world, all horses are white!

1.16. *Show that for any positive integer n the number n^2 is the sum of the first n odd numbers,*

$$n^2 = \sum_{k=1}^{n} (2k-1).$$

1.17. *Give the proof of Proposition 1.2.4.*

1.18. *(a) Show that $A \Leftrightarrow B$ has the same truth table as $(A \Rightarrow B) \wedge (B \Rightarrow A)$.*
(b) Show that $A \vee B$ is equivalent to $(\neg A) \Rightarrow B$ in the same way.

1.19. *The propositional connectives $\wedge$, $\vee$, $\Rightarrow$, $\Leftrightarrow$ are each a function of an ordered pair of truth values, and the value of each of these functions is either true or false. How many distinct logical connectives of this type are possible? Can they all be constructed using the given four if in addition $\neg$ is available to negate one or both of the arguments? As an example consider $f_1(A, B) = (\neg A) \vee B$.*

1.20. *Show that the statement*

$$A \vee B \Rightarrow (C \Rightarrow A \wedge B)$$

is not a tautology, but is a valid consequence of A, B.

1.21. *Consider the following argument.*

Sam will keep his job only if he files a fraudulent corporate tax return. He will avoid jail only if he files a honest tax return. Since Sam must file a corporate return which is either honest or fraudulent, he will either lose his job or go to jail.

Represent the argument using propositional logic, and decide whether or not the argument is sound. Use the letters $A - C$, *to represent the statements*

A: Sam will keep his job.

B: Sam will go to jail.

C: Sam files an honest tax return.

1.22. *The situation in the previous problem becomes a bit more complex. Again, represent the argument using propositional logic, and decide whether the argument is sound.*

Sam will keep his job if he files a fraudulent corporate tax return, or if his boss goes to jail. Sam will go to jail if he files a fraudulent return, or if he files an honest return and the prosecutor is related to his boss. If the prosecutor is related to his boss, his boss will not go to jail. If Sam is lucky enough to keep his job and avoid jail, then he must file an honest return and the prosecutor must be unrelated to his boss.

1.23. *a) Represent the following argument using propositional logic, and determine the soundness the argument. It may be helpful to introduce the symbol* $\sqcup$ *to represent the exclusive "or."*

Jane and Mary each love either William or Harry, but not both. William will marry Jane if she loves him, and William will marry Mary if she loves him. (For the moment we allow the possibility of two wives.) Harry will marry Mary if she does not love William. If William or Harry will not marry, then either Mary loves William, or Jane and Mary love Harry.

b) Find a propositional logic representation for the following premises:

William will marry either Jane or Mary if she loves him, but if he is loved by both he will marry only one.

1.24. *Use truth tables to check that* $A \wedge (B \vee C)$ *is a valid consequence of* $(A \wedge B) \wedge C$.

1.25. Modus ponens *has the form*

$$\begin{array}{c} A \\ A \Rightarrow B \\ \hline B \end{array}$$

Use truth tables to check that the related formula

$$[A \wedge (A \Rightarrow B)] \Rightarrow B$$

is a tautology. Perform a similar analysis of the following rules of inference.

$$\frac{\begin{matrix} A \Rightarrow B \\ A \Rightarrow C \end{matrix}}{A \Rightarrow B \wedge C} \qquad \frac{\begin{matrix} [A \vee B] \Rightarrow C \\ \neg C \end{matrix}}{\neg A \wedge \neg B} \qquad \frac{\begin{matrix} A \vee B \\ A \Rightarrow C \\ B \Rightarrow D. \end{matrix}}{C \vee D}$$

1.26. *Use the predicates*

$$P(x) : x \textit{ is a car}, \quad Q(x) : x \textit{ is a Cadillac},$$

to represent the sentences

all cars are not Cadillacs,

and

not all cars are Cadillacs,

with the predicate calculus. Do these sentences have the same meaning?

1.27. *(a) Suppose the predicate $P(x)$ is $x > 0$ while the predicate $Q(x)$ is $x < 0$. Show that the implication*

$$[\exists x P(x) \wedge \exists x Q(x)] \Rightarrow [\exists x (P(x) \wedge Q(x))]$$

is not valid; consider the domain $\mathcal{D}$ equal to the set of integers.

(b) Find an example showing that the following implication is not valid.

$$[\forall x (P(x) \vee Q(x))] \Rightarrow [\forall x P(x) \vee \forall x Q(x)].$$

1.28. *Suppose $P(x,y)$ denotes the predicate*

$$(x = 0) \vee (y = 0) \vee (x \otimes y \neq 0).$$

Let $\mathcal{D}$ be the set of integers $\{0,1,2\}$, and suppose the product $x \otimes y$ is multiplication modulo 3, so that if xy is normal integer multiplication and $xy = 3n + r$, with $0 \le r < 3$, then $x \otimes y = r$.

(a) Show that $\forall x \forall y P(x,y)$ is correct if the domain is $\mathcal{D}$.

(b) Show that $\forall x \forall y P(x,y)$ is incorrect if $\mathcal{D}$ is the set of integers $\{0,1,2,3\}$, and the product $x \otimes y$ is multiplication modulo 4.

1.29. *Suppose $\mathcal{D}$ is the set of integers. Let $P(x,y)$ be the predicate which is T if $x - y$ is even and F if $x - y$ is odd. Show that P is an equivalence relation. That is, show*

$$P(x,x), \quad P(x,y) \Rightarrow P(y,x),$$

and

$$[P(x,y) \wedge P(y,z)] \Rightarrow P(x,z).$$

1.30. *(a) If m_1/n_1 and m_2/n_2 are rational, show that their sets of equivalent rational forms are either the same or disjoint.*

(b) More generally, suppose $\mathcal{D}$ is a set and $P(x,y)$ is an equivalence relation defined on $\mathcal{D}$. For each element x of $\mathcal{D}$ let

$$S_x = \{z \in \mathcal{D} \mid P(x,z) = T\},$$

that is, S_x is the set of elements equivalent to x. Show that for any choice of x and y in $\mathcal{D}$, either $S_x = S_y$, or $S_x \cap S_y = \emptyset$.

2

Infinite sequences

Polynomials are a delight to calculus students. If

$$p(x) = c_0 + c_1 x + c_2 x^2 + \cdots + c_N x^N, \tag{2.1}$$

then

$$\frac{dp}{dx} = c_1 + 2c_2 x + \cdots + N c_N x^{N-1},$$

and

$$\int_0^x p(t)\, dt = c_0 x + c_1 \frac{x^2}{2} + \cdots + c_N \frac{x^{N+1}}{N+1},$$

so differentiation or integration of $p(x)$ is no challenge. The ease of these computations makes it irresistible to consider functions of the form

$$f(x) = c_0 + c_1 x + c_2 x^2 + c_3 x^3 + \ldots, \tag{2.2}$$

which look like polynomials with infinitely many terms.

The reader may recall examples from calculus. While there is no elementary antiderivative for e^{-x^2}, the formula

$$\int_0^x e^{-t^2}\, dt = \int_0^x \sum_{k=0}^{\infty} \frac{(-t^2)^k}{k!}\, dt = \sum_{k=0}^{\infty} \frac{(-1)^k x^{2k+1}}{(2k+1)k!} = x - \frac{x^3}{3} + \frac{x^5}{10} + \ldots \tag{2.3}$$

is both attractive and computationally efficient for small values of x. One of the author's favorite examples uses the geometric series

$$1 + x + x^2 + \cdots = \sum_{k=0}^{\infty} x^k = \frac{1}{1-x}, \quad |x| < 1, \tag{2.4}$$

to generate the expression

$$\tan^{-1}(x) = \int_0^x \frac{1}{1-(-t^2)}\, dt = \int_0^x \sum_{k=0}^{\infty} (-t^2)^k\, dt = x - \frac{x^3}{3} + \frac{x^5}{5} + \ldots. \tag{2.5}$$

This idea of treating power series as if they were polynomials of "infinite degree" also works well for solving many otherwise recalcitrant differential equations.

Related ideas also arise in the familiar notion that real numbers can be expressed as possibly infinite decimals. Consultation with a calculator indicates that $\pi = 3.14159\ldots$ or $\sqrt{2} = 1.414\ldots$. These decimal representations require an infinite list of digits to achieve actual equality. The truncated decimal expansions for π actually indicate that π is within 10^{-2} of 3.14, and is no more than 10^{-4} from 3.1415, etc.

Despite such apparent successes, the use of infinite processes such as infinite sums requires extra diligence. For example, the geometric series formula (2.4) with $x = -1$ suggests that

$$1 + (-1) + 1 + (-1) + \cdots = \sum_{k=0}^{\infty} (-1)^k = \frac{1}{2}.$$

Even more improbably, if $x = 2$ the formula states that

$$1 + 2 + 4 + \cdots = \sum_{k=0}^{\infty} 2^k = \frac{1}{-1} = -1.$$

These results are certainly false in the sense that the finite sums

$$\sum_{k=0}^{n-1} (-1)^k = \frac{1}{2} - \frac{(-1)^n}{2},$$

and

$$\sum_{k=0}^{n-1} 2^k = \frac{1}{-1} - \frac{2^n}{-1},$$

are not approaching the numbers $1/2$ or -1 respectively.

The pitfalls associated with the use of infinite processes led the ancient Greeks to avoid them [1, p. 13–14], [5, p. 176]. As calculus was developed [5, p. 436–467] the many successful calculations achieved with infinite processes, including infinite series, apparently reduced the influence of the cautious critics. A reconciliation between the success of cavalier calculations and the careful treatment of foundational issues was not achieved until the nineteenth century.

Analyzing the validity of expressions like (2.3) or (2.5) involves several distinct ideas. This chapter takes the first step by introducing infinite sequences. The next chapter will use sequences as the basis for studying infinite sums of numbers. Later chapters will treat the existence and properties of functions defined by infinite series.

2.1 Limits of infinite sequences

2.1.1 Basic ideas

Unless explicitly stated otherwise, assume from now on that $\mathbb{F}$ is an Archimedean ordered field, satisfying the field axioms F.1–F.10 and the order axioms O.1–O.7. The rational numbers are our main example. If $x \in \mathbb{F}$, the *absolute value* of x, denoted $|x|$, is equal to x if $x \geq 0$ and is equal to $-x$ if $x < 0$. To finish our axiomatic description of the real numbers a discussion of infinite sequences is needed. Until the description is completed a number will be an element of an Archimedean ordered field.

Intuitively, an infinite sequence is simply an infinite list of numbers. The $k-$th term of the sequence is denoted c_k or $c(k)$. Examples include the sequences

$$1, \frac{1}{2}, \frac{1}{3}, \frac{1}{4}, \ldots \quad c_k = \frac{1}{k}, \quad k = 1, 2, 3, \ldots,$$

$$1, -1, 1, -1, \ldots, \quad c_k = (-1)^k, \quad k = 0, 1, 2, \ldots,$$

and

$$3, 3.1, 3.14, 3.141, 3.1415, \ldots,$$

where c_k is the first k digits of the decimal expansion of π.

In the usual mathematical language, an *infinite sequence*, or simply a *sequence*, is a function c whose domain is the set $\mathbb{N}$ of positive integers $1, 2, 3, \ldots$. The value $c(k)$, or more commonly c_k, of the function at k is called the $k-th$ *term* of the sequence. For our purposes the values $c(k)$ will typically be numbers, although the idea extends to more complex objects. A slight extension of the idea allows the domain to be the set of nonnegative integers.

Although a sequence is a function, it is common to use a special notation for sequences. As noted above, the terms are often written c_k instead of $c(k)$. The sequence itself is denoted $\{c_k\}$. As an abbreviation, people often write "the sequence c_k," instead of "the sequence $\{c_k\}$," although this can create some confusion between the entire sequence and its k-th term.

The notion of a limit is the most important idea connected with sequences. Say that the sequence of numbers $\{c_k\}$ has the number L as a *limit* if for any $\epsilon > 0$ there is an integer N such that

$$|c_k - L| < \epsilon, \quad \text{whenever} \quad k \geq N.$$

To emphasize the dependence of N on ϵ we may write N_ϵ or $N(\epsilon)$. In mathematical shorthand the existence of a limit is written as

$$\lim_{k \to \infty} c_k = L.$$

An equivalent statement is that the sequence $\{c_k\}$ *converges* to L. This definition has a graphical interpretation which illustrates the utility of the function interpretation of a sequence. The statement that the sequence has the limit L is the same as saying that the graph of the function $c(k)$ has a horizontal asymptote $y = L$, as shown in Figure 2.1 (where $L = 2$).

A substantial amount of both theoretical and applied mathematics is concerned with showing that sequences converge. To pick one elementary example,

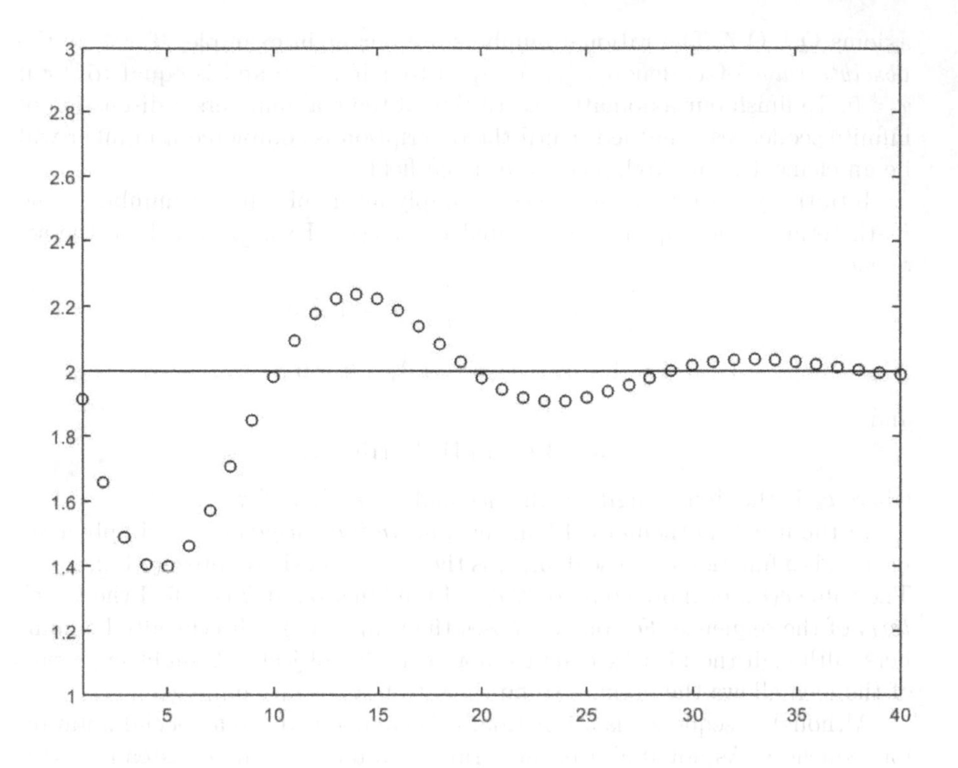

FIGURE 2.1
Graphical representation of a convergent sequence.

suppose that in the course of doing statistical work you need to evaluate an integral of the form

$$I = \int_a^b e^{-x^2}\, dx.$$

As you may recall from calculus, the function e^{-x^2} does not have an elementary antiderivative. To obtain numerical values for the integral one would run to the computer and use a numerical integration scheme. One could use Riemann sums (admittedly a naive and rather inefficient approach), divide the interval $[a, b]$ into k equal subintervals, and estimate the true value of I by a sequence I_k of approximations obtained by summing the areas of rectangles. In this case we may interpret the number ϵ in the definition of limit as describing the error in approximating the integral I by the Riemann sum I_k. The statement that $\lim_{k\to\infty} I_k = I$ simply means that the approximation can be made as accurate as desired if k is chosen sufficiently large. In a practical application some explicit knowledge of the connection between the size of k and the accuracy ϵ of the approximation may also be required.

A few concrete examples will help us to get comfortable with the ideas.

Suppose our goal is to show that

$$\lim_{k\to\infty} 3 + \frac{1}{k} = 3.$$

Pick any $\epsilon > 0$, and ask how large N should be so that all of the numbers $c_k = 3 + 1/k$ will be within ϵ of 3 as long as $k \ge N$. To insure

$$|c_k - L| = |(3 + 1/k) - 3| = 1/k < \epsilon$$

it suffices to take

$$k > \frac{1}{\epsilon}.$$

To make a concrete choice, let N be the smallest integer at least as big as $1/\epsilon + 1$, that is

$$N = \lceil 1/\epsilon + 1 \rceil.$$

As a second example, consider the sequence of numbers

$$s_k = \sum_{j=0}^{k-1} 2^{-j}.$$

This sequence comes from the geometric series and, as previously noted,

$$s_k = \frac{1}{1-1/2} - \frac{2^{-k}}{1-1/2} = 2 - 2^{1-k}.$$

It doesn't strain the imagination to see that the limit is 2. However, we can again ask how large k must be before s_k stays within ϵ of 2. The desired inequality is

$$|s_k - L| = |(2 - 2^{1-k}) - 2| = 2^{1-k} < \epsilon.$$

Assuming for the moment that we know about base two logarithms (see Problem 2.16), this is the same as

$$1 - k < \log_2(\epsilon),$$

or

$$k > 1 - \log_2(\epsilon).$$

In this case N may be chosen to be the smallest integer at least as big as $2 - \log_2(\epsilon)$, that is

$$N = \lceil 2 - \log_2(\epsilon) \rceil.$$

In many cases it is impractical or impossible to find a simple expression for the smallest possible N. Suppose $c_k = 1 + 1/(k^{17} + k^3 + 8k)$, which satisfies $L = \lim_{k\to\infty} c_k = 1$. Rather than struggling with the precise value of c_k, it is convenient to simply note that

$$|c_k - L| = |1/(k^{17} + k^3 + 8k)| < 1/k.$$

As in the first example, it suffices to pick

$$N = \lceil 1/\epsilon + 1 \rceil.$$

For $k \ge N$ it follows that

$$|c_k - 1| < \epsilon.$$

2.1.2 Properties of limits

It is time to turn from the consideration of specific examples to general properties of sequences and limits. The first point to make is that a sequence can have at most one limit. In addition to establishing the result itself, the proof introduces three common techniques. The first is the judicious addition of 0. The second is the use of the triangle inequality

$$|a + b| \le |a| + |b|,$$

which is easily verified for numbers a and b (see Problem 2.1). The third is the observation that if a number r is nonnegative, but $r < \epsilon$ for every positive number ϵ, then $r = 0$.

Theorem 2.1.1. *If*

$$\lim_{k\to\infty} c_k = L_1 \quad \text{and} \quad \lim_{k\to\infty} c_k = L_2,$$

then $L_1 = L_2$.

Proof. Notice that by adding zero and using the triangle inequality,

$$|L_1 - L_2| = |(L_1 - c_k) + (c_k - L_2)| \le |L_1 - c_k| + |c_k - L_2|.$$

Let ϵ be any positive number, and take $\epsilon_1 = \epsilon_2 = \epsilon/2$. Since

$$\lim_{k\to\infty} c_k = L_1 \quad \text{and} \quad \lim_{k\to\infty} c_k = L_2,$$

there are numbers N_1 and N_2 such that $|L_1 - c_k| < \epsilon_1$ for all $k \ge N_1$, and $|L_2 - c_k| < \epsilon_2$ for all $k \ge N_2$.

Now pick k such that $k \ge N_1$ and $k \ge N_2$. Then

$$|L_1 - L_2| \le |L_1 - c_k| + |c_k - L_2| < \epsilon_1 + \epsilon_2 = \epsilon.$$

Since the inequality $|L_1 - L_2| < \epsilon$ holds for any positive number ϵ, it must be the case that $|L_1 - L_2| = 0$, or $L_1 = L_2$. □

A sequence $\{c_k\}$ of numbers is *bounded* if there is a number M such that $|c_k| \le M$ for all positive integers k.

Lemma 2.1.2. *If*

$$\lim_{k\to\infty} c_k = L,$$

then the sequence c_k *is bounded.*

Proof. Choose $\epsilon = 1$. According to the definition, there is an integer N such that

$$|c_k - L| < 1$$

provided that $k \ge N$. This means that for $k \ge N$

$$|c_k| < |L| + 1.$$

If

$$M = \max(|L| + 1, |c_1|, \dots, |c_N|)$$

then $|c_k| \le M$ for all indices k. □

Some sequences which are not bounded still have simple behavior. Say that $\lim_{k\to\infty} c_k = \infty$, or equivalently $c_k \to \infty$, if for any $M > 0$ there is an integer N such that $c_k > M$ for all $k \ge N$. By the previous lemma a sequence satisfying $\lim_{k\to\infty} c_k = \infty$ cannot converge.

The next result describes the interplay between limits and arithmetic operations.

Theorem 2.1.3. *Suppose that*

$$\lim_{k\to\infty} a_k = L, \quad \lim_{k\to\infty} b_k = M,$$

and C is a fixed number. Then

$$\lim_{k\to\infty} Ca_k = CL, \tag{i}$$

$$\lim_{k\to\infty} a_k + b_k = L + M, \tag{ii}$$

$$\lim_{k\to\infty} a_k b_k = LM, \tag{iii}$$

and

$$\lim_{k\to\infty} a_k/b_k = L/M, \quad M \ne 0. \tag{iv}$$

Statement (iv) deserves a comment. It is possible to have $M \ne 0$, but to have many b_k which are 0, in which case some of the terms a_k/b_k are not defined. This problem can be handled by adding the hypothesis that $b_k \ne 0$ for all k. Another option is to notice, as we will in the proof, that if $M \ne 0$ then $b_k \ne 0$ if k is sufficiently large. Thus a_k/b_k is defined for k large enough, which is all that is really required to make sense of limits. The reader is free to choose either point of view.

Before launching into the proof, this may be a good time to recognize that writing a formal proof is usually preceded by some preliminary analysis. Let's start with statement (i). To understand what needs to be done it helps to work backwards. Our goal is to conclude that

$$|Ca_k - CL| < \epsilon$$

for k large enough based on the fact that

$$|a_k - L| < \epsilon_1$$

for k large enough. If $|C| \le 1$ there is no challenge, since we can take $\epsilon_1 = \epsilon$ and get

$$|Ca_k - CL| < |C|\epsilon_1 < \epsilon.$$

If $|C| > 1$, then take $\epsilon_1 = \epsilon/|C|$. As soon as k is large enough that

$$|a_k - L| < \epsilon_1 = \epsilon/|C|,$$

it follows that

$$|Ca_k - CL| < |C|\epsilon_1 = \epsilon.$$

Now let's examine the formal proof.

Proof. (*i*) Take any $\epsilon > 0$. From the definition of

$$\lim_{k\to\infty} a_k = L$$

there is an N_ϵ such that

$$|a_k - L| < \epsilon$$

whenever $k \ge N_\epsilon$. Consider two cases: $|C| \le 1$, and $|C| > 1$.

Suppose first that $|C| \le 1$. Then whenever $k \ge N_\epsilon$ we have the desired inequality

$$|Ca_k - CL| < |C|\epsilon < \epsilon.$$

Next suppose that $|C| > 1$. Let $\epsilon_1 = \epsilon/|C|$. Since

$$\lim_{k\to\infty} a_k = L$$

there is an N_{ϵ_1} such that $k \ge N_{\epsilon_1}$ implies

$$|a_k - L| < \epsilon_1.$$

But this means that

$$|Ca_k - CL| < |C|\epsilon_1 = |C|\epsilon/|C| = \epsilon,$$

concluding the proof of (*i*).

Next consider the proof of (*ii*). If we start by assuming that k is so large that $|a_k - L| < \epsilon$ and $|b_k - M| < \epsilon$, then the triangle inequality gives

$$|(a_k + b_k) - (L + M)| \le |a_k - L| + |b_k - M| < 2\epsilon.$$

On one hand this looks bad, since our goal is to show that $|(a_k + b_k) - (L + M)| < \epsilon$, not 2ϵ. On the other hand the situation looks good since the value of $|(a_k+b_k)-(L+M)|$ can be made as small as we like by making k large enough.

This issue can be resolved by the trick of splitting ϵ in two. (See Problem 2.4 for an alternative.)

Take any $\epsilon > 0$. Now take $\epsilon_1 = \epsilon/3$ and $\epsilon_2 = 2\epsilon/3$. From the limit definitions there are N_1 and N_2 such that if $k \ge N_1$ then

$$|a_k - L| < \epsilon_1,$$

and if $k \ge N_2$ then

$$|b_k - M| < \epsilon_2.$$

Take $N = \max(N_1, N_2)$. If $k \ge N$, then

$$|(a_k + b_k) - (L + M)| \le |a_k - L| + |b_k - M| < \epsilon_1 + \epsilon_2 = \epsilon/3 + 2\epsilon/3 = \epsilon.$$

Of course the choice of $\epsilon_1 = \epsilon/3$ and $\epsilon_2 = 2\epsilon/3$ was largely arbitrary. The choice $\epsilon_1 = \epsilon_2 = \epsilon/2$ was certainly available.

To prove (iii), begin with the algebraic manipulation

$$|a_k b_k - LM| = |a_k[M + (b_k - M)] - LM| \le |(a_k - L)M| + |a_k(b_k - M)|.$$

The plan is to show that $|(a_k - L)M|$ and $|a_k(b_k - M)|$ can be made small.

Take any $\epsilon > 0$, and let $\epsilon_1 = \epsilon/2$. Replacing C by M in part (i), there is an N_1 such that $k \ge N_1$ implies

$$|(a_k - L)M| = |Ma_k - ML| < \epsilon_1.$$

By Lemma 2.1.2 there is a constant C_1 such that $|a_k| \le C_1$. Thus

$$|a_k(b_k - M)| \le |C_1(b_k - M)| = |C_1 b_k - C_1 M|.$$

Again using part (i), there is an N_2 such that $k \ge N_2$ implies

$$|C_1 b_k - C_1 M| < \epsilon_1.$$

Take $N = \max(N_1, N_2)$ so that for $k \ge N$,

$$|a_k b_k - LM| = |a_k[M + (b_k - M)] - LM| \le |(a_k - L)M| + |a_k(b_k - M)|$$

$$< \epsilon_1 + \epsilon_1 = \epsilon.$$

(iv) The proof starts by observing that for k large enough,

$$|b_k| \ge |M|/2.$$

To see this, choose $\epsilon_1 = |M|/2$. Then there is an N_1 such that for $k \ge N_1$,

$$|b_k - M| < |M|/2,$$

or

$$|b_k| \ge |M|/2, \quad k \ge N_1. \tag{2.6}$$

Now an algebraic manipulation is combined with the triangle inequality.

$$|\frac{a_k}{b_k} - \frac{L}{M}| = |\frac{a_k M - b_k L}{b_k M}| = |\frac{M(a_k - L) + L(M - b_k)}{b_k M}|$$

$$\leq |\frac{M(a_k - L)}{b_k M}| + |\frac{L(b_k - M)}{b_k M}|.$$

For $k \geq N_1$ use (2.6) to find

$$|\frac{a_k}{b_k} - \frac{L}{M}| \leq |\frac{M(a_k - L)}{M^2/2}| + |\frac{L(b_k - M)}{M^2/2}| = |\frac{2}{M}(a_k - L)| + |\frac{2L}{M^2}(b_k - M)| \quad (2.7)$$

Now take any $\epsilon > 0$. Use part (i) of this theorem with $C = 2/M$ to conclude that there is an N_2 such that

$$|\frac{2}{M}(a_k - L)| < \epsilon/2$$

if $k \geq N_2$. Similarly, there is an N_3 such that

$$|\frac{2L}{M^2}(b_k - M)| < \epsilon/2$$

if $k \geq N_3$. Finally, if $N = \max(N_1, N_2, N_3)$ and $k \geq N$, then

$$|\frac{a_k}{b_k} - \frac{L}{M}| < \epsilon.$$

□

2.2 Completeness axioms

There are several ways to describe the important property of the real numbers which has been omitted so far. These various descriptions all involve the convergence of sequences. A basic feature distinguishing the real numbers from the rational numbers is that well behaved sequences of rational numbers may fail to converge because the number which should be the limit is not rational. For instance the sequence of truncated decimal expansions for $\sqrt{2}$

$$c_1 = 1, \quad c_2 = 1.4, \quad c_3 = 1.41, \quad c_4 = 1.414, \ldots,$$

is a sequence of rational numbers which wants to converge to $\sqrt{2}$, but, as demonstrated earlier, $\sqrt{2}$ is not in the set of rational numbers. This sequence $\{c_k\}$ has no limit in the rational numbers. This phenomenon does not occur for the real numbers.

The first way to characterize the good behavior of the reals involves

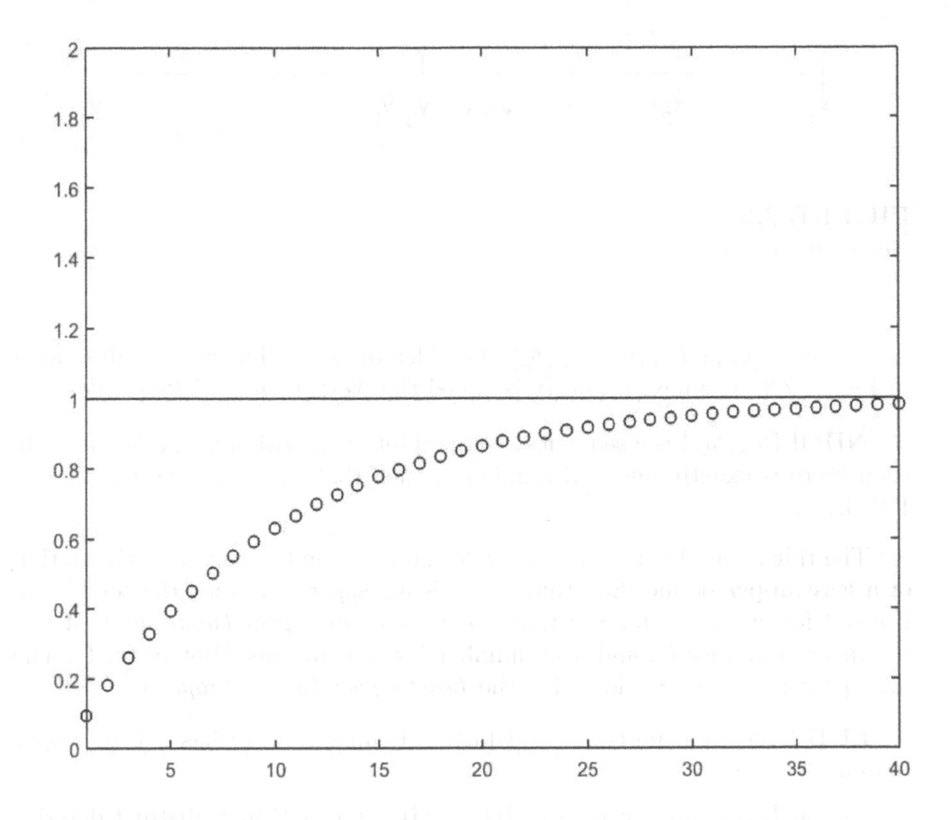

FIGURE 2.2
A bounded monotone sequence.

bounded increasing or decreasing sequences. Such a sequence is illustrated in Figure 2.2. Say that a sequence $c_k \in \mathbb{F}$ is *monotonically increasing* if $c_l \ge c_k$ whenever $l > k$. Similarly c_k is *monotonically decreasing* if $c_l \le c_k$ whenever $l > k$. If c_k is either monotonically increasing or decreasing, the sequence is said to be *monotone.* A set $U \subset \mathbb{F}$ is *bounded* if there is an $M \in \mathbb{F}$ such that $|c| \le M$ for all $c \in U$. A set $U \subset \mathbb{F}$ is *bounded above* if there is an $M \in \mathbb{F}$ such that $c \le M$ for all $c \in U$, while $V \subset \mathbb{F}$ is *bounded below* if there is an $M \in \mathbb{F}$ such that $c \ge M$ for all $c \in V$.

Here is a completeness axiom which gives one description of the good convergence properties of the real numbers. This particular property is called the *Bounded Monotone Sequence Property.*

BMS Every bounded monotone sequence $\{c_k\}$ has a limit L. $(C.1)$

There is a second completeness property of the real numbers which is closely related to C.1. For positive integers k say that the intervals $[a_k, b_k] \subset \mathbb{F}$

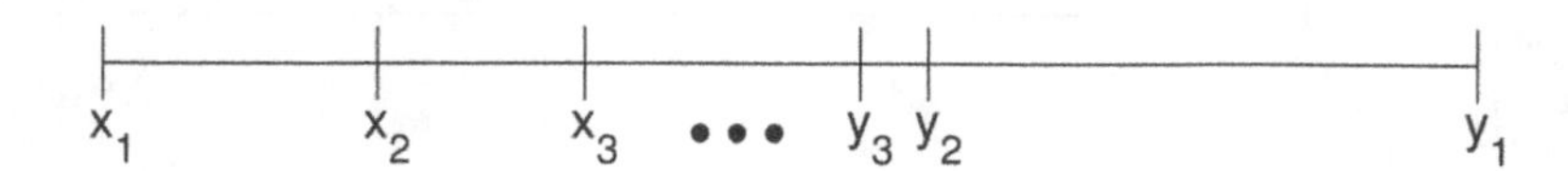

FIGURE 2.3
Nested intervals.

are *nested* if $[a_{k+1}, b_{k+1}] \subset [a_k, b_k]$. The idea of nested intervals is illustrated in Figure 2.3. The next property is called the *Nested Interval Principle.*

NIP If $\{[a_k, b_k]\}$ is a sequence of nested intervals with $\lim_{k\to\infty} b_k - a_k = 0$, then there is exactly one real number L such that $a_k \le L \le b_k$ for all $k = 1, 2, 3, \dots$.

The third completeness property of the real numbers involves the notion of a least upper bound. Say that $b \in \mathbb{F}$ is an *upper bound* for the set $U \subset \mathbb{F}$ if $c \le b$ for every $c \in U$. Say that $z \in \mathbb{F}$ is a *least upper bound* for U if z is an upper bound for U, and if no number $b < z$ is an upper bound for U. The final property to be considered is the *Least Upper Bound Property,*

LUB Every nonempty set U which is bounded above has a least upper bound.

Although the three properties BMS, NIP, and LUB have distinct descriptions, they are in fact equivalent. Axiom C.1 (BMS) will be added to our previous collection of axioms; with this addition our characterization of the real numbers will be finished. That is, the *real numbers* are a set $\mathbb{R}$, together with the functions $+$ and $\cdot$, and the order relation $\le$, which satisfy the field axioms F.1–F.10, the order axioms O.1–O.7, and the completeness axiom C.1. Any two number systems which satisfy the axioms for the real numbers only differ by what amounts to a renaming of the elements, arithmetic functions, and order relation (the proof is omitted). This characterization does not say what the real numbers are; instead, their behavior is described through the axioms, as was our plan.

Before addressing the equivalence of the completeness properties, let us show that a field $\mathbb{F}$ satisfying the Nested Interval Principle contains $\sqrt{2}$. This will demonstrate that a completeness property distinguishes the real numbers from the rationals. The method, called bisection, is a popular computer algorithm.

Theorem 2.2.1. *Suppose the Archimedean ordered field $\mathbb{F}$ satisfies the Nested Interval Principle. Then there is a number $L > 0$ such that $L^2 = 2$.*

Proof. Begin with the interval $[a_0, b_0] = [1, 2]$. Notice that $a_0^2 \le 2$ while $b_0^2 \ge 2$.

Given an interval $[a_k, b_k]$ with $a_k^2 \le 2$ and $b_k^2 \ge 2$, the next interval

$[a_{k+1}, b_{k+1}]$ is constructed as follows. First define $c_k = (a_k + b_k)/2$; notice that $a_k \le c_k \le b_k$ and

$$b_k - c_k = (b_k - a_k)/2 = c_k - a_k.$$

If $c_k^2 > 2$, take $[a_{k+1}, b_{k+1}] = [a_k, c_k]$; otherwise, let $[a_{k+1}, b_{k+1}] = [c_k, b_k]$.

By construction the intervals $[a_k, b_k]$ are nested, and $|b_k - a_k| = 2^{-k}$. The Nested Interval Principle says there is exactly one real number L such that $a_k \le L \le b_k$. Also

$$0 \le b_k^2 - a_k^2 = (b_k + a_k)(b_k - a_k) \le \frac{4}{2^k}.$$

Since x^2 is strictly increasing if $x \in [1, 2]$, and $a_k^2 \le 2$ while $b_k^2 \ge 2$,

$$a_k^2 - b_k^2 \le a_k^2 - 2 \le L^2 - 2 \le b_k^2 - 2 \le b_k^2 - a_k^2.$$

Thus

$$|L^2 - 2| \le b_k^2 - a_k^2 \le \frac{4}{2^k}.$$

Since k is arbitrary, Lemma 1.1.15 gives $L^2 = 2$. □

With a bit more work this result can be generalized. The proof of the resulting intermediate value theorem for polynomials is left to the reader.

Theorem 2.2.2. *Suppose $\mathbb{F}$ satisfies the field and order axioms, and the Nested Interval Principle. Let*

$$p(x) = a_n x^n + \cdots + a_1 x + a_0, \quad a_k \in \mathbb{F},$$

be a polynomial. If $r \in \mathbb{F}$ and there are real numbers x_0 and y_0 such that

$$p(x_0) \le r, \quad p(y_0) \ge r,$$

then there is a number $x \in \mathbb{F}$ satisfying $x_0 \le x \le y_0$ such that $p(x) = r$.

The equivalence of the completeness axiom $C.1$ to the NIP and LUB properties will now be established through a sequence of propositions. The following observation will be useful.

Lemma 2.2.3. *Suppose that $c_k \ge M$ for $k = 1, 2, 3, \dots,$ and that $\lim_{k\to\infty} c_k = L$. Then $L \ge M$.*

Proof. The argument is by contradiction. Suppose that $L < M$, and take $\epsilon = M - L$. The fact that $\lim_{k\to\infty} c_k = L$ means there is an N such that

$$|c_k - L| < \epsilon, \quad k \ge N.$$

Since $c_N \ge M$ and $L < M$, it follows that $c_k - L > 0$ and

$$c_N - L < \epsilon = M - L.$$

This gives $c_N < M$, contradicting the hypotheses. □

Proposition 2.2.4. *The Bounded Monotone Sequence Property implies the Nested Interval Principle.*

Proof. Suppose there is a nested sequence of intervals $[a_k, b_k]$ for $k = 1, 2, 3, \dots$, with $a_k \le a_{k+1} < b_{k+1} \le b_k$ and $\lim_{k\to\infty} b_k - a_k = 0$. The sequence $\{a_k\}$ is increasing, and $a_k \le b_1$ for all k. Since the sequence $\{a_k\}$ is increasing and bounded above, the Bounded Monotone Sequence Property implies that there is a number L_1 such that

$$\lim_{k\to\infty} a_k = L_1.$$

Similarly, the sequence $\{b_k\}$ is decreasing and bounded below, so there is a number L_2 such that

$$\lim_{k\to\infty} b_k = L_2.$$

Notice that

$$L_2 - L_1 = \lim_{k\to\infty} b_k - \lim_{k\to\infty} a_k = \lim_{k\to\infty} b_k - a_k = 0,$$

or $L_2 = L_1$. Let $L = L_1 = L_2$.

For any fixed index j we have $a_k \le b_j$, so by the previous lemma $a_j \le L \le b_j$. Suppose there were a second number $M \in [a_k, b_k]$ for each $k = 1, 2, 3, \dots$. Then $|L - M| \le |b_k - a_k|$. Since $\lim_{k\to\infty} b_k - a_k = 0$, it follows that $|L - M| < \epsilon$ for every $\epsilon > 0$, so $L - M = 0$.

□

Proposition 2.2.5. *The Nested Interval Principle implies the Least Upper Bound Property.*

Proof. Suppose that $U \subset \mathbb{R}$ is a nonempty set which is bounded above by z. Pick a number a_1 which is not an upper bound for U, and a number b_1 which is an upper bound for U. Let $c_1 = (a_1 + b_1)/2$. If c_1 is an upper bound for U define $b_2 = c_1$ and $a_2 = a_1$, otherwise define $a_2 = c_1$ and $b_2 = b_1$. Continuing in this fashion we obtain sequences $\{a_k\}$ and $\{b_k\}$ satisfying $a_k \le a_{k+1} < b_k \le b_{k+1}$ with $b_k - a_k = (b_1 - a_1)/2^{k-1}$. Moreover each point b_k is an upper bound for U, and each point a_k is not an upper bound for U.

By the Nested Interval Principle there is a number L satisfying $a_k \le L \le b_k$ for all $k = 1, 2, 3, \dots$; this implies

$$|L - a_k| \le b_k - a_k, \quad |b_k - L| \le b_k - a_k.$$

If L were not an upper bound for U, then there would be $x \in U$, $x > L$. Write $x = L + (x - L)$. Since $x - L > 0$, the number $L + (x - L) > b_k$ for k sufficiently large. This means $x > b_k$, which is impossible since b_k is an upper bound for U. Thus L is an upper bound for U.

Similarly, if L were not the least upper bound for U there would be some $y < L$ which is an upper bound. Since $y = L - (L - y)$, the number $L - (L - y) < a_k$, or $y < a_k$, for k sufficiently large. This contradiction means that L is the least upper bound for U. □

Another lemma will help complete the chain of logical equivalences for the completeness properties.

Lemma 2.2.6. *Suppose $U \subset \mathbb{R}$ is nonempty and bounded above by L. Then L is the least upper bound for U if and only if for every $\epsilon > 0$ there is an $x \in U$ such that $0 \le L - x < \epsilon$.*

Proof. First suppose that L is the least upper bound. Then for any $\epsilon > 0$ the number $L - \epsilon$ is not an upper bound for U, and so there is an $x \in U$ with $L - \epsilon < x \le L$.

Suppose now that L is an upper bound for U and that for every $\epsilon > 0$ there is an $x \in U$ such that $0 \le L - x < \epsilon$. Suppose $M < L$ is another upper bound for U. Take $\epsilon = L - M$. By assumption there is an $x \in U$ such that

$$0 \le L - x < L - M,$$

or $x > M$. This contradicts the assumption that M is an upper bound for U, so that L is the least upper bound. □

Proposition 2.2.7. *The Least Upper Bound Property implies the Bounded Monotone Sequence Property.*

Proof. Suppose that $\{c_k\}$ is an increasing sequence bounded above by M. The set of numbers in the sequence has a least upper bound L. Since L is the least upper bound, for every $\epsilon > 0$ there is a c_N such that $L - c_N < \epsilon$. Since the sequence is increasing we have $c_k \le c_{k+1} \le L$, so $L - c_k \le L - c_N < \epsilon$ for all $k \ge N$. Thus the sequence $\{c_k\}$ converges to L. □

2.3 Subsequences and compact intervals

A simple example of a sequence that does not converge has the terms

$$c_k = (-1)^k.$$

Notice that while there is no limit for the entire sequence, there are parts of the sequence that do have limits. Looking at the terms with even and odd indices respectively,

$$y_k = c_{2k} = 1, \quad z_k = c_{2k+1} = -1,$$

it is easy to see that the sequences $\{y_k\}$ and $\{z_k\}$ converge. Motivated by this example, one can ask whether this behavior is typical; given a sequence $\{c_k\}$, is there a sequence y_k consisting of some portion of the terms c_k such that $\{y_k\}$ converges?

This question is addressed by considering subsequences; roughly speaking, a subsequence will be an infinite sublist from the sequence $\{c_k\}$. More precisely, if $\{c_k\}$ is a sequence, say that $\{y_j\}$ is a *subsequence* of $\{c_k\}$ if there is a function $k(j) : \mathbb{N} \to \mathbb{N}$ which is strictly increasing, and such that $y_j = c_{k(j)}$. By strictly increasing we mean that $k(j+1) > k(j)$ for all $j = 1, 2, 3, \dots$.

Some examples and comments may help clarify the definition of a subsequence. Suppose the sequence $\{c_k\}$ has terms $c_k = 1/k$, or 1,1/2,1/3 , ... The elements of a subsequence appear in the same order as the corresponding elements of the original sequence. Thus the sequence $1/4, 1/2, 1/3, 1/8, 1/7, \dots$ is not a subsequence of $\{c_k\}$, although a subsequence is obtained if $y_j = c_{3j}$, that is $y_1 = 1/3, y_2 = 1/6, \dots$. Also notice that the sequence with terms $y_j = 1$, or $1, 1, 1, \dots$ is not a subsequence of $\{c_k\}$ since the requirement $k(j+1) > k(j)$ forces $k(2) > 1$, and so $y_2 = c_{k(2)} < 1$.

The sequence with terms $c_k = (-1)^k$ is not convergent, but it does have convergent subsequences. The behavior of the sequence with terms $c_k = k$ is different; this sequence has no convergent subsequences. In general a sequence can have subsequences with many different limits. The behavior of subsequences is simple if the sequence $\{c_k\}$ has a limit.

Lemma 2.3.1. *Suppose $\{c_k\}$ is a sequence with a limit L. Then every subsequence $\{y_j\} = \{c_{k(j)}\}$ of $\{c_k\}$ converges to L.*

Proof. Pick any $\epsilon > 0$. By assumption there is an index N such that $|c_k - L| < \epsilon$ whenever $k \ge N$. Suppose that $j \ge N$. Since $k(j)$ is strictly increasing, $k(j) \ge j \ge N$ and

$$|y_j - L| = |c_{k(j)} - L| < \epsilon.$$

□

It is an important fact that if a sequence of real numbers is bounded, then the sequence has a convergent subsequence.

Theorem 2.3.2. *Suppose $\{c_k\}$ is a bounded sequence of real numbers. Then there is a subsequence $\{c_{k(j)}\}$ of $\{c_k\}$ which converges.*

Proof. Suppose the elements of the sequence satisfy $-M \le c_k \le M$ for some $M > 0$. For convenience we may assume that M is an integer. The plan is to construct a sequence of intervals satisfying the Nested Interval Principle, and arrange that the limit guaranteed by this result is also the limit of a subsequence of $\{c_k\}$.

Break up the interval $[-M, M]$ into subintervals $[n, n+1]$ of length 1 with integer endpoints. Since the set of postive integers is infinite, and the number of intervals $[n, n+1]$ contained in $[-M, M]$ is finite, there must be at least one such interval I_0 containing c_k for an infinite collection of indices k. Let $k(0)$ be the smallest of the indices k for which $c_k \in I_0$.

In a similar way, partition I_0 into 10 nonoverlapping subintervals of length 10^{-1}. At least one of these intervals $I_1 \subset I_0$ contains c_k for an infinite collection of indices k. Let $k(1)$ be the smallest of the indices k for which $c_k \in I_1$ and $k(0) < k(1)$.

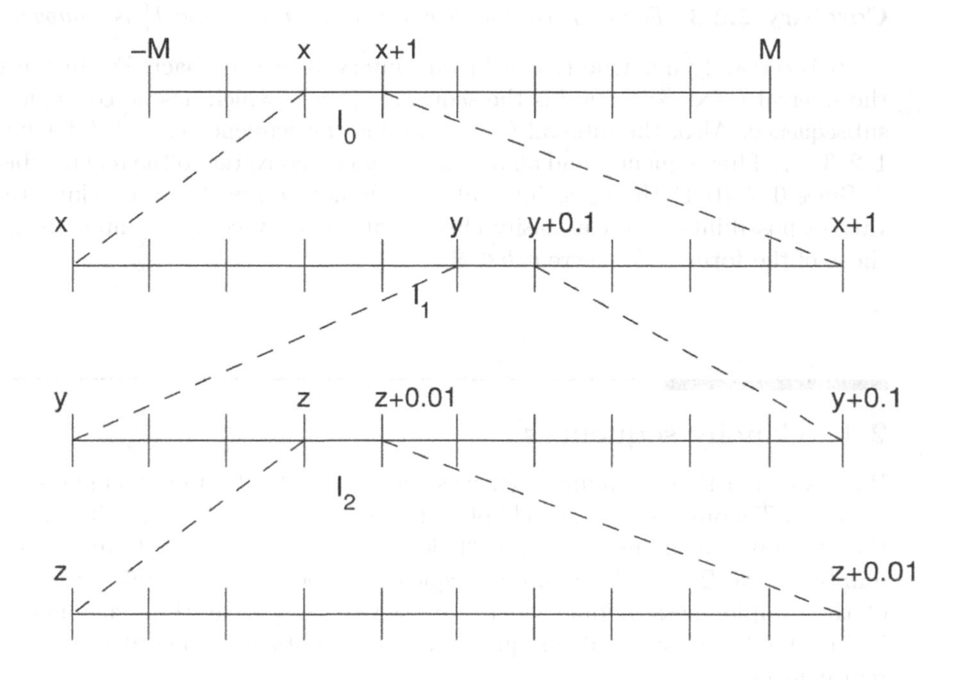

FIGURE 2.4
Constructing a convergent subsequence.

Continue in this fashion (see Figure 2.4) for every positive integer m, partitioning I_{m-1} into 10 nonoverlapping subintervals of length 10^{-m}. At least one of these intervals $I_m \subset I_{m-1}$ contains c_k for an infinite collection of indices k. Let $k(m)$ be the smallest of the indices k for which $c_k \in I_m$ and $k(m-1) < k(m)$.

The intervals I_m are nested by construction, and the length of I_m is 10^{-m}. By the Nested Interval Principle there is a (unique) point z which is in the intersection of all the intervals I_m. Since $c_{k(j)} \in I_m$ if $j \ge m$, it follows that $|z - c_{k(j)}| \le 10^{-m}$ for all $j \ge m$. Thus the subsequence $c_{k(j)}$ converges to z. □

This proof actually shows that if $\{c_k\}$ is any sequence from the set $[-M, M]$, then there is a subsequence $\{c_{k(j)}\}$ which converges to a point $z \in [-M, M]$, since $I_m \subset [-M, M]$. With no essential change the argument shows that this observation may be extended to any closed interval $[a, b]$. Say that $K \subset \mathbb{R}$ is *compact* if every sequence with terms $c_k \in K$ has a subsequence which converges to a point of K. The next result is a consequence of the previous theorem.

Corollary 2.3.3. *For any real numbers $a \le b$ the interval $[a, b]$ is compact.*

It is certainly not true that arbitrary intervals are compact. For instance the interval $(-\infty, \infty)$ contains the sequence $c_k = k$, which has no convergent subsequence. Also, the interval $(0, 1)$ contains the sequence $c_k = 1/k$ for $k = 1, 2, 3, \dots$. This sequence, and all its subsequences, converge to the real number 0. Since $0 \notin (0, 1)$ the open interval $(0, 1)$ is not compact. By checking the various possibilities we can easily check that the only compact intervals are those of the form $[a, b]$ where $a, b \in \mathbb{R}$.

2.4 Cauchy sequences

We saw earlier that the number $\sqrt{2}$ was not in the field of rational numbers. As shown by Theorem 2.2.1, the field of real numbers does include $\sqrt{2}$. It appears that the rational numbers have a "hole" at $\sqrt{2}$ which is "filled" in the real numbers. The Bounded Monotone Sequence Property and the other versions of the completeness axiom, which are used to show that the real numbers include the limits of suitable sequences, take advantage of the ordering of the real numbers.

It is possible to address the completeness of the real numbers without referring to ordering. This is done by introducing Cauchy sequences, which are sequences that behave as if they should have limits. (Augustin-Louis Cauchy (1789–1857) was a prominent mathematician of the nineteenth century.) A sequence $\{c_k\}$ of numbers is a *Cauchy sequence* if for any $\epsilon > 0$ there is an index N such that $|c_j - c_k| < \epsilon$ for all $j, k \ge N$. Cauchy sequences are quite important when completeness is studied in more general settings.

Proposition 2.4.1. *If the sequence $\{c_k\}$ has a limit L, then $\{c_k\}$ is a Cauchy sequence.*

Proof. For any $\epsilon > 0$ there is an index N such that $|c_k - L| < \epsilon/2$ whenever $k \ge N$. If in addition $j \ge N$, then by the triangle inequality

$$|c_j - c_k| = |(c_j - L) + (L - c_k)| \le |c_j - L| + |L - c_k| < \epsilon.$$

□

Suppose $\{q_k\}$ is a sequence of rational numbers with limit $\sqrt{2}$. Since the sequence has a limit in the real numbers, the sequence of rational numbers is a Cauchy sequence. In the rational numbers there are thus Cauchy sequences without rational limits. The next theorem shows that the real numbers do not have this defect. The following lemma is helpful.

Lemma 2.4.2. *If $\{c_k\}$ is a Cauchy sequence of real numbers, then $\{c_k\}$ is bounded.*

Proof. If $\epsilon = 1$, there is an index N such that $|c_j - c_k| < 1$ for all $j, k \geq N$. In particular $|c_k - c_N| < 1$, so by the triangle inequality

$$|c_k| = |(c_k - c_N) + c_N| \leq 1 + |c_N|, \quad k \geq N.$$

Since there are only finitely many indices k with $k < N$, the sequence $\{c_k\}$ is bounded. □

Theorem 2.4.3. *If $\{c_k\}$ is a Cauchy sequence of real numbers, then there is a real number L such that*

$$\lim_{k \to \infty} c_k = L.$$

Proof. Since $\{c_k\}$ is bounded, it has a subsequence $\{c_{k(j)}\}$ with a limit $L \in \mathbb{R}$ by Theorem 2.3.2. In fact the whole Cauchy sequence $\{c_k\}$ will have limit L. To see this, pick $\epsilon > 0$ and let N_1 be an index such that $|c_j - c_k| < \epsilon/2$ for all $j, k \geq N_1$. There is also an index N_2 such that $|c_{k(j)} - L| < \epsilon/2$ for all $j \geq N_2$. Take $N \geq \max(N_1, N_2)$. Using the triangle inequality for $k \geq N$,

$$|c_k - L| = |(c_k - c_{k(N)}) + (c_{k(N)} - L)| \leq |c_k - c_{k(N)}| + |c_{k(N)} - L|.$$

Recall that the subsequence index $k(N)$ is strictly increasing, so $k(N) \geq N \geq N_1$. This implies $|c_k - c_{k(N)}| < \epsilon/2$. Similarly, $N \geq N_2$, so $|c_{k(N)} - L| < \epsilon/2$, finishing the proof. □

2.5 Continued fractions

One interesting way in which sequences can be generated is by iteration of a function. A function $f(x)$ and an initial value x_0 are given. For $n \geq 0$ the sequence is then defined by

$$x_{n+1} = f(x_n).$$

As an example, consider the sequence defined by

$$x_{n+1} = 2 + \frac{1}{x_n}, \quad x_0 = 2.$$

The first few terms in this sequence are

$$x_0 = 2, \quad x_1 = 2 + \frac{1}{2}, \quad x_2 = 2 + \frac{1}{2 + \frac{1}{2}}, \quad x_3 = 2 + \frac{1}{2 + \frac{1}{2 + \frac{1}{2}}}, \quad \ldots.$$

Of course these fractions x_n are rational numbers, so could be written as a quotient of integers. A more interesting approach explores the limiting process suggested by the representation of numbers as such continued fractions.

Continued fractions are more commonly encountered in number theory, where they play a role in the study of the approximation of real numbers by rational numbers.

Suppose first that $a_1,\dots,a_N$ is a finite sequence of positive real numbers. A *finite continued fraction* is the expression

$$a_0 + \frac{1}{a_1 + \frac{1}{a_2 + \cdots + \frac{1}{a_N}}}. \tag{2.8}$$

The first term a_0 is not required to be positive. Since the expression (2.8) is so awkward, the continued fraction is usually denoted $[a_0, a_1, \dots, a_N]$.

If $a_1, a_2, a_3, \dots$ is an infinite sequence of positive numbers, it is possible to consider the *infinite continued fraction*

$$[a_0, a_1, a_2, \dots] = a_0 + \frac{1}{a_1 + \frac{1}{a_2 + \dots}}. \tag{2.9}$$

For $n = 0, 1, 2, \dots$ let x_n denote the real number represented by the finite continued fraction $[a_0, a_1, \dots, a_n]$. The infinite continued fraction $[a_0, a_1, a_2, \dots]$ is said to be convergent if the sequence of numbers $\{x_n\}$ has a limit.

For $n \le N$ the continued fractions $[a_0, \dots, a_n]$ are said to be *convergents* of $[a_0, \dots, a_n, \dots, a_N]$; the terminology is the same in the case of an infinite continued fraction. When the numbers $a_0, a_1, \dots$ are further restricted to be integers, the continued fractions are called *simple*. Simple continued fractions provide an alternative to the decimal representation for real numbers. They are particularly important for studying approximations of real numbers by rational numbers.

Our first goal is to try to understand a finite continued fraction when it is expressed as a simple ratio. Let

$$\frac{p_n}{q_n} = [a_0, \dots, a_n].$$

Evaluating the first few cases gives

$$\frac{p_0}{q_0} = \frac{a_0}{1}, \quad \frac{p_1}{q_1} = \frac{a_0 a_1 + 1}{a_1},$$

$$\frac{p_2}{q_2} = \frac{a_0(a_1 a_2 + 1) + a_2}{a_1 a_2 + 1} = \frac{(a_0 a_1 + 1)a_2 + a_0}{a_1 a_2 + 1}$$

$$\frac{p_3}{q_3} = \frac{(a_0 a_1 a_2 + a_2 + a_0)a_3 + a_0 a_1 + 1}{(a_1 a_2 + 1)a_3 + a_1} = \frac{[(a_0 a_1 + 1)a_2 + a_0)]a_3 + [a_0 a_1 + 1]}{(a_1 a_2 + 1)a_3 + a_1}.$$

There is a recursive pattern which holds in general.

Theorem 2.5.1. *Suppose $a_k \in \mathbb{R}$ and $a_k > 0$ for $k \ge 1$. If*

$$p_0 = a_0, \quad q_0 = 1, \quad p_1 = a_0 a_1 + 1, \quad q_1 = a_1,$$

and

$$p_n = a_n p_{n-1} + p_{n-2}, \quad q_n = a_n q_{n-1} + q_{n-2}, \quad n \ge 2, \tag{2.10}$$

then $p_n/q_n = [a_0, \dots, a_n]$.

Proof. The proof is by induction, with cases $n = 0, \dots, 3$ already established. We will make use of the observation that

$$[a_0, \dots, a_n] = [a_0, \dots, a_{n-2}, a_{n-1} + 1/a_n],$$

so here it is important that the a_k not be restricted to integer values.

Assuming the identity holds for all partial fractions with $m \le n$ terms $a_0, \dots, a_{m-1}$, it follows that

$$[a_0, \dots, a_n] = [a_0, \dots, a_{n-2}, a_{n-1} + 1/a_n] = \frac{(a_{n-1} + 1/a_n)p_{n-2} + p_{n-3}}{(a_{n-1} + 1/a_n)q_{n-2} + q_{n-3}}$$

$$= \frac{(a_n a_{n-1} + 1)p_{n-2} + a_n p_{n-3}}{(a_n a_{n-1} + 1)q_{n-2} + a_n q_{n-3}} = \frac{a_n p_{n-1} - a_n p_{n-3} + p_{n-2} + a_n p_{n-3}}{a_n q_{n-1} - a_n q_{n-3} + q_{n-2} + a_n q_{n-3}}$$

$$= \frac{a_n p_{n-1} + p_{n-2}}{a_n q_{n-1} + q_{n-2}} = \frac{p_n}{q_n}.$$

□

The relations (2.10) give

$$p_n q_{n-1} - q_n p_{n-1} = (a_n p_{n-1} + p_{n-2})q_{n-1} - p_{n-1}(a_n q_{n-1} + q_{n-2})$$

$$= -[p_{n-1}q_{n-2} - q_{n-1}p_{n-2}], \quad n \ge 2.$$

Repeated use of this identity to reduce the index leads to

$$p_n q_{n-1} - q_n p_{n-1} = (-1)^{n-1}[p_1 q_0 - q_1 p_0]$$

$$= (-1)^{n-1}[(a_0 a_1 + 1) - (a_0 a_1)] = (-1)^{n-1}.$$

This gives the next result, which expresses the difference between two consecutive convergents.

Theorem 2.5.2.

$$\frac{p_n}{q_n} - \frac{p_{n-1}}{q_{n-1}} = (-1)^{n-1}\frac{1}{q_{n-1}q_n}.$$

Theorem 2.5.3. *Suppose a_k is a positive integer for $k = 1, 2, 3, \dots$. Then for $n \ge 0$, the integers p_n and q_n have no common integer factors $m \ge 2$.*

Proof. For $n = 0$, an appeal to the definitions of p_0 and q_0 is sufficient. For $n \ge 1$,

$$p_n q_{n-1} - q_n p_{n-1} = (-1)^{n-1}. \tag{2.11}$$

If an integer $m \ge 2$ divides p_n and q_n, then m divides $(-1)^{n-1}$, which is impossible. □

The Nested Interval Principle will now be used to analyze the convergence of infinite continued fractions.

Theorem 2.5.4. *Suppose* $a_k \in \mathbb{R}$ *and* $a_k > 0$ *for* $k = 1, 2, 3 \dots.$ *Then*

$$\frac{p_{n+2}}{q_{n+2}} > \frac{p_n}{q_n}, \quad n \text{ even},$$

$$\frac{p_{n+2}}{q_{n+2}} < \frac{p_n}{q_n}, \quad n \text{ odd}.$$

In addition the odd convergents are greater than the even convergents.

Proof. From Theorem 2.5.2,

$$\frac{p_{n+2}}{q_{n+2}} - \frac{p_n}{q_n} = \Big(\frac{p_{n+2}}{q_{n+2}} - \frac{p_{n+1}}{q_{n+1}}\Big) + \Big(\frac{p_{n+1}}{q_{n+1}} - \frac{p_n}{q_n}\Big)$$

$$= (-1)^{n+1}\frac{1}{q_{n+1}q_{n+2}} + (-1)^n\frac{1}{q_nq_{n+1}}.$$

By (2.10) the q_n are positive and increasing, so

$$\frac{1}{q_{n+1}q_{n+2}} < \frac{1}{q_nq_{n+1}},$$

and the first part of the result follows by checking the signs.

To show that the odd convergents are greater than the even convergents, first use Theorem 2.5.2 to show that

$$\frac{p_n}{q_n} > \frac{p_{n-1}}{q_{n-1}}, \quad n \text{ odd},$$

$$\frac{p_n}{q_n} < \frac{p_{n-1}}{q_{n-1}}, \quad n \text{ even}.$$

Then it suffices to note that the magnitudes of the differences between successive convergents

$$|\frac{p_n}{q_n} - \frac{p_{n-1}}{q_{n-1}}| = \frac{1}{q_{n-1}q_n}.$$

are strictly decreasing. □

Theorem 2.5.5. *Every simple continued fraction is convergent.*

Proof. In this case $a_k \ge 1$, so (2.10) gives $q_n \ge n$. The Nested Interval Principle may be immediately applied to the intervals

$$[\frac{p_n}{q_n}, \frac{p_{n+1}}{q_{n+1}}], \quad n \text{ even}.$$

□

Consider the representation of real numbers by continued fractions. Suppose $x \in \mathbb{R}$, and $a_0 = \lfloor x \rfloor$ is the greatest integer less than or equal to x. Let e_0 be the difference between x and a_0, or

$$x = a_0 + e_0, \quad 0 \le e_0 < 1.$$

If $e_0 \ne 0$ let

$$a_1' = \frac{1}{e_0}, \quad a_1 = \lfloor a_1' \rfloor.$$

Since $0 < e_0 < 1$, it follows that the integer a_1 satisfies $a_1 \ge 1$. The process may be continued if a_1' is not an integer, with

$$a_1' = a_1 + e_1, \quad a_2' = \frac{1}{e_1}, \quad a_2 = \lfloor a_1' \rfloor,$$

and generally if $e_n \ne 0$,

$$a_n' = a_n + e_n, \quad a_n' = \frac{1}{e_{n-1}}, \quad a_n = \lfloor a_n' \rfloor.$$

If a term $e_n = 0$ is encountered, the algorithm simply terminates with the sequence $a_0, \dots, a_n$. Of course if the algorithm terminates, then x is rational.

Notice that

$$x = [a_0, a_1'] = [a_0, a_1 + \frac{1}{a_2'}] = [a_0, a_1, a_2'] = [a_0, a_1, a_2 + \frac{1}{a_3'}] = \dots.$$

If x is irrational the algorithm cannot terminate, and so an infinite (convergent) simple continued fraction is obtained. Now apply Theorem 2.5.1 with

$$x = [a_0, a_1, \dots, a_{n-1}, a_n']$$

to get

$$x = \frac{p_{n+1}'}{q_{n+1}'} = \frac{a_{n+1}' p_n + p_{n-1}}{a_{n+1}' q_n + q_{n-1}}.$$

This leads to

$$x - \frac{p_n}{q_n} = \frac{q_n[a_{n+1}' p_n + p_{n-1}] - p_n[a_{n+1}' q_n + q_{n-1}]}{q_n[a_{n+1}' q_n + q_{n-1}]} = \frac{q_n p_{n-1} - p_n q_{n-1}}{q_n[a_{n+1}' q_n + q_{n-1}]}.$$

The identity (2.11), the inequality $a_{n+1}' > a_{n+1} \ge 1$, and the definition of q_n combine to give

$$|x - \frac{p_n}{q_n}| \le \frac{1}{q_n[a_{n+1} q_n + q_{n-1}]} = \frac{1}{q_n q_{n+1}} \le \frac{1}{q_n^2} \le \frac{1}{n^2}.$$

This estimate and the associated algorithm prove the next result.

Theorem 2.5.6. *Every real number has a simple continued fraction representation.*

If x is irrational, the inequality

$$|x - \frac{p_n}{q_n}| \le \frac{1}{q_n^2}$$

holds for infinitely many distinct rationals p_n/q_n in lowest terms. This result is the first step in quantifying the approximation of real numbers by rational numbers; further developments are in [4].

2.6 Problems

2.1. *a) Show that in an ordered field, the triangle inequality*

$$|a+b| \le |a| + |b|,$$

holds.

b) Show that

$$|a-b| \ge |a| - |b|.$$

2.2. *Suppose* $\{a_k\}$ *is a sequence with limit* L. *If* $b_k = a_{k+1}$, *show that* $\lim_{k\to\infty} b_k = L$.

2.3. *Suppose* $\{a_k\}$ *is a sequence with limit* L. *If there is an integer* M *with* $b_k = a_k$ *for* $k \ge M$, *show that* $\lim_{k\to\infty} b_k = L$.

2.4. *Let* $C > 0$ *be a fixed number. Suppose that for any* $\epsilon > 0$ *there is an* N *such that*

$$|a_k - L| < C\epsilon$$

whenever $k \ge N$. *Show that*

$$\lim_{k\to\infty} a_k = L.$$

2.5. *a) Find an example of a bounded sequence without a limit.*
b) Find an example of a monotone sequence without a limit.

2.6. *Find an example of a pair of sequences* $\{x_k\}$ *and* $\{y_k\}$ *such that the intervals* $[x_k, y_k]$ *are nested, but there are two distinct numbers* L_1 *and* L_2 *satisfying* $x_k \le L_1 < L_2 \le y_k$.

2.7. *Find the least upper bound for the following sets:*
a) $S_1 = \{x \in \mathbb{R} \mid -2 < x < 1\}$,
b) $S_2 = \{x \in \mathbb{R} \mid |x-3| \le 5\}$,
c) S_3 *is the set of rational numbers less than* π.
d) $S_4 = \{1 - 1/k,\ k = 2, 3, 4, \dots\}$.

2.8. *Suppose that* $x_{k+1} \ge x_k$, $y_{k+1} \le y_k$, *and that* $x_k \le y_k$ *for each positive integer* k. *Show that if* j, k *are any two positive integers, then* $x_j \le y_k$.

2.9. *Suppose that* $\lim_{n\to\infty} b_n = b$. *Define*

$$a_n = \frac{1}{n} \sum_{k=1}^{n} b_k.$$

Show that $\lim_{n\to\infty} a_n = b$.

2.10. *Suppose that* $\lim_{n\to\infty} x_n = L$ *and* $\lim_{n\to\infty} y_n = L$. *Define the sequence* $\{z_n\}$ *by interleaving these sequences,*

$$z_{2n-1} = x_n, \quad z_{2n} = y_n, \quad n = 1, 2, 3, \ldots .$$

Show that $\lim_{n\to\infty} z_n = L$.

2.11. *Suppose that* $S \subset \mathbb{R}$ *is a set with least upper bound* L. *Show that if* $L \notin S$ *then there is a strictly increasing infinite sequence* $x_k \in S$ *that converges to* L. *Show by example that this conclusion may be false if* $L \in S$.

2.12. *Suppose that* A *and* B *are subsets of* $\mathbb{R}$ *with least upper bounds* L *and* M *respectively. Prove or give a counterexample:*
a) The least upper bound of $A \cup B$ *is the maximum of* L *and* M.
b) The least upper bound of $A \cap B$ *is the minimum of* L *and* M.

2.13. *Suppose you do not know the Nested Interval Principle. Show directly that the Bounded Monotone Sequence property implies the Least Upper Bound property.*

2.14. *a) Suppose that* x, y *are in a field. Prove that for* $n = 1, 2, 3, \ldots,$

$$x^n - y^n = (x - y)\sum_{k=0}^{n-1} x^k y^{n-1-k}.$$

b) Prove that if $\mathbb{F}$ *satisfies the field and order axioms, and the Nested Interval Principle, then for every* $x \geq 0$ *and every positive integer* n *there is a* $z \in \mathbb{F}$ *with* $z^n = x$.

2.15. *Give an example of an unbounded sequence which has a convergent subsequence.*

2.16. *Without using logarithms, show that in an ordered field,*

$$\lim_{n\to\infty} 1/2^n = 0.$$

2.17. *Without using logarithms, show that in an Archimedean ordered field,* $\lim_{n\to\infty} x^n = 0$ *if* $0 < x < 1$.

2.18. *(a) Suppose that* $z_1, \ldots, z_K$ *is a finite collection of real numbers. Construct a sequence* $\{x_n\}$ *such that each* z_k *is the limit of a subsequence of the sequence* $\{x_n\}$.
(b) Construct a sequence $\{x_n\}$ *such that each real number* $z \in [0, 1]$ *is the limit of a subsequence of the sequence* $\{x_n\}$.

2.19. *Show that any sequence* $x_k \in \mathbb{R}$ *has a monotone subsequence. (Hint: Handle the cases when* $\{x_k\}$ *is bounded and* $\{x_k\}$ *is unbounded separately. If* $\{x_k\}$ *is bounded it has a subsequence* $\{x_{k(n)}\}$ *converging to* x_0. *Either there are infinitely many* n *with* $x_{k(n)} \leq x_0$ *or infinitely many* n *with* $x_{k(n)} \geq x_0$.)

2.20. *Show that a compact set is bounded.*

The next series of problems will make use of the following terminology. Let $B \subset \mathbb{R}$. A point $z \in \mathbb{R}$ is an *accumulation point* of B if there is a sequence $\{x_k\}$ with $x_k \in B$, $x_k \neq z$ for all k, and $\lim_{k\to\infty} x_k = z$. A point $z \in \mathbb{R}$ is an *limit point* of B if there is a sequence $\{x_k\}$ with $x_k \in B$ and $\lim_{k\to\infty} x_k = z$. A set $B \subset \mathbb{R}$ is *closed* if every limit point of B is an element of B.

2.21. *Let $B = (0,1)$. What is the set of accumulation points for B? What is the set of limit points for B?*

2.22. *Let $\mathbb{Z}$ denote the set of integers. What is the set of accumulation points for $\mathbb{Z}$? What is the set of limit points for $\mathbb{Z}$?*

2.23. *Suppose that z is an accumulation point of B. Show that there is a sequence $\{x_k\}$ of distinct points ($j \neq k$ implies $x_j \neq x_k$) with $x_k \in B$, and $\lim_{k\to\infty} x_k = z$.*

2.24. *Let $B = \{1/k, k = 1, 2, 3, \dots\}$. What is the set of accumulation points for B? What is the set of limit points for B?*

2.25. *Show that the set $(0,1)$ is not closed, but the set $[0,1]$ is closed.*

2.26. *Show that the set $\mathbb{R}$ is closed.*

2.27. *Suppose that for $n = 1, \dots, N$ the real numbers a_n, b_n satisfy $a_n < b_n$. Show that*

$$K = \bigcup_{n=1}^{N} [a_n, b_n]$$

is compact.

2.28. *Show that a compact set is closed.*

2.29. *Assume that the sets $K_n \subset \mathbb{R}$ are compact, $n = 1, 2, 3, \dots$. Show that*

$$K = \bigcap_{n=1}^{\infty} K_n$$

is compact. (Assume K is not the empty set.)

2.30. *Use the Nested Interval Principle to show that every Cauchy sequence from the real numbers has a limit. (Hint: Define $a_n = \inf\{x_k, k \geq n\}$ and $b_n = \sup\{x_k, k \geq n\}$.)*

2.31. *Prove that if $K \subset \mathbb{R}$ is compact and $\{x_k\}$ is a Cauchy sequence from K, then $\{x_k\}$ has a limit in K.*

2.32. *Suppose $\{x_k\}$ is any sequence of real numbers. Define*

$$U_k = \mathbb{R} \setminus x_k = \{x \neq x_k\}.$$

Show that $\bigcap_k U_k$ is not empty. That is, there is no sequence which includes all real numbers.

2.33. *What number x is represented by the continued fraction*

$$x = [2, 2, 2, \dots]?$$

Recall that this continued fraction is generated by the recursion formula

$$x_{n+1} = 2 + \frac{1}{x_n}.$$

2.34. *What number x is represented by the continued fraction*

$$x = [2, 3, 2, 3, \dots]?$$

(Hint: find a recursion formula.)

2.35. *A continued fraction $[a_0, a_1, a_2, \dots]$ is said to be periodic if there is a positive integer K such that $a_{n+K} = a_n$ for $n = 0, 1, 2, \dots$ (or more generally for all $n \ge N$). Prove that a periodic simple continued fraction satisfies a quadratic polynomial [4, p. 144].*

2.36. *Construct a number x satisfying*

$$|x - \frac{p_n}{q_n}| < \frac{1}{q_n^3}$$

for infinitely many distinct rationals p_n/q_n in lowest terms. (Hint: Let $q_n = 10^n$ and consider decimal expansions of x with only zeroes and ones.)

3

Infinite series

The theory of infinite sequences and their limits is now applied to the problem of making sense of infinite sums, and the representation of functions by means of power series. Consider the formal sum

$$\sum_{k=0}^{\infty} c_k = c_0 + c_1 + c_2 + \dots.$$

Depending on the values of the numbers c_k such a sum may or may not make sense. For instance, our previous work with the geometric series suggests that

$$\sum_{k=0}^{\infty} 2^{-k} = 1 + 1/2 + 1/4 + \cdots = 2,$$

but the sum

$$1 + 1 + 1 + \dots$$

is unlikely to represent any real number.

3.1 Basics

The key to analyzing infinite series is to convert the problem to one involving sequences. The terms of the sequences will be the finite sums

$$s_n = \sum_{k=1}^{n} c_k = c_1 + c_2 + \cdots + c_n, \quad n = 1, 2, 3, \dots,$$

which are called the n-th *partial sum* of the series

$$\sum_{k=1}^{\infty} c_k.$$

In many cases it is convenient to start the index at $k = 0$, in which case the n-th partial sum is

$$s_n = \sum_{k=0}^{n-1} c_k.$$

The series $\sum_{k=1}^{\infty} c_k$ is said to *converge* if there is a number S such that

$$S = \lim_{n\to\infty} s_n.$$

In this case S is said to be the *sum of the series* $\sum_{k=1}^{\infty} c_k$. A series is said to diverge if it does not converge.

A particularly important example is the geometric series. Given a real number x and $c_k = x^k$, the geometric series is

$$\sum_{k=0}^{\infty} x^k = 1 + x + x^2 + \dots.$$

The partial sums of the geometric series are

$$s_n(x) = \sum_{k=0}^{n-1} x^k = \frac{1-x^n}{1-x}$$

if x is any number other than 1. If $x = 1$ the partial sums are $s_n = n$, which is an unbounded sequence, so the series cannot converge in this case.

Since the denominator $1 - x$ for $s_n(x)$ is independent of n, the series converges (for $x \ne 1$) if $\lim_{n\to\infty} 1 - x^n$ exists. If $|x| < 1$ then

$$\lim_{n\to\infty} x^n \to 0.$$

If $|x| > 1$ then $|x|^n \to \infty$. Since the sequence $\{s_n\}$ is unbounded, it has no limit. Finally, if $x = -1$ then $x^n = (-1)^n = -1, 1, -1, 1, \dots$, which again has no limit. We conclude that the geometric series converges (to $S(x) = 1/(1-x)$) if and only if $|x| < 1$.

A different approach helps with the next example. Let

$$c_k = \frac{1}{k(k+1)}, \qquad k = 1, 2, 3, \dots.$$

Using the algebraic identity

$$c_k = \frac{1}{k(k+1)} = \frac{1}{k} - \frac{1}{k+1},$$

the partial sums are seen to telescope, with

$$s_n = \sum_{k=1}^{n} c_k = \sum_{k=1}^{n} (\frac{1}{k} - \frac{1}{k+1})$$

$$= (1 - \frac{1}{2}) + (\frac{1}{2} - \frac{1}{3}) + (\frac{1}{3} - \frac{1}{4}) + \dots + (\frac{1}{n} - \frac{1}{n+1}) = 1 - \frac{1}{n+1}.$$

Clearly

$$\lim_{n\to\infty} s_n = 1,$$

so that

$$1 = \sum_{k=1}^{\infty} \frac{1}{k(k+1)}.$$

A new idea is needed for the harmonic series

$$\sum_{k=1}^{\infty} \frac{1}{k} = 1 + \frac{1}{2} + \frac{1}{3} + \frac{1}{4} + \dots. \tag{3.1}$$

Since the terms of the series are positive, the partial sums

$$s_n = \sum_{k=1}^{n} \frac{1}{k}$$

obtained by adding the first n terms form an increasing sequence. The size of s_n can be estimated by grouping the terms of the sums into a sequence of blocks of size 2^j,

$$s_n = 1 + [\frac{1}{2}] + [\frac{1}{3} + \frac{1}{4}] + [\frac{1}{5} + \dots + \frac{1}{8}]$$

$$+ [\frac{1}{9} + \dots + \frac{1}{16}] + \dots + [(\frac{1}{2^j} + 1) + \dots + \frac{1}{2^{j+1}}].$$

In each block of size 2^j the last term 2^{-j-1} is the smallest, so the sum of the terms in each block is at least $2^j 2^{-j-1} = 1/2$. If n is large enough to include M blocks, then $s_n \ge M/2$. The infinite series (3.1) diverges.

Theorem 2.1.3 immediately gives us a way to generate new convergent series from old ones.

Lemma 3.1.1. *If α_1 and α_2 are any real numbers, and the infinite series*

$$\sum_{k=1}^{\infty} c_k, \quad \text{and} \quad \sum_{k=1}^{\infty} b_k,$$

converge, so does

$$\sum_{k=1}^{\infty} (\alpha_1 c_k + \alpha_2 b_k).$$

3.2 Positive series

The following general problem confronts us. Given an infinite series of numbers,

$$\sum_{k=0}^{\infty} c_k = c_0 + c_1 + \dots,$$

what general procedures are available to determine if the series converges? Rather than considering the most general series, let's simplify the discussion by considering series whose terms are nonnegative, $c_k \geq 0$. Notice that the sequence of partial sums is *increasing*, that is $s_n \geq s_m$ if $n \geq m$. This observation suggests using the Bounded Monotone Sequence (BMS) Property. With the aid of this important property it is easy to obtain several results about infinite series whose terms are nonnegative.

Theorem 3.2.1. *If $c_k \geq 0$ for $k = 0, 1, 2, \dots$, then the infinite series*

$$\sum_{k=0}^{\infty} c_k = c_0 + c_1 + \dots$$

converges if and only if the sequence of partial sums $\{s_n\}$ is bounded.

Proof. If the series converges, then the sequence of partial sums has a limit. By Lemma 2.1.2 the sequence of partial sums is bounded.

Suppose the sequence of partial sums is bounded. Then $\{s_n\}$ is a bounded increasing sequence, so has a limit by the BMS property. □

Theorem 3.2.2. *(Comparison Test) Suppose that $0 \leq a_k \leq c_k$ for $k = 0, 1, 2, \dots$. If the infinite series $\sum_{k=0}^{\infty} c_k = c_0 + c_1 + \dots$ converges, so does the series $\sum_{k=0}^{\infty} a_k = a_0 + a_1 + \dots$. If the infinite series $\sum_{k=0}^{\infty} a_k = a_0 + a_1 + \dots$ diverges, so does the series $\sum_{k=0}^{\infty} c_k = c_0 + c_1 + \dots$.*

Proof. Look at the partial sums

$$s_n = \sum_{k=0}^{n-1} c_k = c_0 + c_1 + \dots + c_{n-1}, \quad \sigma_n = \sum_{k=0}^{n-1} a_k = a_0 + a_1 + \dots + a_{n-1}.$$

Since $0 \leq a_k \leq c_k$, the inequality $\sigma_n \leq s_n$ holds for all positive integers n. If the series $\sum c_k$ converges, then the sequence of partial sums $\{s_n\}$ is bounded above, and so is the sequence of partial sums $\{\sigma_n\}$. By the BMS property the series $\sum a_k$ converges.

Suppose that the series $\sum a_k$ diverges. By the BMS property the sequence of partial sums $\{\sigma_n\}$ is not bounded above. Consequently, the sequence of partial sums $\{s_n\}$ is not bounded above, and the series $\sum c_k$ diverges. □

As an example, consider the two series whose terms are

$$a_k = \frac{1}{k^2 + k}, \quad c_k = \frac{1}{2k^2}.$$

Since $k^2 \geq k$ for $k \geq 1$,

$$0 \leq \frac{1}{2k^2} \leq \frac{1}{k^2 + k},$$

and the hypotheses of the Comparison Test are satisfied. The earlier calculation,

$$\sum_{k=1}^{\infty} \frac{1}{k^2+k} = \sum_{k=1}^{\infty} (\frac{1}{k} - \frac{1}{k+1}) = \lim_{n\to\infty} 1 - \frac{1}{n+1} = 1,$$

in combination with the Comparison Test then implies that the series

$$\sum_{k=1}^{\infty} \frac{1}{2k^2}$$

converges. The convergence of

$$\sum_{k=1}^{\infty} \frac{1}{k^2}$$

follows by an application of Lemma 3.1.1.

Theorem 3.2.3. *(Ratio Test) Suppose that $c_k > 0$ for $k = 0, 1, 2, \dots,$ and that*

$$\lim_{k\to\infty} \frac{c_{k+1}}{c_k} = L.$$

Then the series $\sum c_k$ converges if $L < 1$ and diverges if $L > 1$.

Proof. First suppose that $L < 1$, and let L_1 be another number satisfying $L < L_1 < 1$. Since

$$\lim_{k\to\infty} \frac{c_{k+1}}{c_k} = L,$$

there is an integer N such that

$$0 < \frac{c_{k+1}}{c_k} < L_1, \quad k \ge N.$$

This implies that for $k \ge N$ we have

$$c_k = c_N \frac{c_{N+1}}{c_N} \frac{c_{N+2}}{c_{N+1}} \cdots \frac{c_k}{c_{k-1}} \le c_N L_1^{k-N}.$$

A comparison with the geometric series is now helpful. For $m > N$,

$$s_m = \sum_{k=1}^{m} c_k = \sum_{k=1}^{N-1} c_k + \sum_{k=N}^{m} c_k$$

$$\le \sum_{k=1}^{N-1} c_k + c_N \sum_{k=N}^{m} L_1^{k-N} \le \sum_{k=1}^{N-1} c_k + c_N \sum_{j=0}^{\infty} L_1^j$$

$$= \sum_{k=1}^{N-1} c_k + c_N \frac{1}{1-L_1}.$$

Since the sequence of partial sums is bounded, the series converges.

If instead $L > 1$, then a similar argument can be made with $1 < L_1 < L$. This time the comparison with the geometric series shows that the sequence of partial sums is unbounded, and so the series diverges. □

The ratio test provides an easy means of checking convergence of the usual Taylor series for e^x when $x > 0$. The series is

$$\sum_{k=0}^{\infty} \frac{x^k}{k!},$$

with $c_k = x^k/k!$ and

$$\frac{c_{k+1}}{c_k} = \frac{x^{k+1}}{(k+1)!} \frac{k!}{x^k} = \frac{x}{k+1}.$$

Clearly $\lim_{k\to\infty} c_{k+1}/c_k = 0$ for any fixed $x > 0$, so the series converges.

As another example consider the series

$$\sum_{k=0}^{\infty} kx^k, \quad x > 0.$$

The ratios are

$$\frac{c_{k+1}}{c_k} = \frac{(k+1)x^{k+1}}{kx^k} = x\frac{(k+1)}{k}.$$

Since

$$\lim_{k\to\infty} \frac{c_{k+1}}{c_k} = x,$$

the series converges when $0 < x < 1$.

3.3 General series

For a series whose terms c_k are not all positive, other methods are needed to show that the sequence of partial sums

$$s_n = \sum_{k=1}^{n} c_k$$

has a limit. To begin, note that if a series converges, then the terms c_k must have 0 as their limit.

Lemma 3.3.1. *If* $\sum c_k$ *converges then*

$$\lim_{k\to\infty} c_k = 0.$$

Proof. The individual terms of the series can be expressed as a difference of partial sums,

$$c_k = s_k - s_{k-1}.$$

The sequences $\{s_k\}$ and $\{s_{k-1}\}$ have the same limit, so

$$\lim_{k\to\infty} c_k = \lim_{k\to\infty} s_k - \lim_{k\to\infty} s_{k-1} = 0.$$

□

3.3.1 Absolute convergence

The main technique for establishing convergence of a general series $\sum c_k$ is to study the related positive series $\sum_{k=1}^{\infty} |c_k|$. Say that a series $\sum_{k=1}^{\infty} c_k$ *converges absolutely* if $\sum_{k=1}^{\infty} |c_k|$ converges.

Theorem 3.3.2. *If a series $\sum_{k=1}^{\infty} c_k$ converges absolutely, then it converges.*

Proof. If

$$s_n = \sum_{k=1}^{n} |c_k|$$

is the n-th partial sum for the series $\sum_{k=1}^{\infty} |c_k|$, then the sequence $\{s_n\}$ increases to its limit S. This means in particular that $s_n \le S$.

Let $\{a_j\}$ and $\{-b_j\}$ be respectively the sequences of nonnegative and negative terms from the sequence $\{c_k\}$, as illustrated below. (One of these sequences may be a finite list rather than an infinite sequence.)

$$c_1, c_2, c_3, \cdots = |c_1|, |c_2|, -|c_3|, |c_4|, -|c_5|, |c_6|, |c_7|, \dots,$$

$$a_1 = c_1, a_2 = c_2, a_3 = c_4, a_4 = c_6, a_5 = c_7, \quad b_1 = |c_3|, b_2 = |c_5|, \dots.$$

If σ_m is a partial sum of the series $\sum_j a_j$, then for some $n \ge m$ we have

$$\sigma_m \le s_n \le S.$$

Since the partial sums of the positive series $\sum_j a_j$ are bounded, the series converges to a number A. By a similar argument the positive series $\sum_j b_j$ converges to a number B.

Let's show that

$$\sum_{k=1}^{\infty} c_k = A - B.$$

Pick $\epsilon > 0$, and as usual let $\epsilon_1 = \epsilon/2$. There are numbers N_1 and N_2 such that

$$|\sum_{j=1}^{n} a_j - A| < \epsilon_1, \quad n \ge N_1, \quad |\sum_{j=1}^{n} b_j - B| < \epsilon_1, \quad n \ge N_2.$$

Find a number N_3 such that $n \ge N_3$ implies that the finite list $c_1, \dots, c_{N_3}$ contains at least the first N_1 positive terms a_k and the first N_2 negative terms $-b_k$. Then for $n \ge N_3$ we have

$$|\sum_{k=1}^{n} c_k - (A - B)| \le |\sum_{j=1}^{n_1} a_j - A| + |B - \sum_{j=1}^{n_2} b_j| < \epsilon_1 + \epsilon_1 = \epsilon,$$

since $n_1 \ge N_1$ and $n_2 \ge N_2$. □

This theorem may be used in conjunction with the tests for convergence of positive series. As an example, reconsider the usual power series for e^x. The series is

$$\sum_{k=0}^{\infty} \frac{x^k}{k!}.$$

Replace the terms in this series by their absolute values,

$$\sum_{k=0}^{\infty} \frac{|x|^k}{k!}.$$

If $x = 0$ the series converges, and if $x \ne 0$ we may apply the ratio test with $c_k = |x|^k/k!$ to get

$$\frac{c_{k+1}}{c_k} = \frac{|x|^{k+1}}{(k+1)!}\frac{k!}{|x|^k} = \frac{|x|}{k+1}.$$

Clearly $\lim_{k\to\infty} c_{k+1}/c_k = 0$ for any fixed x, so the original series for e^x converges since it converges absolutely.

3.3.2 Alternating series

There is a more specialized convergence test for some series which may not be absolutely convergent.

Theorem 3.3.3. *(Alternating series test) Suppose that $c_k > 0$, $c_{k+1} \le c_k$, and $\lim_{k\to\infty} c_k = 0$. Then the series*

$$\sum_{k=1}^{\infty} (-1)^{k+1} c_k = c_1 - c_2 + c_3 - c_4 + \dots$$

converges. Furthermore, if S is the sum of the series, and $\{s_n\}$ is its sequence of partial sums, then

$$s_{2m} \le S \le s_{2m-1}, \quad m \ge 1.$$

Proof. For $m \ge 1$ define new sequences $e_m = s_{2m}$ and $o_m = s_{2m-1}$. The proof is based on some observations about these sequences of partial sums with even and odd indices. Notice that

$$e_{m+1} = s_{2m+2} = \sum_{k=1}^{2m+2} (-1)^{k+1} c_k = s_{2m} + c_{2m+1} - c_{2m+2} = e_m + [c_{2m+1} - c_{2m+2}].$$

The assumption $c_{k+1} \le c_k$ means that $c_{2m+1} - c_{2m+2} \ge 0$, so that

$$e_{m+1} \ge e_m.$$

Essentially the same argument shows that

$$o_{m+1} \le o_m.$$

Since $c_k \geq 0$,

$$e_m = s_{2m-1} - c_{2m} = o_m - c_{2m} \leq o_m \leq o_1.$$

Similarly,

$$o_{m+1} = s_{2m} + c_{2m+1} = e_m + c_{2m+1} \geq e_m \geq e_1.$$

Thus the sequence e_m is increasing and bounded above, while the sequence o_m is decreasing and bounded below. By the BMS theorem the sequences e_m and o_m have limits E and O respectively.

Now

$$O - E = \lim_{m\to\infty} o_m - \lim_{m\to\infty} e_m = \lim_{m\to\infty} o_m - e_m$$

$$= \lim_{m\to\infty} s_{2m-1} - s_{2m} = \lim_{m\to\infty} c_{2m} = 0,$$

or $O = E$. Take $S = O = E$. Since the even partial sums s_{2m} increase to S, and the odd partial sums s_{2m-1} decrease to S, the conclusion

$$s_{2m} \leq S \leq s_{2m-1}, \quad m \geq 1$$

is established.

Finally, to see that $S = \lim_{n\to\infty} s_n$, let $\epsilon > 0$. Since the even partial sum sequence converges to S, there is a number M_1 such that

$$|e_m - S| = |s_{2m} - S| < \epsilon, \quad \text{if} \quad m \geq M_1.$$

There is a corresponding number M_2 for the odd partial sum sequence,

$$|o_m - S| = |s_{2m-1} - S| < \epsilon, \quad \text{if} \quad m \geq M_2.$$

Consequently if $N = \max(2M_1, 2M_2 - 1)$, then whenever $n \geq N$ we have

$$|s_n - S| < \epsilon.$$

□

When a series is alternating it is possible to be quite precise about the speed of convergence. Start with the inequality

$$s_{2m} \leq S \leq s_{2m-1}, \quad m \geq 1.$$

This implies that

$$S - s_{2m} \leq s_{2m-1} - s_{2m} = c_{2m}$$

and similarly

$$s_{2m-1} - S \leq s_{2m-1} - s_{2m} = c_{2m}.$$

On the other hand

$$(s_{2m-1} - S) + (S - s_{2m}) = s_{2m-1} - s_{2m} = c_{2m},$$

so either $|s_{2m-1} - S| \geq c_{2m}/2$ or $|S - s_{2m}| \geq c_{2m}/2$.

As an example of an alternating series, take $c_k = 1/\log_{10}(k+1)$. The alternating series $\sum_{k=1}^{\infty}(-1)^k/\log_{10}(k+1)$ converges, but since $c_k \geq 1/(k+1)$, comparison with the harmonic series shows that the series does not converge absolutely. The difference between the partial sum s_n and the sum S is on the order of the last term c_n, so to insure that $|s_n - S| < 10^{-6}$, for instance, we would take $\log_{10}(k+1) > 10^6$, or $k+1 > 10^{1,000,000}$.

3.4 Power series

The class of elementary functions (including rational, trigonometric, and exponential functions) is too meager for the evaluation of many integrals such as $\int_0^x \sin(t)/t\,dt$, or the solution of differential equations like

$$y'' - xy = 0. \tag{3.2}$$

The collection of functions at our disposal is greatly enlarged by the addition of power series, which are functions having the form

$$f(x) = \sum_{k=0}^{\infty} a_k x^k,$$

or more generally

$$g(x) = \sum_{k=0}^{\infty} a_k (x - x_0)^k.$$

Our treatment of infinite series has already alerted us to concerns about when such series make sense. The next theorem guarantees that the domain of a function defined by a power series has a relatively simple form.

Theorem 3.4.1. *Suppose the power series*

$$\sum_{k=0}^{\infty} a_k (x - x_0)^k$$

converges for $x = x_1 \neq x_0$. *Then the series converges absolutely for* $|x - x_0| < |x_1 - x_0|$.

Proof. Since the series $\sum_{k=0}^{\infty} a_k (x_1 - x_0)^k$ converges, Lemma 3.3.1 says that

$$\lim_{k \to \infty} a_k (x_1 - x_0)^k = 0.$$

Therefore the sequence with terms $a_k (x_1 - x_0)^k = 0$ is bounded, and there must be a number M such that

$$|a_k (x_1 - x_0)^k| \leq M, \quad k = 0, 1, 2, \ldots.$$

Testing for absolute convergence, suppose $|x - x_0| < |x_1 - x_0|$. Let

$$r = \frac{|x - x_0|}{|x_1 - x_0|} < 1.$$

Then

$$|a_k(x - x_0)^k| = |a_k(x_1 - x_0)^k|\left|\frac{(x - x_0)^k}{(x_1 - x_0)^k}\right| \le Mr^k.$$

Since $0 \le r < 1$ the power series converges by comparision with the geometric series. □

This theorem justifies defining the *radius of convergence* of a power series, which is the largest number R such that the power series converges for all $|x - x_0| < R$. If the power series converges for all x we say the radius of convergence is ∞. In some cases the radius of convergence of a power series may be readily computed.

Theorem 3.4.2. *Suppose that* $|a_k| \ne 0$, *and*

$$\lim_{k\to\infty} \frac{|a_k|}{|a_{k+1}|} = L.$$

Then the radius of convergence of the power series $\sum a_k(x - x_0)^k$ *is* L. *(The case* $L = \infty$ *is included.)*

Proof. When $L \ne 0$ and $L \ne \infty$ the ratio test may be applied to

$$\sum_{k=0}^{\infty} |a_k||x - x_0|^k.$$

In this case

$$\lim_{k\to\infty}\left|\frac{a_{k+1}(x - x_0)^{k+1}}{a_k(x - x_0)^k}\right| = \frac{|x - x_0|}{L},$$

so by the ratio test the power series converges absolutely if $|x - x_0| < L$. On the other hand if $|x - x_0| > L$, the power series does not converge absolutely, so by the previous theorem it cannot converge at all for any $|x - x_0| > L$.

The same argument only requires slight modifications in case $L = 0$ or $L = \infty$. □

Example 1: The series for e^x centered at $x = 0$ is

$$\sum_{k=0}^{\infty} \frac{x^k}{k!}.$$

In this case

$$\frac{|a_k|}{|a_{k+1}|} = k + 1 \to \infty$$

and the series converges for all x.

Example 2: Suppose that our series has the form

$$\sum_{k=0}^{\infty} k^m x^k$$

for some positive integer m. Then

$$\frac{|a_k|}{|a_{k+1}|} = \frac{k^m}{(k+1)^m} = \frac{1}{(1+1/k)^m} \to 1,$$

so the power series has radius of convergence 1.

3.5 Problems

3.1. *Show that the series*

$$\sum_{k=0}^{\infty} 2^{-k} \frac{1}{1+k^2}$$

converges.

3.2. *Suppose that $c_k \ge 0$ for $k = 1,2,3,\dots$, and that $\sum_{k=1}^{\infty} c_k$ converges. Show that if $0 \le b_k \le M$ for some number M, then $\sum_{k=1}^{\infty} b_k c_k$ converges.*

3.3. *Show that the series $\sum_{k=1}^{\infty} k2^{-k}$ converges. Show that if m is a fixed positive integer then the series $\sum_{k=1}^{\infty} k^m 2^{-k}$ converges.*

3.4. *Prove Lemma 3.1.1.*

3.5. *a) Show that the series*

$$\sum_{k=0}^{\infty} \frac{2^k}{k!}$$

converges.

b) Show that the series

$$\sum_{k=0}^{\infty} \frac{k^k}{k!}$$

diverges.

3.6. *Suppose that $c_k \ge 0$ and*

$$\lim_{k\to\infty} c_k = r > 0.$$

Show that $\sum_{k=1}^{\infty} c_k$ diverges.

3.7. *Assume that $c_k \ge 0$ and $\sum_{k=1}^{\infty} c_k$ converges. Suppose that there is a sequence $\{a_k\}$, a positive integer N, and a positive real number r such that $0 \le a_k \le rc_k$ for $k \ge N$. Show that $\sum_{k=1}^{\infty} a_k$ converges.*

3.8. *Assuming that $k^2 + ak + b \ne 0$ for $k = 1,2,3,\dots$, show that the series*

$$\sum_{k=1}^{\infty} \frac{1}{k^2 + ak + b}$$

converges.

3.9. *Find an example of a divergent series $\sum_{k=1}^{\infty} c_k$ for which $\lim_{k\to\infty} c_k = 0$.*

3.10. *(Root Test) Suppose that $c_k \ge 0$ for $k = 1,2,3,\dots$, and that*

$$\lim_{k\to\infty} c_k^{1/k} = L.$$

Show that the series $\sum_{k=1}^{\infty} c_k$ converges if $L < 1$ and diverges if $L > 1$.

3.11. *a) Establish the convergence of the series*

$$\sum_{k=1}^{\infty} \frac{k+1}{k^3+6}.$$

b) Suppose that $p(k)$ and $q(k)$ are polynomials. State and prove a theorem about the convergence of the series

$$\sum_{k=1}^{\infty} \frac{p(k)}{q(k)}.$$

3.12. *Prove convergence of the series*

$$\sum_{k=1}^{\infty} \frac{(-1)^k}{k} + \frac{1}{k^2}.$$

3.13. *Give an alternative proof of Theorem 3.3.2 along the following lines. For $n > m$ the partial sums s_n of a series satisfy*

$$|s_n - s_m| = |(s_n - s_{n-1}) + (s_{n-1} - s_{n-2}) + \cdots + (s_{m+1} - s_m)| \le \sum_{k=m}^{n-1} |c_k|.$$

Now appeal to Cauchy sequences.

3.14. *For $k = 0, 1, 2, \dots$ suppose*

$$b(k) = \begin{Bmatrix} 1, & k = 0,1 \mod 3, \\ -1, & k = 2 \mod 3 \end{Bmatrix}.$$

Does the series

$$\sum_{k=1}^{\infty} b(k-1)\frac{1}{k} = 1 + \frac{1}{2} - \frac{1}{3} + \frac{1}{4} + \frac{1}{5} - \frac{1}{6} + \dots$$

converge or diverge?

3.15. *A series is said to converge conditionally if $\sum c_k$ converges, but $\sum |c_k|$ diverges. Suppose that the series $\sum c_k$ converges conditionally.*

(a) Show that there are infinitely many positive and negative terms c_k.

(b) Let a_j be the j-th nonnegative term in the sequence $\{c_k\}$, and let b_j be the j-th negative term in the sequence $\{c_k\}$. Show that both series $\sum a_j$ and $\sum b_j$ diverge.

3.16. *In the paragraph following the alternating series test theorem we showed that*

$$S - s_{2m} \le s_{2m-1} - s_{2m} = c_{2m}, \quad s_{2m-1} - S \le s_{2m-1} - s_{2m} = c_{2m},$$

and either $|s_{2m-1} - S| \geq c_{2m}/2$ *or* $|S - s_{2m}| \geq c_{2m}/2$. *Starting with*

$$s_{2m+1} - s_{2m} = c_{2m+1},$$

develop similar estimates comparing the differences $|s_{2m+1} - S|$ *and* $|s_{2m} - S|$ *to* c_{2m+1}.

3.17. *For which values of* x *does the power series*

$$\sum_{k=1}^{\infty} \frac{(x-1)^k}{k}$$

converge, and for which does it diverge?

3.18. *For which values of* x *does the series*

$$\sum_{k=1}^{\infty} (2x+3)^k$$

converge, and for which does it diverge?

3.19. *Show that if* $p(k)$ *is a (nontrivial) polynomial, then the power series*

$$\sum_{k=1}^{\infty} p(k)x^k$$

converges if $|x| < 1$, *but diverges if* $|x| > 1$.

3.20. *Find the radius of convergence of the following series:*

$$a) \sum_{k=1}^{\infty} (2^k + 10)x^k, \quad b) \sum_{k=1}^{\infty} k!(x-5)^k, \quad c) \sum_{k=1}^{\infty} \frac{k^{k/2}}{k!} x^k.$$

3.21. *Suppose* $|c_k| > 0$ *and*

$$\lim_{k \to \infty} \frac{|c_{k+1}|}{|c_k|} = L.$$

Show that if $L > 1$, *then*

$$\lim_{k \to \infty} |c_k| = \infty,$$

and so the series $\sum_k c_k$ *diverges.*

3.22. *(Products of series) If power series are formally multiplied, then collecting terms with equal powers of* x *gives*

$$(\sum_{k=0}^{\infty} a_k x^k)(\sum_{k=0}^{\infty} b_k x^k) = (a_0 + a_1 x + a_2 x^2 + \dots)(b_0 + b_1 x + b_2 x^2 + \dots)$$

$$= a_0 + (a_1 + b_1)x + (a_0 b_2 + a_1 b_1 + a - 2b_0)x^2 + \dots .$$

This suggests defining the product of two power series by

$$(\sum_{k=0}^{\infty} a_k x^k)(\sum_{k=0}^{\infty} b_k x^k) = (\sum_{k=0}^{\infty} c_k x^k), \quad c_k = \sum_{j=0}^{k} a_j b_{k-j}.$$

Setting $x = 1$ leads to the definition

$$(\sum_{k=0}^{\infty} a_k)(\sum_{k=0}^{\infty} b_k) = (\sum_{k=0}^{\infty} c_k).$$

Prove that if the series $\sum_{k=0}^{\infty} a_k$ and $\sum_{k=0}^{\infty} b_k$ converge absolutely, with sums A and B respectively, then the series $\sum_{k=0}^{\infty} c_k$ converges absolutely, and its sum is AB.

4

More sums

Several additional series topics are treated in this chapter. The first section addresses rearrangements and other alterations of infinite series. Absolutely convergent series can tolerate a wide variety of manipulations without a change of their convergence properties or the actual sum. Conditionally convergent series are much more delicate; careless rearrangements can change convergent series to divergent ones, or alter the sum when convergence is preserved.

After consideration of rearrangements, the subsequent sections consider finite sums such as

$$\sum_{k=1}^{n} \frac{1}{k(k+1)} = 1 - \frac{1}{n+1},$$

or

$$\sum_{k-0}^{n} x^k = \frac{1-x^n}{1-x}, \quad x \neq 1,$$

for which exact formulas are available. This work starts with the development of a calculus of sums and differences which parallels the usual calculus of integrals and derivatives. After establishing this calculus, the final section treats the problem of extending the formulas

$$\sum_{k=1}^{n} k = \frac{n(n+1)}{2}, \quad \sum_{k=1}^{n} k^2 = \frac{n(n+1)(2n+1)}{6},$$

to the more general case

$$\sum_{k=1}^{n} k^m.$$

4.1 Grouping and rearrangement

To motivate some further developments in the theory of infinite series, consider power series solutions of Airy's differential equation

$$y'' - xy = 0, \quad -\infty < x < \infty. \tag{4.1}$$

Simply assume for now that this equation has series solutions

$$y(x) = \sum_{n=0}^{\infty} c_n x^n,$$

with derivatives

$$y'(x) = \sum_{n=1}^{\infty} n c_n x^{n-1}, \quad y''(x) = \sum_{n=2}^{\infty} n(n-1) c_n x^{n-2}.$$

Putting these series into the equation gives

$$\sum_{n=2}^{\infty} n(n-1) c_n x^{n-2} - x \sum_{n=0}^{\infty} c_n x^n = 0.$$

When equal powers of x are collected the resulting expression is

$$2c_2 + \sum_{n=1}^{\infty} [(n+2)(n+1) c_{n+2} - c_{n-1}] x^n = 0.$$

The desired constants c_n can be found by requiring the coefficient for each power of x to be zero. The resulting equations are

$$c_2 = 0, \quad c_{n+2} = \frac{c_{n-1}}{(n+2)(n+1)}, \quad n = 1, 2, 3, \dots.$$

These algebraic manipulations (see Problem 4.1) suggest that the coefficients c_0 and c_1 are arbitrary, while the subsequent coefficients c_k, for $k = 0, 1, 2, \dots$, come in three types:

$$c_{3k} = \frac{c_0}{2 \cdot 3 \cdot 5 \cdot 6 \cdots (3k-1) \cdot (3k)}, \tag{4.2}$$

$$c_{3k+1} = \frac{c_1}{3 \cdot 4 \cdot 6 \cdot 7 \cdots (3k) \cdot (3k+1)},$$

$$c_{3k+2} = 0.$$

A basic issue is whether the resulting power series converge for all x. Given the structure of this series, it is tempting to split it into three parts, looking at

$$\sum_{k=0}^{\infty} c_{3k} x^{3k}, \quad \text{and} \quad \sum_{k=0}^{\infty} c_{3k+1} x^{3k+1}$$

separately. Of course the third part is simply 0. Rewrite the first of these constituent series as

$$\sum_{k=0}^{\infty} c_{3k} x^{3k} = \sum_{j=0}^{\infty} \alpha_j (x^3)^j.$$

It is easy to check that $|\alpha_j| \le 1/j!$, so by the comparison and ratio tests this series will converge absolutely for all values of x^3, which means for all values of x. The second constituent series can be treated in the same fashion. What is still missing is a license to shuffle the three constituent convergent series together, so that convergence of the original series may be determined.

For more elementary examples, start with the conditionally convergent alternating harmonic series,

$$\Sigma_0 = \sum_{n=1}^{\infty} (-1)^{n+1}\frac{1}{n} = 1 - \frac{1}{2} + \frac{1}{3} - \frac{1}{4} + \dots .$$

As variations on this series, consider first the rearrangement

$$\Sigma_1 = 1 + \frac{1}{3} - \frac{1}{2} + \frac{1}{5} + \frac{1}{7} - \frac{1}{4} + \frac{1}{9} + \frac{1}{11} - \frac{1}{6} + \dots, \tag{4.3}$$

where the order of the terms has been altered. A second example maintains the original order, but organizes the terms into groups which are summed first,

$$\Sigma_2 = 1 + (-\frac{1}{2} + \frac{1}{3}) + (-\frac{1}{4} + \frac{1}{5} - \frac{1}{6}) + (\frac{1}{7} - \frac{1}{8} + \frac{1}{9} - \frac{1}{10}) + \dots . \tag{4.4}$$

For the final variation the signs of the terms are modified,

$$\Sigma_3 = 1 + \frac{1}{2} - \frac{1}{3} - \frac{1}{4} + \frac{1}{5} + \frac{1}{6} - \frac{1}{7} - \frac{1}{8} + \dots, \tag{4.5}$$

These series are closely related to a previously treated series; it would be nice to have some guidelines for analyzing their convergence.

As a first step toward this end, let's begin with the notion of a rearrangment of an infinite series. Suppose that $\{p(k),\ k = 1, 2, 3, \dots\}$ is a sequence of positive integers which includes each positive integer exactly once. The infinite series

$$\sum_{k=1}^{\infty} c_{p(k)}$$

is said to be a *rearrangment* of the series $\sum_k c_k$. If the series $\sum_k c_k$ converges absolutely, the story about rearrangements is simple and satisfying.

Theorem 4.1.1. *If the series*

$$\sum_{k=1}^{\infty} c_k$$

converges absolutely, then any rearrangement of the series also converges absolutely, with the same sum.

Proof. Suppose

$$A = \sum_{k=1}^{\infty} |c_k|. \tag{4.6}$$

For any m-th partial sum of the rearranged series there is an n such that

$$\sum_{k=1}^{m} |c_{p(k)}| \le \sum_{k=1}^{n} |c_k| \le A,$$

since for n sufficiently large every term in the left sum will also appear in the right sum. The partial sums of the rearranged series form a bounded increasing sequence, so the rearranged series is absolutely convergent.

To check that the sum of the rearranged series agrees with the sum of the original series, let

$$S = \sum_{k=1}^{\infty} c_k.$$

If $\{\sigma_n\}$ denotes the sequence of partial sums for the series (4.6), then for any $\epsilon > 0$ there is an N such that $n > N$ implies

$$\sigma_n - \sigma_N = \sum_{k=N+1}^{n} |c_k| < \epsilon/2,$$

and

$$|S - \sum_{k=1}^{N} c_k| < \epsilon/2.$$

Find M such that each of the terms c_k for $k = 1, \dots, N$ appears in the partial sum

$$\sum_{k=1}^{M} c_{p(k)}.$$

For any $m \ge M$, and for n sufficiently large,

$$|S - \sum_{k=1}^{m} c_{p(k)}| = |(S - \sum_{k=1}^{N} c_k) + (\sum_{k=1}^{N} c_k - \sum_{k=1}^{m} c_{p(k)})| \le |S - \sum_{k=1}^{N} c_k| + \sum_{k=N+1}^{n} |c_k| < \epsilon.$$

□

If a series converges, but does not converge absolutely, the series is said to *converge conditionally*. The situation is less satisfactory for rearrangements of conditionally convergent series. In fact, as Riemann discovered [7, p. 67], given any real number x, there is a rearrangement of a conditionally convergent series which converges to x.

Conditionally convergent series can be rearranged without changing the sum if a tight rein is kept on the rearrangement. A rearrangement will be called *bounded* if there is a number C such that

$$|p(k) - k| \le C.$$

The following result is mentioned in [8] along with a variety of more sophisticated tests.

Theorem 4.1.2. *If the series*

$$\sum_{k=1}^{\infty} c_k$$

converges, then any bounded rearrangement of the series converges to the same sum.

Proof. Consider the partial sums for the original series and the rearranged series,

$$s_n = \sum_{k=1}^{n} c_k, \quad \sigma_n = \sum_{k=1}^{n} c_{p(k)}.$$

Compare these sums by writing

$$\sigma_n = s_n - o_n + i_n,$$

where o_n is the sum of terms from s_n omitted in σ_n, while i_n is the sum of terms from σ_n which are not in s_n. Importantly, if c_k is a term from the original series included in o_n or i_n, then $k \ge n - C$ because the rearrangement is bounded.

The condition $|p(k) - k| \le C$ implies that the sums o_n and i_n contain no more than C terms c_k each, with $k \ge n - C$. Since $\lim_{k\to\infty} c_k = 0$

$$\lim_{n\to\infty} o_n = 0, \quad \lim_{n\to\infty} i_n = 0.$$

It follows that $\lim_n \sigma_n = \lim_n s_n$. □

There are several circumstances in which convergent series may be altered without changing their sum if the original ordering of the terms is respected. As in (4.4), blocks of terms from a convergent series may be added first, without changing the sum. Recall that $\mathbb{N}$ denotes the set of positive integers $1, 2, 3, \dots$.

Theorem 4.1.3. *Suppose that $m(k) : \mathbb{N} \to \mathbb{N}$ is a strictly increasing function, with $m(1) = 1$. If the series*

$$\sum_{k=1}^{\infty} c_k$$

converges, then so does

$$\sum_{k=1}^{\infty} \Big(\sum_{j=m(k)}^{m(k+1)-1} c_j \Big),$$

and the sums are the same.

Proof. Let s_n be the n-th partial sum for the original series, and let t_n be the n-th partial sum for the series with grouped terms. Then

$$t_n = \sum_{k=1}^{n} \Big(\sum_{j=m(k)}^{m(k+1)-1} c_j \Big) = \sum_{k=1}^{m(n+1)-1} c_k = s_{m(n+1)-1}.$$

Since the sequence $\{t_n\}$ of partial sums of the series with grouped terms is a subsequence of the sequence $\{s_n\}$, the sequences of partial sums have the same limits. □

It is also possible to shuffle several convergent series without losing convergence. Suppose $M \ge 2$ is an integer. Partition the set $\mathbb{N}$ of positive integers into M infinite subsets

$$\mathbb{N} = U_1 \cup \cdots \cup U_M, \quad U_i \cap U_j = \emptyset, i \ne j,$$

so that each positive integer lies in exactly one of the sets U_m. For $m = 1, \dots, M$ and $j = 1, 2, 3, \dots$, let $\{k(m,j)\}$ be the sequence of all elements of U_m in increasing order. For any positive integer k there will be exactly one pair m, j such that $k = k(m,j)$. Given M series

$$\sum_{j=1}^{\infty} c_j(m), \quad m = 1, \dots, M,$$

say that the series $\sum_k a_k$ is a *shuffle* of the M series $\sum_j c_j(m)$ if

$$a_{k(m,j)} = c_j(m), \quad m = 1, \dots, M, \quad j = 1, 2, 3, \dots.$$

For example, the series (4.5) is a shuffle of the two alternating series

$$1 - \frac{1}{3} + \frac{1}{5} - \frac{1}{7} + \dots, \quad \text{and} \quad \frac{1}{2} - \frac{1}{4} + \frac{1}{6} - \frac{1}{8} + \dots,$$

with $k(1,j) = 2j - 1$ and $k(2,j) = 2j$.

Theorem 4.1.4. *Suppose that the series*

$$\sum_{j=1}^{\infty} c_j(m), \quad m = 1, \dots, M,$$

are convergent, with sums S_m. *If the series*

$$\sum_{k=1}^{\infty} a_k$$

is a shuffle of the series $\sum_j c_j(m)$, *then* $\sum_k a_k$ *converges to* $S_1 + \cdots + S_M$.

Proof. Given $\epsilon > 0$ there is an integer N such that

$$|S_m - \sum_{j=1}^{n} c_j(m)| < \epsilon/M, \quad n \ge N, \quad m = 1, \dots, M.$$

If L is large enough that all terms $c_j(m)$ for $j \le N$ and $m = 1, \dots, M$ appear in the sum

$$\sum_{k=1}^{L} a_k,$$

then there are numbers $N_1 \geq N, \ldots, N_M \geq N$ such that

$$|S_1 + \cdots + S_M - \sum_{k=1}^{L} a_k| = |[S_1 - \sum_{j=1}^{N_1} c_j(1)] + \cdots + [S_M - \sum_{j=1}^{N_M} c_j(M)]| < \epsilon.$$

□

4.2 A calculus of sums and differences

A basic idea in traditional calculus is that integrals are like sums and derivatives are slopes. Motivated in part by the problem of understanding the sum of powers

$$\sum_{k=1}^{n} k^m = 1^m + 2^m + 3^m + \cdots + n^m,$$

a bit of discrete calculus will be developed for functions $f(n)$ which are defined for nonnegative integers n. Functions with integer arguments are common: in particular they arise in digital communications, and when computer algorithms are designed to perform mathematical calculations. Examples of such functions include polynomials like $f(n) = 5n^3 + n^2 + 3n$, or rational functions like $g(n) = n/(1+n^2)$, which are defined only when n is a nonnegative integer. The graphs of the functions $1/(n+1)$ and $\sin(\pi n/10)$ are shown in Figure 4.1 and Figure 4.2.

Of course the function $a(n)$ whose domain is the set $\mathbb{N}_0$ of nonnegative integers is the same as an infinite sequence $a_0, a_1, a_2, \ldots$. Functional notation has simply replaced subscript notation for the index.

Functions $f(n)$ of an integer variable have a calculus similar to the better known calculus of functions of a real variable. However most of the limit problems that arise in real variable calculus are missing in the study of discrete calculus. As a consequence, many of the proofs are greatly simplified.

In discrete calculus the role of the conventional integral

$$\int_a^x f(t)\,dt, \quad a \leq x,$$

will be replaced by the sum or *discrete integral*

$$\sum_{k=m}^{n} f(k), \quad m \leq n.$$

The problem of finding simple expressions for (signed) areas such as

$$\int_0^x t^3\,dt$$

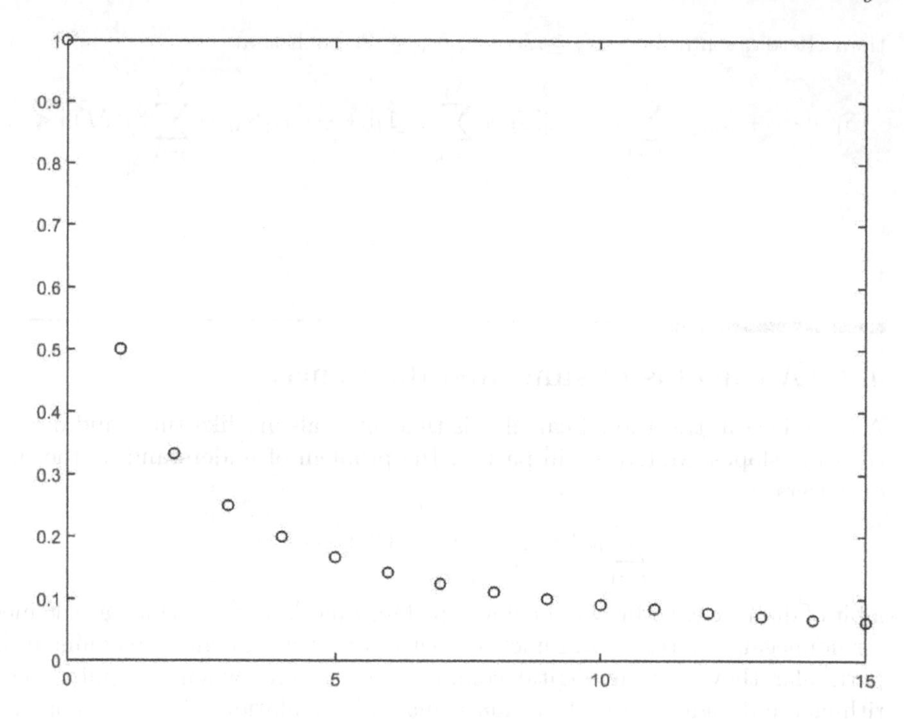

FIGURE 4.1
Graph of $f(n) = \frac{1}{n+1}$.

is replaced by the analogous problem of finding a simple expression for a sum

$$\sum_{k=1}^{n} k^3.$$

To find a replacement for the derivative

$$f'(x) = \frac{df(x)}{dx} = \lim_{h \to 0} \frac{f(x+h) - f(x)}{h},$$

recall the approximation of the derivative by the slope of a secant line,

$$f'(n) \simeq \frac{f(n+1) - f(n)}{1} = f(n+1) - f(n).$$

Since the difference between consecutive integers is 1, the denominator conveniently drops out. It is helpful to introduce two notations for these differences, which will play the role of the derivative:

$$f^+(n) = \Delta^+ f(n) = f(n+1) - f(n).$$

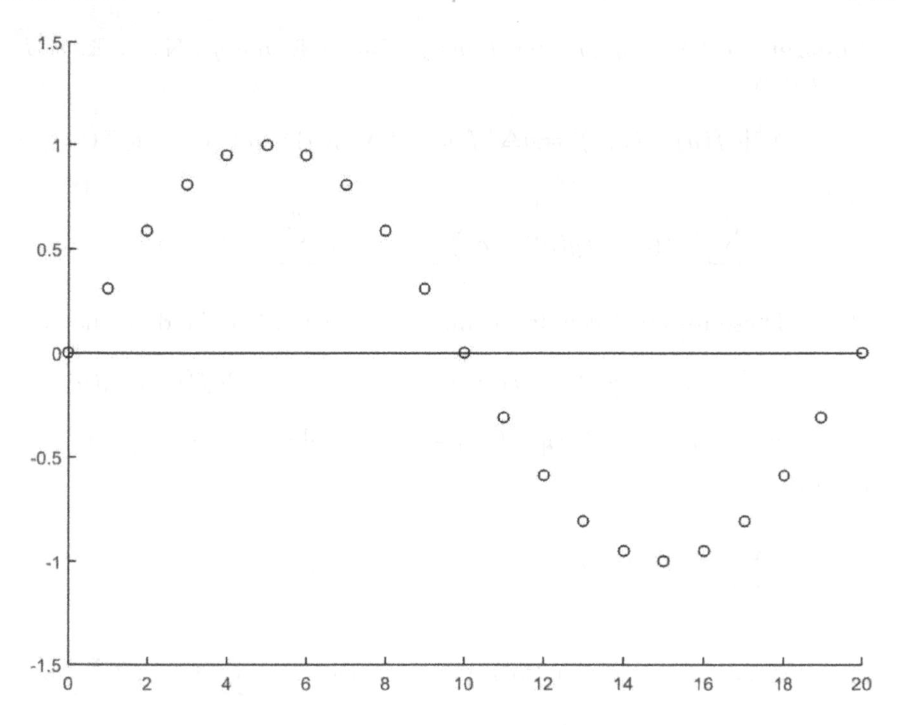

FIGURE 4.2
Graph of $g(n) = sin(\pi n/10)$

This new function $f^+(n)$ is called the forward difference of $f(n)$.

Here are forward differences for a few simple functions.

$$f(n) = n, \quad f^+(n) = (n+1) - n = 1,$$

$$g(n) = n^2, \quad g^+(n) = (n+1)^2 - n^2 = 2n + 1.$$

Notice that $g^+(n)$ is not $2n$, as we might expect from derivative calculations. Another example is

$$h(n) = 3^{-n}, \quad h^+(n) = 3^{-(n+1)} - 3^{-n} = [3^{-1} - 1]3^{-n} = -\frac{2 \cdot 3^{-n}}{3}.$$

In fact for any fixed number x,

$$\Delta^+ x^n = x^{n+1} - x^n = (x-1)x^n. \tag{4.7}$$

There are several theorems from real variable calculus with close parallels in discrete calculus. The first says that sums and differences are linear. To simplify notation, recall that $\mathbb{N}_0$ denotes the set of nonnegative integers $0, 1, 2, \ldots$. The notation $f : \mathbb{N}_0 \to \mathbb{R}$ means that the function f takes real values, and has domain $\mathbb{N}_0$.

Theorem 4.2.1. *For any functions $f : \mathbb{N}_0 \to \mathbb{R}$ and $g : \mathbb{N}_0 \to \mathbb{R}$, and any real numbers a, b*

$$\Delta^+[af(n) + bg(n)] = a\Delta^+ f(n) + b\Delta^+ g(n) = af^+(n) + bg^+(n),$$

and

$$\sum_{k=m}^{n} [af(k) + bg(k)] = a \sum_{k=m}^{n} f(k) + b \sum_{k=m}^{n} g(k), \quad m \le n.$$

Proof. These results follow from simple arithmetic. For the differences,

$$\Delta^+[af(n) + bg(n)] = [af(n+1) + bg(n+1)] - [af(n) + bg(n)]$$

$$= a[f(n+1) - f(n)] + b[g(n+1) - g(n)] = af^+(n) + bg^+(n).$$

For the sums,

$$\sum_{k=m}^{n} [af(k) + bg(k)] = af(m) + bg(m) + \cdots + af(n) + bg(n)$$

$$= a[f(m) + \cdots + f(n)] + b[g(m) + \cdots + g(n)] = a \sum_{k=m}^{n} f(k) + b \sum_{k=m}^{n} g(k).$$

□

There is a product rule only slightly different from what one might expect.

Theorem 4.2.2. *For any functions $f : \mathbb{N}_0 \to \mathbb{R}$ and $g : \mathbb{N}_0 \to \mathbb{R}$,*

$$[f(n)g(n)]^+ = f^+(n)g(n+1) + f(n)g^+(n) = g^+(n)f(n+1) + g(n)f^+(n).$$

Proof. The proof, whose first step is the judicious addition of 0, is again easy.

$$[f(n)g(n)]^+ = f(n+1)g(n+1) - f(n)g(n)$$

$$= f(n+1)g(n+1) - f(n)g(n+1) + f(n)g(n+1) - f(n)g(n)$$

$$= [f(n+1) - f(n)]g(n+1) + f(n)[g(n+1) - g(n)] = f^+(n)g(n+1) + f(n)g^+(n).$$

Since $f(n)g(n) = g(n)f(n)$, we may change the order to get

$$[f(n)g(n)]^+ = g^+(n)f(n+1) + g(n)f^+(n).$$

□

One of the important steps in calculus is the introduction of the indefinite integral, or antiderivative,

$$F(x) = \int_a^x f(t)\, dt.$$

The Fundamental Theorem of Calculus shows that differentiation and antidifferentiation are essentially inverse operations, in the sense that

$$F'(x) = f(x), \quad \int_a^x F'(t)\,dt = F(x) - F(a).$$

The analogous idea for sums is developed by introducing the indefinite sum

$$F(n) = \sum_{k=0}^{n} f(k), \quad n = 0, 1, 2, \ldots.$$

Notice that the indefinite sum is a way of producing new functions, rather than numbers. Here are a few examples which take advantage of our previous computations:

$$f(n) = n, \quad F(n) = \sum_{k=0}^{n} k = 0 + 1 + \cdots + n = \frac{n(n+1)}{2},$$

$$f(n) = n^2, \quad F(n) = \sum_{k=0}^{n} k^2 = \frac{n(n+1)(2n+1)}{6},$$

$$f(n) = 1, \quad F(n) = \sum_{k=0}^{n} 1 = n + 1.$$

The main result relates indefinite sums and differences much as integrals and derivatives are related.

Theorem 4.2.3. (The Fundamental Theorem of Discrete Calculus) *For any function $f : \mathbb{N}_0 \to \mathbb{R}$, and $n \geq 0$,*

$$\sum_{k=0}^{n} f^+(k) = f(n+1) - f(0),$$

and

$$\Delta^+ \sum_{k=0}^{n} f(k) = f(n+1).$$

Proof. For the first part we have

$$\sum_{k=0}^{n} f^+(k) = [f(n+1) - f(n)] + [f(n) - f(n-1)] + \cdots + [f(1) - f(0)].$$

Adjacent terms cancel, leaving only the difference of the first and last terms.

For the second part,

$$\Delta^+ \sum_{k=0}^{n} f(k) = \sum_{k=0}^{n+1} f(k) - \sum_{k=0}^{n} f(k) = f(n+1).$$

□

There are times when it is convenient to start additions from a number $m > 0$. A simple corollary of the Fundamental Theorem of Discrete Calculus is the formula

$$\sum_{k=m}^{n} f^+(k) = \sum_{k=0}^{n} f^+(k) - \sum_{k=0}^{m-1} f^+(k) \tag{4.8}$$
$$= [f(n+1) - f(0)] - [f(m) - f(0)] = f(n+1) - f(m).$$

An important consequence of the Fundamental Theorem of Discrete Calculus is that every difference formula has a corresponding sum formula. For example, the calculation

$$\Delta^+ \frac{1}{n+1} = \frac{1}{n+2} - \frac{1}{n+1} = \frac{-1}{(n+1)(n+2)}$$

gives

$$\sum_{k=0}^{n} \frac{1}{(k+1)(k+2)} = 1 - \frac{1}{n+2}.$$

Recall from real variable calculus that the product rule and the Fundamental Theorem of Calculus combine to give integration by parts. Here is a similar result.

Theorem 4.2.4. (Summation by parts) *For any functions $f : \mathbb{N}_0 \to \mathbb{R}$, $g : \mathbb{N}_0 \to \mathbb{R}$, and for $m \le n$,*

$$\sum_{k=m}^{n} f(k)g^+(k) = [f(n+1)g(n+1) - f(m)g(m)] - \sum_{k=m}^{n} f^+(k)g(k+1).$$

Proof. As with derivatives and integrals we start with the product rule.

$$[f(n)g(n)]^+ = f^+(n)g(n+1) + f(n)g^+(n).$$

Applying (4.8) yields

$$[f(n+1)g(n+1) - f(m)g(m)] = \sum_{k=m}^{n} f^+(k)g(k+1) + \sum_{k=m}^{n} f(k)g^+(k),$$

which is equivalent to the result. □

As an application of summation by parts, consider

$$\sum_{k=0}^{n} \frac{k}{2^k},$$

or more generally

$$\sum_{k=0}^{n} kx^k, \quad x \ne 1,$$

for a fixed number x. It will prove fruitful to take

$$f(n) = n, \quad g(n) = x^n/(x-1).$$

The earlier computation (4.7) showed that $g^+(n) = x^n$.

The summation by parts formula and the previous computation

$$\sum_{k=0}^{n} x^k = \frac{1-x^{n+1}}{1-x}$$

now give

$$\sum_{k=0}^{n} kx^k = \sum_{k=0}^{n} f(k)g^+(k) = [f(n+1)g(n+1) - f(0)g(0)] - \sum_{k=0}^{n} f^+(k)g(k+1)$$

$$= \frac{(n+1)x^{n+1}}{x-1} - \sum_{k=0}^{n} \frac{x^{k+1}}{x-1} = \frac{(n+1)x^{n+1}}{x-1} - \frac{x}{x-1}\sum_{k=0}^{n} x^k$$

so that

$$\sum_{k=0}^{n} kx^k = \frac{(n+1)x^{n+1}}{x-1} + \frac{x(1-x^{n+1})}{(1-x)^2}. \tag{4.9}$$

4.3 Computing the sums of powers

The Fundamental Theorem of Discrete Calculus mimics the version from real variable calculus, showing in particular that if

$$F(n) = \sum_{k=0}^{n} f(k),$$

then

$$\Delta^+ F(n) = f(n+1).$$

This does not quite answer the question of whether for any function $f(n)$ there is a function $F_1(n)$ such that $\Delta^+ F_1(n) = f(n)$, and to what extent the "antidifference" is unique. Following the resolution of these problems we consider whether the indefinite sum of a polynomial function is again a polynomial. This question will lead us back to the sums of powers formulas which appeared at the beginning of the chapter.

Lemma 4.3.1. *Suppose the function $f : \mathbb{N}_0 \to \mathbb{R}$ satisfies*

$$\Delta^+ f(n) = 0$$

for all $n \ge 0$. Then $f(n)$ is a constant.

Proof. Write

$$f(n) = f(0) + [f(1) - f(0)] + [f(2) - f(1)] + \cdots + [f(n) - f(n-1)],$$

to see that $f(n) = f(0)$. □

Theorem 4.3.2. *If $f : \mathbb{N}_0 \to \mathbb{R}$ is any function, then there is a function $F_1 : \mathbb{N}_0 \to \mathbb{R}$ such that*

$$\Delta^+ F_1(n) = f(n).$$

Moreover, if there are two functions $F_1(n)$ and $F_2(n)$ such that

$$\Delta^+ F_1(n) = f(n) = \Delta^+ F_2(n),$$

then for some constant C

$$F_1(n) = F_2(n) + C.$$

Proof. To show the existence of F_1 take

$$F_1(n) = \left\{ \begin{matrix} \sum_{k=0}^{n-1} f(k), & n \ge 1, \\ 0, & n = 0. \end{matrix} \right\}$$

The forward difference of this function is

$$\Delta^+ F_1(n) = F_1(n+1) - F_1(n) = \sum_{k=0}^{n} f(k) - \sum_{k=0}^{n-1} f(k) = f(n), \quad n \ge 1.$$

If $n = 0$ then

$$\Delta^+ F_1(0) = F_1(1) - F_1(0) = f(0).$$

If there are two functions $F_1(n)$ and $F_2(n)$ whose forward differences agree at all points n, then

$$\Delta^+ [F_1(n) - F_2(n)] = 0.$$

By the previous lemma there is a constant C such that

$$F_1(n) - F_2(n) = C.$$

□

Discrete calculus avoids limits, but a drawback is that the formulas are more complex than the corresponding derivative formulas. Looking in particular at the power functions n^m, the Binomial Theorem gives

$$\Delta^+ n^m = (n+1)^m - n^m = \sum_{k=0}^{m} \binom{m}{k} n^k - n^m$$

$$= \sum_{k=0}^{m-1} \binom{m}{k} n^k = mn^{m-1} + \frac{m(m-1)}{2} n^{m-2} + \cdots + mn + 1$$

This calculation is worth highlighting in the following lemma.

Lemma 4.3.3. *If m is a nonnegative integer then*

$$\Delta^+ n^m = \sum_{k=0}^{m-1} \binom{m}{k} n^k.$$

Since forward differences of the power functions are polynomials in n, Theorem 4.2.1 implies that the forward difference of a polynomial is always a polynomial. It is natural to then ask whether indefinite sums of polynomials are again polynomials, and if so, are there convenient formulas. This question was already considered by Euler [3, pp. 36–42]. It suffices to consider the sum of powers

$$\sum_{k=0}^{n} k^m.$$

Theorem 4.3.4. *For every nonnegative integer m there is a polynomial $p_m(n)$ of degree $m+1$ such that*

$$\sum_{k=0}^{n} k^m = p_m(n).$$

Moreover, these polynomials satisfy the recursion formula

$$(m+1)p_m(n) = (m+1)\sum_{k=0}^{n} k^m = (n+1)^{m+1} - \sum_{j=0}^{m-1} \binom{m+1}{j} p_j(n).$$

Proof. The proof is by induction on m. The first case is $m = 0$, where

$$\sum_{k=0}^{n} k^0 = \sum_{k=0}^{n} 1 = n+1.$$

Note that 0^0 is interpreted as 1 in this first case. If $m > 0$ then $0^m = 0$.

For $m \ge 1$ consider two evaluations of the sum

$$\sum_{k=0}^{n} \Delta^+ k^{m+1}. \tag{4.10}$$

On one hand, the Fundamental Theorem of Discrete Calculus gives

$$\sum_{k=0}^{n} \Delta^+ k^{m+1} = (n+1)^{m+1} - 0^{m+1} = (n+1)^{m+1}.$$

On the other hand, Lemma 4.3.3 shows that

$$\Delta^+ k^{m+1} = (k+1)^{m+1} - k^{m+1} = \sum_{j=0}^{m} \binom{m+1}{j} k^j.$$

Putting this expression for $\Delta^+ k^{m+1}$ into the sum in (4.10) gives

$$\sum_{k=0}^{n} \Delta^+ k^{m+1} = \sum_{k=0}^{n}\sum_{j=0}^{m} \binom{m+1}{j} k^j.$$

Equating the two expressions for

$$\sum_{k=0}^{n} \Delta^+ k^{m+1}$$

gives

$$(n+1)^{m+1} = \sum_{k=0}^{n}\sum_{j=0}^{m} \binom{m+1}{j} k^j.$$

Interchanging the order of summation leads to

$$(n+1)^{m+1} = \sum_{j=0}^{m}\sum_{k=0}^{n} \binom{m+1}{j} k^j = \sum_{j=0}^{m} \Big[\binom{m+1}{j} \sum_{k=0}^{n} k^j\Big] \tag{4.11}$$

$$= (m+1)\sum_{k=0}^{n} k^m + \sum_{j=0}^{m-1} \Big[\binom{m+1}{j} \sum_{k=0}^{n} k^j\Big]$$

$$= (m+1)\sum_{k=0}^{n} k^m + \sum_{j=0}^{m-1} \binom{m+1}{j} p_j(n).$$

By virtue of the induction hypothesis

$$p_m(n) = \sum_{k=0}^{n} k^m$$

is a polynomial in n of degree $m+1$ satisfying the given recursion formula. □

We make two comments about ideas arising in this proof. The first comment has to do with changing the order of summation in (4.11). Suppose we have any function $F(j,k)$ of the integer variables j,k. For the case in question $F(j,k) = \binom{m+1}{j} k^j$, and the values $F(j,k)$ are added for all j,k in a rectangle in the j,k plane (see Figure 4.3). For a finite sum, the sum does not depend on the order of summation, so we may add rows first, or columns first, whichever proves more convenient.

The second comment concerns a slight variation on induction which entered in this last proof. Rather than counting on the truth of the m-th statement to imply the truth of the $(m+1)$-st statement, we actually assumed the truth of all statements with index less than or equal to m. A review of the logic behind induction shows that this variant is equally legitimate.

The recursion formula for the functions $p_m(n)$ means that in principle we

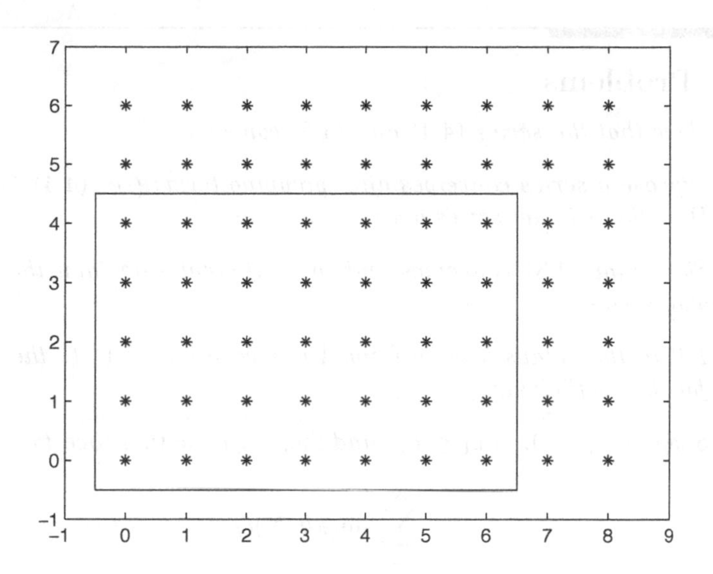

FIGURE 4.3
Adding $F(j,k)$ for $j = 0,\ldots,6$, and $k = 0,\ldots,4$.

can write down arbitrary sums of powers formulas, although they immediately look pretty messy. For instance the previously established formulas

$$p_0(n) = n+1, \quad p_1(n) = \frac{n(n+1)}{2}, \quad p_2(n) = \frac{n(n+1)(2n+1)}{6},$$

lead to

$$4p_3(n) = 4\sum_{k=0}^{n} k^3 = (n+1)^4 - \sum_{j=0}^{2} \binom{4}{j} p_j(n)$$
$$= (n+1)^4 - p_0(n) - 4p_1(n) - 6p_2(n)$$
$$= (n+1)^4 - (n+1) - 2n(n+1) - n(n+1)(2n+1) = (n+1)^2 n^2.$$

That is,

$$\sum_{k=0}^{n} k^3 = \frac{(n+1)^2 n^2}{4}.$$

4.4 Problems

4.1. *Show that the series (4.4) and (4.5) converge.*

4.2. *Suppose a series converges after grouping terms (see (4.4) for an example). Does the original series always converge?*

4.3. *Show that (4.3) converges, but to a different sum than the alternating harmonic series.*

4.4. *Fill in the details leading from Airy's equation (4.1) to the expressions (4.2) for the coefficients.*

4.5. *Suppose* $c_k > 0$, $c_{k+1} \le c_k$, *and* $\lim_{k\to\infty} c_k = 0$. *Prove the convergence of*

$$\sum_{k=1}^{\infty} \sin(\pi k/N) c_k$$

for any positive integer N.

4.6. *State and prove a version of Theorem 4.1.2 which allows for certain rearrangements which are not bounded.*

4.7. *Suppose that for nonnegative integers* m *the function* T *satisfies the recurrence formula*

$$T(2^m) \le aT(2^{m-1}) + 2^m b, \quad m \ge 1,$$

$$T(1) \le b.$$

Here a *and* b *are nonnegative numbers.*

Use induction to show that for every positive integer m,

$$T(2^m) \le 2^m b \sum_{k=0}^{m} (a/2)^k = 2^m b \frac{1-(a/2)^{m+1}}{1-(a/2)}.$$

Such recurrence formulas are often encountered in studying the execution time T *of computer algorithms as a function of the size* 2^m *of a set of inputs.*

4.8. *Use trigonometric identities to show that*

$$\Delta^+ \sin(an) = \sin(an)[\cos(a) - 1] + \sin(a)\cos(an) = 2\cos(a[n+1/2])\sin(a/2).$$

4.9. *Find* $f^+(n)$ *if*

$$a)\ f(n) = n^3, \quad b)\ g(n) = \frac{1}{(n+1)^2}.$$

4.10. *Use the results of Problem 4.9 to find*

$$a) \sum_{k=0}^{n-1}(3k^2+3k+1), \quad b) \sum_{k=0}^{n-1}\frac{2k+3}{(k+1)^2(k+2)^2}.$$

4.11. *Verify the quotient rule*

$$\Delta^+\frac{f(n)}{g(n)} = \frac{f^+(n)g(n)-f(n)g^+(n)}{g(n)g(n+1)}.$$

4.12. *Use the quotient rule to evaluate* $f^+(n)$ *if*

$$a) \ f(n) = \frac{n^2}{2n^2+n+1}, \quad b) \ f(n) = \frac{n^2}{3^n}.$$

Use the Fundamental Theorem to derive summation formulas from these two calculations.

4.13. *For an integer* $m \geq 2$, *compute*

$$\sum_{k=0}^{n-1}\frac{m}{(k+1)(k+m+1)}.$$

(Hint: compute $\Delta^+[\frac{1}{n+1}+\frac{1}{n+2}]$, *and generalize.)*

4.14. *For fixed* n *let*

$$f(k) = \binom{n}{k}.$$

Find $f^+(k)$ *and determine when* f *is an increasing, respectively decreasing, function of* k.

4.15. *Use the summation by parts formula to find*

$$\sum_{k=0}^{n-1} k^2x^k.$$

(Hint: Follow the method used to derive (4.9).*)*

4.16. *Show that*

$$\sum_{k=0}^{n-1}(\sin(k)[\cos(1)-1]+\sin(1)\cos(k)) = \sin(n).$$

4.17. *Express the function*

$$p_4(n) = \sum_{k=0}^{n} k^4$$

as a polynomial in n.

4.18. *Show that if $p(n)$ is a polynomial, then so is*

$$\sum_{k=0}^{n} p(k).$$

4.19. *For integers $k \ge 1$ define the function*

$$q_k(n) = n(n+1)\cdots(n+k-1).$$

a) Show that $nq_k(n+1) = (n+k)q_k(n)$.
b) Show that $nq_k^+(n) = kq_k(n)$.
c) Show that

$$\sum_{j=0}^{n-1} q_k(j) = \frac{n}{k}q_k(n) - \frac{1}{k}\sum_{j=0}^{n-1} q_k(j+1) = \frac{n}{k}q_k(n) - \frac{1}{k}\sum_{j=1}^{n} q_k(j),$$

so that

$$\sum_{j=0}^{n-1} q_k(j) = \frac{n-1}{k+1}q_k(n).$$

4.20. *(a) For integer $m \ge 1$ show that*

$$\sum_{k=0}^{n-1} k^m x^k = \frac{n^m x^n}{x-1} - \sum_{k=0}^{n-1}[\sum_{j=0}^{m-1}\binom{m}{j}k^j]x^k.$$

(b) Use part (a) to compute

$$\sum_{k=0}^{n-1} k^2 x^k.$$

5

Functions

One of the cornerstones of analysis is the study of real valued functions of a real variable. To the extent that functions appear in elementary mathematics, they tend to appear either as narrow classes related to arithmetic, such as polynomials, rational functions, or roots, or as specific examples of "transcendental" functions. Thus a function of a real variable x might be defined by operating on x by the elementary arithmetic operations addition, subtraction, multiplication and division, yielding such examples as

$$p(x) = x^2 + 3x + 7, \quad r(x) = \frac{x-1}{x^3 + 7x^2},$$

or by the use of such particular functions as

$$\sin(x), \quad e^x, \quad \log(x).$$

This procedural view of functions was shared to a considerable extent by researchers when the concept of a function was developed during the seventeenth century [5, p. 403–406] and [2].

The expectation that a function will be defined by an explicit procedure runs into trouble almost immediately. For instance, a polynomial

$$p(x) = a_n x^n + \cdots + a_1 x + a_0, \quad a_n \neq 0$$

will have between 0 and n real solutions, and n complex solutions (when counted with appropriate multiplicity). Is it appropriate to say that the roots are a function of the coefficients $a_0, \dots, a_n$? If the degree n is two, the quadratic formula provides a procedure for explicitly expressing the roots in terms of the coefficients. Such explicit elementary formulas are not available when $n \geq 5$, and more sophisticated procedures, possibly involving infinite processes like power series, are needed.

Some bold attempts were made to extend the available procedures for defining functions. Early in the historical development of functions the infinite repetition of arithmetic operations was allowed, providing function descriptions using power series, infinite products, and continued fractions. While these methods were productive, the use of infinite processes adds complexity and new issues; as an illustration, a function defined by a power series may have a limited domain of convergence.

Some conceptual problems persisted for hundreds of years. Properties of familiar examples were assumed valid for "general" functions. Thus it was assumed that functions would have derivatives of all orders except for occasional singular points. It was shocking when Weierstrass pointed out that a continuous function might fail to have a derivative at every point.

In response to these issues, the modern view is to initially downplay the importance of any procedure when discussing functions. In principle, one can simply consider the elements x of the domain of the function, and the corresponding values of $f(x)$. The constructive procedure is completely removed, having been replaced by a generalized version of a list or table of function values. Functions are often described as "any rule which produces a single output from any permissible input".

The main emphasis of this chapter is on the properties of functions that make them susceptible to mathematical study, and useful for applications to science and engineering. Starting with the existence of limits, the development continues with such important properties as continuity, uniform continuity, and differentiability. Consequences of these properties will include important elements of calculus such as the Extreme Value Theorem, the Intermediate Value Theorem, and the Mean Value Theorem, along with the various theorems facilitating the calculation of derivatives.

5.1 Basics

Suppose that A and B are two sets. To define a function f, we first identify a subset $D_f \subset A$ called the *domain* of f. It is traditional, at least at an elementary level, to use the following definition: a *function* is a rule for assigning a unique element $f(x) \in B$ to each element $x \in D_f$. The *range* of f, denoted R_f, is the set of all $y \in B$ such that $y = f(x)$ for some $x \in D_f$.

While this definition is useful in practice, there are a few fine points worthy of attention. Suppose f and g are two functions with the same domain, $D_f = D_g$ Suppose too that f and g are defined by distinct rules, which happen to always give the same value for every $x \in D_f$. For example, the domain might be the set $\mathbb{R}$ of all real numbers, and the rules could be

$$f(x) = (x+1)^3, \quad g(x) = x^3 + 3x^2 + 3x + 1.$$

The rules are obviously different, but the result is always the same. In this case we agree to declare the functions equal.

To handle this technical detail, as well as to emphasize the generality of allowed "rules", functions may also be defined in terms of sets. To define a function f, consider a set G_f of ordered pairs (a, b) with $a \in A$, $b \in B$, having the property that if $(a, b_1) \in G_f$ and $(a, b_2) \in G_f$, then $b_1 = b_2$. That is, the second element of the pair is uniquely determined by the first element. The set G_f is sometimes called the graph of f, which is supposed to be the implicitly defined "rule". Notice that in this definition there is no explicit mention of the rule which produces b from a.

For those who have some familiarity with computing, it may help to describe functions with that vocabulary. Functions have inputs and outputs. The inputs are elements of the domain. In programming, the type of the input must usually be specified, and we can think of A as defining the type (Problem 5.1). Similarly, the collection of all outputs is the range of the function, and the type of the output is given by B. Two functions, or procedures, are said to be the same as long as the allowed inputs are the same, and the outputs agree whenever the inputs agree. The notation $f : D_f \to B$ is often used to name a function, its domain, and the type of output. The same notation $f : A \to B$ may also be used to merely specify the type of the inputs and outputs, leaving implicit the exact domain. For example, one might define the rational function $r : \mathbb{R} \to \mathbb{R}$ by $r(x) = 1/x$.

In elementary analysis the domain and range of our functions are usually subsets of the real numbers $\mathbb{R}$, so we may take $A = \mathbb{R}$ and $B = \mathbb{R}$. In fact the domain of a function is often an *interval* I. A set $I \subset \mathbb{R}$ is an interval if for every pair $a, b \in I$, the number $x \in I$ if $a \le x \le b$. Often one is concerned with *open intervals*

$$(a, b) = \{x \mid a < x < b\}$$

or *closed intervals*

$$[a, b] = \{x \mid a \le x \le b\}.$$

A function f is a *polynomial* if it can be written in the form

$$f(x) = \sum_{k=0}^{n} c_k x^k.$$

The *coefficients* c_k will usually be real numbers, although even in elementary algebra it is not uncommon to allow the c_k to be complex numbers. A function g is a *rational function* if it can be written in the form

$$g(x) = \frac{p(x)}{q(x)},$$

where p and q are polynomials, and q is not everywhere 0. The value of a polynomial may be computed whenever x is a real number, and the value of a rational function may be computed whenever x is a real number and $q(x) \neq 0$. When using familiar functions whose domain may be defined by virtue of the operations in the function's rule, the explicit description of the domain is often omitted, with the understanding that the "natural" domain is implied.

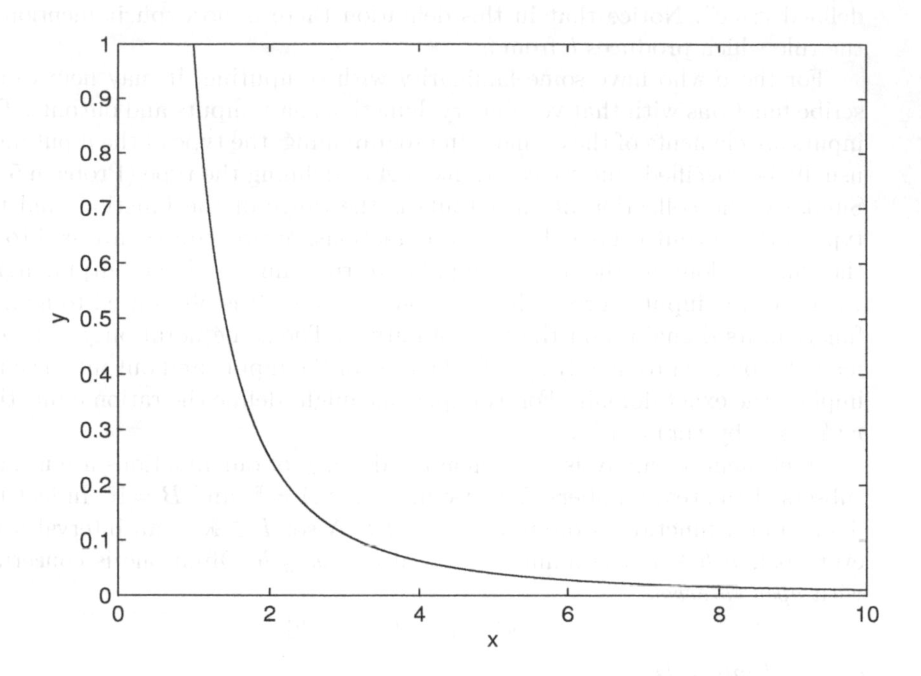

FIGURE 5.1
The graph of $1/x^2$.

5.2 Limits and continuity

5.2.1 Limits

5.2.1.1 Limit as $x \to \infty$

One context in which limits seem natural is when the behavior of a function $f(x)$ is considered for large values of x. Starting with a very simple example, let

$$f(x) = \frac{1}{x^2}.$$

If the graph in Figure 5.1 is to be trusted, it is obvious that $f(x) \to 0$ as $x \to \infty$, or $\lim_{x\to\infty} f(x) = 0$. The challenge is to develop techniques which will apply when the answer is not simply obvious.

If L is a real number, say that $\lim_{x\to\infty} f(x) = L$ (respectively $\lim_{x\to-\infty} f(x) = L$) if for every $\epsilon > 0$ there is a $N > 0$ such that

$$|f(x) - L| < \epsilon$$

whenever $x \geq N$ (resp. $x \leq -N$).

In the case $f(x) = 1/x^2$, some algebraic manipulation is involved to establish that $\lim_{x\to\infty} f(x) = 0$. Pick any number $\epsilon > 0$. The inequality

$$|f(x) - 0| = \left|\frac{1}{x^2}\right| < \epsilon,$$

is the same as requiring

$$x^2 > \frac{1}{\epsilon},$$

or

$$|x| > \frac{1}{\sqrt{\epsilon}}.$$

One possible choice is

$$N = \frac{2}{\sqrt{\epsilon}}.$$

In this case it was productive to work backwards, starting with the desired conclusion, and converting it to an equivalent statement about x. Having understood how big to take x, it is easy to find an N such that whenever $x \geq N$, it follows that $|f(x) - L| < \epsilon$. In fact we have found an explicit formula for N as a function of ϵ. It will not always be possible to obtain such a convenient or explicit expression.

Here is a second example. Let

$$f(x) = \frac{x^2 + 1}{2x^2 + 3}.$$

We claim that $\lim_{x\to\infty} f(x) = 1/2$. Pick any real number $\epsilon > 0$. Write

$$|f(x) - L| = \left|\frac{x^2 + 1}{2x^2 + 3} - \frac{1}{2}\right|$$

$$= \left|\frac{x^2 + 3/2}{2x^2 + 3} - \frac{1/2}{2x^2 + 3} - \frac{1}{2}\right| = \left|-\frac{1/2}{2x^2 + 3}\right| = \frac{1}{2}\left|\frac{1}{2x^2 + 3}\right|.$$

Suppose that $N = 1/\sqrt{\epsilon}$, and $x > N$. Then

$$|f(x) - L| = \frac{1}{2}\left|\frac{1}{2x^2 + 3}\right| < \frac{1}{x^2} \leq \frac{1}{1/\epsilon} = \epsilon.$$

Whenever $x \geq N$, it follows that $|f(x) - 1/2| < \epsilon$, so that $\lim_{x\to\infty} f(x) = 1/2$. Notice that there was some flexibility in our choice of N.

5.2.1.2 Limit as $x \to x_0$

Suppose x_0, a, and b are real numbers, with $a < x_0 < b$. Assume that f is a real valued function defined on the set $(a, x_0) \cup (x_0, b)$; that is, f is defined on some open interval which contains the number x_0, except that f may not be defined at x_0 itself. Say that

$$\lim_{x\to x_0} f(x) = L, \quad L \in \mathbb{R}$$

if for every $\epsilon > 0$ there is a $\delta > 0$ such that

$$|f(x) - L| < \epsilon$$

whenever $0 < |x - x_0| < \delta$.

To amplify on the possible omission of x_0, consider the function

$$f(x) = \frac{\sin(x)}{x}.$$

This formula does not provide a value for f at $x = 0$. Nonetheless we can consider

$$\lim_{x\to 0} \frac{\sin(x)}{x}.$$

In fact this limit exists, and turns out to be 1.

Of course limits arise in the definition of derivatives. Suppose that x_0 is fixed. The following limit problem amounts to computing the derivative of x^2 at the point x_0.

$$\lim_{x\to x_0} \frac{x^2 - x_0^2}{x - x_0} = \lim_{x\to x_0} \frac{(x - x_0)(x + x_0)}{x - x_0} = \lim_{x\to x_0} (x + x_0) = 2x_0.$$

Notice that the function

$$f(x) = \frac{x^2 - x_0^2}{x - x_0}$$

is not defined at x_0 because division by 0 is not defined.

When considering limits as x approaches x_0, it is sometimes convenient to restrict x to $x > x_0$ or $x < x_0$. The definition for this version of limits simply reflects the restriction. To describe a limit from above, say that

$$\lim_{x\to x_0^+} f(x) = L, \quad L \in \mathbb{R}$$

if for every $\epsilon > 0$ there is a $\delta > 0$ such that

$$|f(x) - L| < \epsilon$$

whenever $0 < x - x_0 < \delta$. Similarly, to describe a limit from below, say that

$$\lim_{x\to x_0^-} f(x) = L, \quad L \in \mathbb{R}$$

if for every $\epsilon > 0$ there is a $\delta > 0$ such that

$$|f(x) - L| < \epsilon$$

whenever $0 < x_0 - x < \delta$.

It is also convenient to talk about functions growing without bound. The statement

$$\lim_{x \to x_0} f(x) = \infty$$

means that for every $M > 0$ there is a number $\delta > 0$ such that

$$f(x) > M \quad \text{whenever} \quad 0 < |x - x_0| < \delta.$$

The statement

$$\lim_{x \to \infty} f(x) = \infty$$

means that for every $M > 0$ there is a number $N > 0$ such that

$$f(x) > M \quad \text{whenever} \quad x > M.$$

5.2.1.3 Limit rules

Limits are well behaved with respect to arithmetic operations. The next theorem makes this point precise while also providing a good illustration of the use of existence statements. Since the theorem is a general assertion about limits, rather than an analysis of a particular case, the proof makes use of the general properties, not the details of an example. The proof for this theorem is quite similar to the proof of the analogous theorem for limits of sequences, so only part of the proof is provided. (See Problem 5.6.)

Theorem 5.2.1. *Suppose that L, M and c are real numbers, and that*

$$\lim_{x \to x_0} f(x) = L, \qquad \lim_{x \to x_0} g(x) = M.$$

Then

$$\lim_{x \to x_0} cf(x) = cL, \tag{i}$$

$$\lim_{x \to x_0} [f(x) + g(x)] = L + M, \tag{ii}$$

$$\lim_{x \to x_0} [f(x)g(x)] = LM, \tag{iii}$$

and, if $M \neq 0$,

$$\lim_{x \to x_0} f(x)/g(x) = L/M. \tag{iv}$$

Proof. To prove (i), take any $\epsilon > 0$. From the definition of

$$\lim_{x \to x_0} f(x) = L$$

there is a $\delta > 0$ such that

$$|f(x) - L| < \epsilon$$

whenever $0 < |x - x_0| < \delta$. We consider two cases: $|c| \le 1$, and $|c| > 1$.

Suppose first that $|c| \le 1$. Then whenever $0 < |x - x_0| < \delta$ the desired inequality holds, since

$$|cf(x) - cL| < |c|\epsilon < \epsilon.$$

Next suppose that $|c| > 1$. Let $\epsilon_1 = \epsilon/|c|$. Since

$$\lim_{x \to x_0} f(x) = L$$

there is a $\delta > 0$ such that $0 < |x - x_0| < \delta$ implies

$$|f(x) - L| < \epsilon_1.$$

But this means that

$$|cf(x) - cL| < |c|\epsilon_1 = |c|\epsilon/|c| = \epsilon.$$

To prove (ii), take any $\epsilon > 0$ and define $\epsilon_1 = \epsilon/2$. From the limit definitions there are δ_1 and δ_2 such that $|f(x) - L| < \epsilon_1$ if $0 < |x - x_0| < \delta_1$, and $|g(x) - M| < \epsilon_1$ if $0 < |x - x_0| < \delta_2$.

Take $\delta = \min(\delta_1, \delta_2)$. If $0 < |x - x_0| < \delta$, then

$$|(f(x) + g(x)) - (L + M)| \le |f(x) - L| + |g(x) - M| < \epsilon_1 + \epsilon_1 = \epsilon/2 + \epsilon/2 = \epsilon.$$

□

In Theorem 5.2.1 the statement

$$\lim_{x \to x_0} f(x)/g(x) = L/M, \quad M \ne 0,$$

deserves a comment. One can show (see Problem 5.5) that if $\lim_{x \to x_0} g(x) = M$, and $M \ne 0$, then there is some $\delta > 0$ such that $g(x) \ne 0$ for $0 < |x - x_0| < \delta$. In this set the quotient $f(x)/g(x)$ will be defined, and the limit may be considered.

The limit rules of Theorem 5.2.1 for

$$\lim_{x \to x_0} f(x) = L$$

also apply in the cases of

$$\lim_{x \to \infty} f(x) = L,$$

and

$$\lim_{x \to x_0^{\pm}} f(x) = L.$$

5.2.2 Continuity

Suppose that $I \subset \mathbb{R}$ is an interval. A function $f : I \to \mathbb{R}$ is *continuous at* $x_0 \in I$ if

$$\lim_{x \to x_0} f(x) = f(x_0).$$

If x_0 is the left or right endpoint of the interval I, this limit is taken to be the limit from above or below, as appropriate. The function f is said to be *continuous*, or continuous on I, if f is continuous at every point of I. When $I = [a, b]$ is a closed interval, saying that f is continuous on I means that

$$\lim_{x \to x_0} f(x) = f(x_0), \quad a < x_0 < b,$$

$$\lim_{x \to a^+} f(x) = f(a), \quad \lim_{x \to b^-} f(x) = f(b).$$

Notice that if f is a continuous function on an interval I, then f is also continuous on every interval $I_1 \subset I$.

Theorem 5.2.1 shows that if f and g are continuous at x_0, so are $f + g$ and fg. If $g(x_0) \ne 0$ the function f/g is also continuous at x_0. It is easy to check that the function $f(x) = x$ is continuous for any interval I. It follows immediately that polynomials are continuous on any interval I, and rational functions are continuous at each point where the denominator is not 0. Problem 5.9 and Problem 5.11 describe other ways of exhibiting continuous functions.

Before diving into the next theorem, it will help to make an observation. Suppose that f is continuous at x_0. This means that $\lim_{x \to x_0} f(x)$ exists, that $f(x_0)$ is defined, and that these two numbers are the same. For a real valued function f defined on an open interval I, the definition of continuity of f at x_0 can thus be written as follows: for every $\epsilon > 0$ there is a $\delta > 0$ such that

$$|f(x) - f(x_0)| < \epsilon$$

whenever

$$|x - x_0| < \delta.$$

In the definition of limit the inequality $0 < |x - x_0| < \delta$ appeared; for the case of continuity the possibility $x = x_0$ is included.

It is sometimes convenient to use the alternate characterization of continuity provided by the next result. The theorem is stated for functions defined on open intervals, but the same result holds for arbitrary intervals if appropriate one-sided limits are used.

Theorem 5.2.2. *Suppose that I is an open interval, $x_0 \in I$, and $f : I \to \mathbb{R}$. Then f is continuous at x_0 if and only if*

$$\lim_{n \to \infty} f(x_n) = f(x_0)$$

whenever $\{x_n\}$ is a sequence in I with limit x_0.

Proof. First suppose that f is continuous at x_0, that is

$$\lim_{x \to x_0} f(x) = f(x_0).$$

Picking any $\epsilon > 0$, there is a $\delta > 0$ such that

$$|f(x) - f(x_0)| < \epsilon$$

whenever

$$0 \le |x - x_0| < \delta.$$

Now assume that $\{x_n\}$ is a sequence in I with limit x_0. Using δ from above, there is a positive integer N such that

$$|x_n - x_0| < \delta$$

whenever $n \ge N$. Of course this means that when $n \ge N$ the inequality

$$|f(x_n) - f(x_0)| < \epsilon,$$

holds, showing that

$$\lim_{n \to \infty} f(x_n) = f(x_0).$$

To show the converse implication, assume that

$$\lim_{x \to x_0} f(x) \ne f(x_0).$$

Either the limit fails to exist, or the limit exists, but its value is different from $f(x_0)$. In either case there is some $\epsilon_1 > 0$ such that for any $\delta > 0$

$$|f(z) - f(x_0)| \ge \epsilon_1$$

for some z satisfying

$$0 < |z - x_0| < \delta.$$

Since I is an open interval, there is a number $r > 0$ such that

$$(x_0 - r, x_0 + r) \subset I.$$

For $k = 1, 2, 3, \dots$ let $\delta_k = \min(1/k, r)$. Pick x_k such that

$$0 < x_k - x_0 < \delta_k$$

and

$$|f(x_k) - f(x_0)| \ge \epsilon_1.$$

By construction the sequence $\{x_k\}$ has limit x_0, but

$$\lim_{k \to \infty} f(x_k) \ne f(x_0).$$

□

Theorem 5.2.3. *(Extreme Value Theorem) Suppose that $f : [a,b] \to \mathbb{R}$ is a continuous function on the compact interval $[a,b]$. Then there are points $x_{min}, x_{max} \in [a,b]$ such that*

$$f(x_{min}) \le f(x), \quad a \le x \le b,$$

$$f(x_{max}) \ge f(x), \quad a \le x \le b.$$

Proof. If the range of f is bounded above, let y_{max} denote the least upper bound of the range. If the range of f is not bounded above, write $y_{max} = \infty$. Let x_n be a sequence of points in $[a,b]$ such that

$$\lim_{n\to\infty} f(x_n) = y_{max}.$$

Since the interval $[a,b]$ is compact, the sequence x_n has a subsequence $x_{n(k)}$ which converges to $z \in [a,b]$. Since f is continuous at z,

$$\lim_{k\to\infty} f(x_{n(k)}) = f(z) = y_{max}.$$

Thus $y_{max} \in \mathbb{R}$ and we may take $x_{max} = z$. The existence of x_{min} has a similar proof. □

Theorem 5.2.4. *(Intermediate Value Theorem) Suppose $f : [a,b] \to \mathbb{R}$ is a continuous function, and $f(a) < f(b)$. For every number $y \in [f(a), f(b)]$ there is an $x \in [a,b]$ such that $f(x) = y$.*

Proof. The set $J = \{x \in [a,b] \mid f(x) \le y\}$ is nonempty, and has a least upper bound $z \le b$. Pick a sequence of points $x_n \in J$ converging to z. Since $f(x_n) \le y$ for each n, and f is continuous,

$$\lim_{n\to\infty} f(x_n) = f(z) \le y. \tag{5.1}$$

It is possible that $z = b$. In this case $y \le f(b)$ by assumption, and $f(b) = f(z) \le y$ by (5.1), so $y = f(b)$ and the desired point x is b.

If $z < b$, then $f(x) > y$ for every $x \in (z,b]$ by the definition of J. Pick a sequence of points $w_n \in (z,b]$ such that $\{w_n\}$ converges to z. Since $f(w_n) > y$,

$$f(z) = \lim_{n\to\infty} f(w_n) \ge y.$$

Now $f(z) \ge y$ and $f(z) \le y$, so it follows that $f(z) = y$. □

The sequential characterization of continuity is also useful for establishing the next result, which says that the composition of continuous functions is continuous.

Theorem 5.2.5. *Suppose that I_0 and I_1 are open intervals, that $f : I_0 \to \mathbb{R}$, $g : I_1 \to \mathbb{R}$, and that $f(I_0) \subset I_1$. Assume that f is continuous at $x_0 \in I_0$, and g is continuous at $f(x_0) \in I_1$. Then $g(f(x))$ is continuous at x_0.*

Proof. Suppose that $\{x_k\}$ is any sequence in I_0 with limit x_0. Since f is continuous at x_0, the sequence $y_k = f(x_k)$ is a sequence in I_1 with limit $y_0 = f(x_0)$. Since g is continuous at y_0, we also have

$$\lim_{k \to \infty} g(y_k) = g(y_0),$$

or

$$\lim_{k \to \infty} g(f(x_k)) = g(f(x_0)),$$

as desired. □

5.2.2.1 Rootfinding 1

Suppose $f : I \to \mathbb{R}$ is a function defined on the interval I. The number x is said to be a root of f if $f(x) = 0$. The accurate approximation of roots is a common problem of computational mathematics. As an example one might consider finding solutions of the equations

$$\tan(x) - x - 1 = 0, \quad 0 \le x < \pi/2,$$

or

$$x^7 + 3x^6 + 17x^2 + 4 = 0.$$

The Intermediate Value Theorem can be used to justify a simple computational technique called the bisection method.

Suppose that f is continuous on I, and there are points $a, b \in I$ such that $a < b$ and $f(a)f(b) < 0$. This last condition simply means that f is positive at one of the two points, and negative at the other point. By the Intermediate Value Theorem there must be a root in the interval $[a, b]$. There is no loss of generality in assuming that $f(a) < 0$ and $f(b) > 0$ since if the signs are switched we can simply replace f by the function $g = -f$. The functions f and g have the same roots.

We will now define two sequences of points $\{a_n\}$, $\{b_n\}$, starting with $a_0 = a$ and $b_0 = b$. The definition of the subsequent points in the sequence is given recursively. The points a_n and b_n will always be chosen so that $f(a_n)f(b_n) \le 0$, and if $f(a_n)f(b_n) = 0$ then either a_n or b_n is a root.

Let c_n be the midpoint of the interval $[a_n, b_n]$, or

$$c_n = \frac{a_n + b_n}{2}.$$

If $f(c_n) = 0$, we have a root and can stop. If $f(c_n) < 0$, define $a_{n+1} = c_n$, and $b_{n+1} = b_n$. If $f(c_n) > 0$, define $a_{n+1} = a_n$, and $b_{n+1} = c_n$. Since $f(a_{n+1}) < 0$ and $f(b_{n+1}) > 0$, the Intermediate Value Theorem implies that a root lies in the interval $[a_{n+1}, b_{n+1}]$.

Finally, notice that

$$|b_{n+1} - a_{n+1}| = \frac{1}{2}|b_n - a_n|.$$

By induction this means that

$$|b_n - a_n| = 2^{-n}|b - a|.$$

The intervals $[a_n, b_n]$ are nested, and the lengths $|b_n - a_n|$ have limit 0, so by the Nested Interval Principle there is a number r such that

$$\lim_{n\to\infty} a_n = r = \lim_{n\to\infty} b_n.$$

Since $f(a_n) < 0$,

$$f(r) = \lim_{n\to\infty} f(a_n) \le 0.$$

On the other hand, $f(b_n) > 0$, so

$$f(r) = \lim_{n\to\infty} f(b_n) \ge 0,$$

and $f(r) = 0$. Moreover

$$|b_n - r| \le |b_n - a_n| = 2^{-n}|b - a|,$$

so an accurate estimate of the root is obtained rapidly.

5.2.3 Uniform continuity

To begin discussing uniform continuity, it will be helpful to review the definition of continuity at a point x_0, which could have been phrased as follows. The function $f : I \to \mathbb{R}$ is continuous at a point $x_0 \in I$ if for every $\epsilon > 0$ there is a $\delta(\epsilon, x_0) > 0$ such that $|f(x) - f(x_0)| < \epsilon$ whenever $|x - x_0| < \delta(\epsilon, x_0)$. The new emphasis is on the possible dependence of δ on both ϵ and x_0.

To illustrate this point, consider the example $f(x) = 1/x$ on the interval $(0, \infty)$. For this function, the value of δ will depend on both ϵ and x_0. To see this, note that the condition

$$|\frac{1}{x} - \frac{1}{x_0}| < \epsilon,$$

for $x \in (0, \infty)$ means that

$$\frac{1}{x_0} - \epsilon < \frac{1}{x} < \frac{1}{x_0} + \epsilon,$$

or

$$\frac{1 - \epsilon x_0}{x_0} < \frac{1}{x} < \frac{1 + \epsilon x_0}{x_0}.$$

This requires

$$\frac{x_0}{1 + \epsilon x_0} < x < \frac{x_0}{1 - \epsilon x_0}.$$

If a fixed $\epsilon > 0$ is chosen, then the size of the interval

$$|x - x_0| < \delta(\epsilon, x_0)$$

where

$$|\frac{1}{x} - \frac{1}{x_0}| < \epsilon$$

shrinks to 0 as x_0 goes to 0.

For other functions or other intervals it may be possible to choose δ independent of the value of x_0. This leads to the notion of a uniformly continuous function. Say that $f : I \to \mathbb{R}$ is *uniformly continuous* on I if for every $\epsilon > 0$ there is a $\delta(\epsilon) > 0$, such that $|f(y)-f(x)| < \epsilon$ whenever $x, y \in I$ and $|y-x| < \delta$.

Theorem 5.2.6. *If f is continuous on a compact interval I, then f is uniformly continuous on I.*

Proof. The argument is by contradiction. If f is not uniformly continuous then there is some $\epsilon_1 > 0$ such that for every $\delta > 0$ there are points x and y in I satisfying $|y - x| < \delta$, but for which $|f(y) - f(x)| \ge \epsilon_1$. For this ϵ_1 consider $\delta_n = 1/n$, and let x_n and a_n be points such that $|x_n - a_n| < 1/n$, but $|f(x_n) - f(a_n)| \ge \epsilon_1$.

Since the interval I is compact, the sequence a_n has a convergent subsequence $a_{n(k)}$. Suppose that the limit of this subsequence is c. By the triangle inequality,

$$|x_{n(k)} - c| \le |x_{n(k)} - a_{n(k)}| + |a_{n(k)} - c| \le 1/n(k) + |a_{n(k)} - c|,$$

so that c is also the limit of the subsequence $x_{n(k)}$.

The function f is assumed to be continuous at c. Let $\epsilon = \epsilon_1/2$. There is a $\delta(\epsilon_1/2, c)$ such that $|f(x) - f(c)| < \epsilon_1/2$ whenever $|x - c| < \delta(\epsilon_1/2, c)$.

Now use the triangle inequality again to get

$$|f(x_{n(k)}) - f(a_{n(k)})| \le |f(x_{n(k)}) - f(c)| + |f(c) - f(a_{n(k)})|. \tag{5.2}$$

For k large enough, (5.2) will lead to a contradiction. When both $|x_{n(k)} - c|$ and $|a_{n(k)} - c|$ are smaller than $\delta(\epsilon_1/2, c)$, the inequalities $|f(x_{n(k)}) - f(c)| < \epsilon_1/2$ and $|f(c) - f(a_{n(k)})| < \epsilon_1/2$ are valid. (5.2) then gives $|f(x_{n(k)}) - f(a_{n(k)})| < \epsilon_1$, while the sequences $\{x_n\}$ and $\{a_n\}$ were chosen so that $|f(x_n)-f(a_n)| \ge \epsilon_1$. This contradiction implies that f must have been uniformly continuous. □

Of course it is possible to have a uniformly continuous function on a noncompact interval. Suppose f is uniformly continuous on the compact interval $[a, b]$. Then f is also uniformly continuous on every interval $(c, d) \subset [a, b]$.

One striking consequence of Theorem 5.2.6 involves the approximation of a continuous function f on a compact interval $[a, b]$ by functions of a particularly simple type. Say that a function $g : [a, b] \to \mathbb{R}$ is a *step function* if there is a finite collection of points $x_0, \dots, x_N$ such that

$$a = x_0 < x_1 < \cdots < x_N = b,$$

and $g(x)$ is constant on each of the intervals (x_n, x_{n+1}). The approximation of continuous functions by step functions is extremely important for developing the theory of integration using Riemann sums.

One way of getting a step function g from an arbitrary function f is to define g using samples of f from the intervals $[x_n, x_{n+1}]$, where $n = 0, \dots, N - 1$. For instance the left endpoint Riemann sums commonly seen in calculus use the function

$$g_l(x) = \begin{Bmatrix} f(x_n), & x_n \le x < x_{n+1} \\ f(x_{N-1}), & x = b. \end{Bmatrix}$$

More generally, consider the sample points ξ_n, where $x_n \le \xi_n \le x_{n+1}$, and define

$$g(x) = \begin{Bmatrix} f(\xi_n), & x_n \le x < x_{n+1} \\ f(\xi_{N-1}), & x = b. \end{Bmatrix} \tag{5.3}$$

A corollary of Theorem 5.2.6 is that it is always possible to approximate a continuous function f on a compact interval as well as you like with a step function.

Corollary 5.2.7. *Suppose that f is continuous on a compact interval $[a, b]$, and g is a step function, defined as in (5.3). Then for any $\epsilon > 0$ there is a $\delta > 0$ such that*

$$|f(x) - g(x)| < \epsilon, \quad x \in [a, b]$$

if

$$0 < x_{n+1} - x_n < \delta, \quad n = 0, \dots, N - 1.$$

Proof. By Theorem 5.2.6 the function f is uniformly continuous on $[a, b]$. Given $\epsilon > 0$, let δ be chosen so that

$$|f(y) - f(x)| < \epsilon \quad \text{if} \quad |y - x| < \delta.$$

Pick a finite collection of points $a = x_0 < x_1 < \cdots < x_N = b$ from $[a, b]$ and suppose that $x_n \le \xi_n \le x_{n+1}$.

Assume that $g(x)$ is defined as in (5.3), and

$$|x_{n+1} - x_n| < \delta, \quad n = 0, \dots, N - 1.$$

For each $x \in [a, b)$, we have $x_n \le x < x_{n+1}$ for some n. Because

$$|\xi_n - x| \le |x_{n+1} - x_n| < \delta,$$

it follows that

$$|f(x) - g(x)| = |f(x) - f(\xi_n)| < \epsilon.$$

The argument is essentially the same for $x = b$. □

5.3 Derivatives

The notion of the derivative of a function is essential for the study of various basic problems: how to make sense of velocity and acceleration for objects in motion, how to define and compute tangents to curves, and how to minimize or maximize functions. These problems were studied with some success in the seventeenth century by a variety of researchers, including Pierre de Fermat (1601–65), René Descartes (1596–1650), and Isaac Barrow (1630–77). In the later part of the seventeenth century derivatives became a central feature of the calculus developed by Isaac Newton (1642–1727) and Gottfried Wilhelm Leibniz (1646–1716) [5, pp. 342–390].

Suppose that (a,b) is an open interval and $f:(a,b)\to\mathbb{R}$. The *derivative* of f at $x_0\in(a,b)$ is

$$f'(x_0)=\lim_{x\to x_0}\frac{f(x)-f(x_0)}{x-x_0}$$

if this limit exists. If the derivative exists, the function f is said to be *differentiable* at x_0. The function f is differentiable on the interval (a,b) if it has a derivative at each $x\in(a,b)$.

When a function f is differentiable on an open interval (a,b), then the derivative function $f'(x)$ may itself have derivatives at $x_0\in(a,b)$. If f is differentiable on (a,b), the *second derivative* of f at $x_0\in(a,b)$ is

$$f''(x_0)=\lim_{x\to x_0}\frac{f'(x)-f'(x_0)}{x-x_0}$$

if this limit exists. Denoting repeated differentiation of f with more $'$s leads to an unwieldy notation. As an alternative, write $f^{(1)}(x_0)$ for $f'(x_0)$, and $f^{(2)}(x_0)$ for $f''(x_0)$. Continuing in this fashion, if $f,f^{(1)},\dots,f^{(n-1)}$ are differentiable on (a,b), the *n-th derivative* of f at $x_0\in(a,b)$ is

$$f^{(n)}(x_0)=\lim_{x\to x_0}\frac{f^{(n-1)}(x)-f^{(n-1)}(x_0)}{x-x_0}$$

if this limit exists.

It is often desirable to talk about a function which is differentiable on a closed interval $[a,b]$. This will mean that there is some open interval (c,d) such that $[a,b]\subset(c,d)$, and the function $f:(c,d)\to\mathbb{R}$ is differentiable. (Alternatively, we could ask for the existence of limits from above and below at a and b respectively.)

By defining $h=x-x_0$, which gives $x=x_0+h$, the derivative may be defined in the equivalent form

$$f'(x_0)=\lim_{h\to 0}\frac{f(x_0+h)-f(x_0)}{h}.$$

When it is notationally convenient the derivative is written as

$$\frac{df}{dx}(x_0) = f'(x_0),$$

and higher derivatives are

$$\frac{d^n f}{dx^n}(x_0) = f^{(n)}(x_0).$$

Occasionally one also encounters the notation

$$D^n f(x_0) = f^{(n)}(x_0).$$

5.3.1 Computation of derivatives

By virtue of Theorem 5.2.1, sums and constant multiples of differentiable functions are differentiable.

Lemma 5.3.1. *Suppose that $f : (a,b) \to \mathbb{R}$, $g : (a,b) \to \mathbb{R}$, and $c \in \mathbb{R}$. If f and g have derivatives at $x_0 \in (a,b)$, so do cf and $f+g$, with*

$$(f+g)'(x_0) = f'(x_0) + g'(x_0),$$

$$(cf)'(x_0) = cf'(x_0).$$

Proof. An application of Theorem 5.2.1 gives

$$(f+g)'(x_0) = \lim_{x \to x_0} \frac{(f(x)+g(x)) - (f(x_0)+g(x_0))}{x - x_0}$$

$$= \lim_{x \to x_0} \Big(\frac{f(x) - f(x_0)}{x - x_0} + \frac{g(x)) - g(x_0)}{x - x_0} \Big)$$

$$= \lim_{x \to x_0} \frac{f(x) - f(x_0)}{x - x_0} + \lim_{x \to x_0} \frac{g(x)) - g(x_0)}{x - x_0} = f'(x_0) + g'(x_0),$$

and

$$(cf)'(x_0) = \lim_{x \to x_0} \frac{cf(x) - cf(x_0)}{x - x_0} = \lim_{x \to x_0} c\frac{f(x) - f(x_0)}{x - x_0}$$

$$= c \lim_{x \to x_0} \frac{f(x) - f(x_0)}{x - x_0} = cf'(x_0).$$

□

Having a derivative at x_0 is a stronger requirement than being continuous at x_0.

Theorem 5.3.2. *If f has a derivative at x_0, then f is continuous at x_0.*

Proof. Write $x = x_0 + h$, and consider the following calculation.

$$\lim_{x \to x_0} f(x) - f(x_0) = \lim_{h \to 0} f(x_0 + h) - f(x_0) = \lim_{h \to 0} \frac{f(x_0 + h) - f(x_0)}{h} h$$

$$= \lim_{h \to 0} \frac{f(x_0 + h) - f(x_0)}{h} \lim_{h \to 0} h = f'(x_0) \cdot 0 = 0.$$

Thus

$$\lim_{x \to x_0} f(x) = f(x_0).$$

□

It is also possible to differentiate products and quotients, with rules familiar from calculus.

Theorem 5.3.3. *If $f : (a,b) \to \mathbb{R}$ and $g : (a,b) \to \mathbb{R}$ have derivatives at $x_0 \in (a,b)$, so does fg, with*

$$(fg)'(x_0) = f'(x_0)g(x_0) + f(x_0)g'(x_0).$$

If in addition $g(x_0) \neq 0$, then f/g has a derivative at x_0, with

$$\left(\frac{f}{g}\right)'(x_0) = \frac{f'(x_0)g(x_0) - f(x_0)g'(x_0)}{g^2(x_0)}.$$

Proof. The addition of 0 is helpful. First,

$$(fg)'(x_0) = \lim_{x \to x_0} \frac{f(x)g(x) - f(x_0)g(x_0)}{x - x_0}$$

$$= \lim_{x \to x_0} \frac{f(x)g(x) - f(x_0)g(x) + f(x_0)g(x) - f(x_0)g(x_0)}{x - x_0}$$

$$= \lim_{x \to x_0} g(x)\frac{f(x) - f(x_0)}{x - x_0} + \lim_{x \to x_0} f(x_0)\frac{g(x) - g(x_0)}{x - x_0}$$

$$= g(x_0) \lim_{x \to x_0} \frac{f(x) - f(x_0)}{x - x_0} + f(x_0) \lim_{x \to x_0} \frac{g(x) - g(x_0)}{x - x_0}$$

$$= f'(x_0)g(x_0) + f(x_0)g'(x_0),$$

since the limit of a product is the product of the limits by Theorem 5.2.1, and g is continuous at x_0.

A similar technique establishes the quotient rule.

$$(f/g)'(x_0) = \lim_{x \to x_0} \frac{1}{x - x_0}\left(\frac{f(x)}{g(x)} - \frac{f(x_0)}{g(x_0)}\right) = \lim_{x \to x_0} \frac{1}{x - x_0} \frac{f(x)g(x_0) - g(x)f(x_0)}{g(x)g(x_0)}$$

$$= \lim_{x \to x_0} \frac{1}{x - x_0} \frac{(f(x) - f(x_0))g(x_0) - (g(x) - g(x_0)f(x_0)}{g(x)g(x_0)}$$

$$= \lim_{x\to x_0} \frac{g(x_0)}{g(x)g(x_0)} \frac{f(x)-f(x_0)}{x-x_0} - \lim_{x\to x_0} \frac{f(x_0)}{g(x)g(x_0)} \frac{g(x)-g(x_0)}{x-x_0}$$
$$= \frac{f'(x_0)g(x_0)}{g^2(x_0)} - \frac{f(x_0)g'(x_0)}{g^2(x_0)}.$$

□

Thanks to Theorem 5.3.3 any rational function is differentiable as long as the denominator is not zero. Some additional ways of exhibiting differentiable functions are explored in Problem 5.28 and Problem 5.37.

Another important differentiation rule is the chain rule, which tells us how to differentiate the composition of two functions. Recall the alternate notations for composition,

$$f(g(x)) = (f \circ g)(x).$$

The chain rule says roughly that if f and g are differentiable, then

$$(f \circ g)'(x_0) = f'(g(x_0))g'(x_0).$$

In preparation for the proof of the chain rule, we establish a series of lemmas which follow quickly from the definition of the derivative.

Lemma 5.3.4. *Suppose that g has a derivative at x_0. If $g'(x_0) \neq 0$, then there is a $\delta > 0$ such that $0 < |x - x_0| < \delta$ implies*

$$|g'(x_0)||x - x_0|/2 < |g(x) - g(x_0)| < 2|g'(x_0)||x - x_0|.$$

If $g'(x_0) = 0$, then for any $\epsilon > 0$ there is a $\delta > 0$ such that $0 < |x - x_0| < \delta$ implies

$$|g(x) - g(x_0)| \le \epsilon |x - x_0|.$$

Proof. Assume that $g'(x_0) \neq 0$. There is no loss of generality in assuming that $g'(x_0) > 0$, since if $g'(x_0) < 0$ the function $-g$ can be considered instead.

Take $\epsilon = g'(x_0)/2$. From the limit definition there is a $\delta > 0$ such that $0 < |x - x_0| < \delta$ implies

$$\left|\frac{g(x) - g(x_0)}{x - x_0} - g'(x_0)\right| < g'(x_0)/2,$$

which is the same as

$$g'(x_0) - g'(x_0)/2 < \frac{g(x) - g(x_0)}{x - x_0} < g'(x_0) + g'(x_0)/2.$$

Since the middle term is positive, and $3/2 < 2$,

$$g'(x_0)/2 < \left|\frac{g(x) - g(x_0)}{x - x_0}\right| < 2g'(x_0).$$

Multiply by $|x - x_0|$ to get the first result.

In case $g'(x_0) = 0$ the limit definition says that for any $\epsilon > 0$ there is a $\delta > 0$ such that $0 < |x - x_0| < \delta$ implies

$$|\frac{g(x) - g(x_0)}{x - x_0}| < \epsilon.$$

Multiply by $|x - x_0|$ to get the desired inequality. □

Lemma 5.3.5. *Suppose that $g'(x_0) \neq 0$. Then there is a $\delta > 0$ such that $0 < |x - x_0| < \delta$ implies $g(x) \neq g(x_0)$.*

Proof. Since $g'(x_0) \neq 0$, the previous lemma says there is a $\delta > 0$ such that $0 < |x - x_0| < \delta$ implies

$$|g(x) - g(x_0)| > |x - x_0||g'(x_0)|/2.$$

Since $|x - x_0| \neq 0$, it follows that $g(x) - g(x_0) \neq 0$. □

The last lemma develops an estimate valid for any value of $g'(x_0)$.

Lemma 5.3.6. *Suppose that g has a derivative at x_0. Then there is a $\delta > 0$ such that $0 < |x - x_0| < \delta$ implies*

$$|g(x) - g(x_0)| \leq \big[1 + 2|g'(x_0)|\big]\ |x - x_0|.$$

Proof. One conclusion of Lemma 5.3.4 is that for x close to x_0 either

$$|g(x) - g(x_0)| \leq 2|g'(x_0)||x - x_0|,$$

or for any $\epsilon > 0$,

$$|g(x) - g(x_0)| \leq \epsilon|x - x_0|,$$

depending on the value of $g'(x_0)$. In this last inequality take $\epsilon = 1$. In any case $|g(x) - g(x_0)|$ will be smaller than the sum of the right-hand sides, which is the claim. □

Theorem 5.3.7. *(Chain rule) Suppose that $f : (a, b) \to \mathbb{R}$, $g : (c, d) \to \mathbb{R}$, g is differentiable at $x_0 \in (c, d)$, and f is differentiable at $g(x_0) \in (a, b)$. Then $f \circ g$ is differentiable at x_0 and*

$$(f \circ g)'(x_0) = f'(g(x_0))g'(x_0).$$

Proof. First suppose that $g'(x_0) \neq 0$. Lemma 5.3.5 assures us that $g(x) \neq g(x_0)$ for x close to x_0, so

$$\lim_{x \to x_0} \frac{f(g(x)) - f(g(x_0))}{x - x_0}$$

$$= \lim_{x \to x_0} \frac{f(g(x)) - f(g(x_0))}{g(x) - g(x_0)} \frac{g(x) - g(x_0)}{x - x_0}.$$

The existence of the limit

$$\lim_{x \to x_0} \frac{g(x) - g(x_0)}{x - x_0} = g'(x_0)$$

was part of the hypotheses. Let $y_0 = g(x_0)$ and let $h : (a, b) \to \mathbb{R}$ be the function

$$h(y) = \left\{ \begin{matrix} \frac{f(y) - f(y_0)}{y - y_0}, & y \neq y_0, \\ f'(y_0), & y = y_0 \end{matrix} \right\}.$$

The assumption that f is differentiable at y_0 is precisely the assumption that h is continuous at y_0. Since g is continuous at x_0, Theorem 5.2.5 says that $h(g(x))$ is continuous at x_0, or

$$\lim_{x \to x_0} \frac{f(g(x)) - f(g(x_0))}{g(x) - g(x_0)} = f'(g(x_0)).$$

Since the product of the limits is the limit of the product, the pieces may be put together to give

$$(f \circ g)'(x_0) = \lim_{x \to x_0} \frac{f(g(x)) - f(g(x_0))}{g(x) - g(x_0)} \lim_{x \to x_0} \frac{g(x) - g(x_0)}{x - x_0} = f'(g(x_0))g'(x_0).$$

Now the case $g'(x_0) = 0$ is considered. Take any $\epsilon > 0$. Since g is continuous at x_0, and f has a derivative at $g(x_0)$, Lemma 5.3.6 shows that for x close enough to x_0

$$|f(g(x)) - f(g(x_0))| \le \left[1 + 2|f'(g(x_0))|\right] \; |g(x) - g(x_0)|,$$

and

$$|g(x) - g(x_0)| \le \epsilon |x - x_0|.$$

Putting these estimates together yields

$$|f(g(x)) - f(g(x_0))| \le \left[1 + 2|f'(g(x_0))|\right]\epsilon |x - x_0|.$$

This is the same as

$$|\frac{f(g(x)) - f(g(x_0))}{x - x_0} - 0| \le \left[1 + 2|f'(g(x_0))|\right]\epsilon,$$

or

$$0 = \lim_{x \to x_0} \frac{f(g(x)) - f(g(x_0))}{x - x_0} = (f \circ g)'(x_0) = f'(g(x_0))g'(x_0).$$

□

5.3.2 The Mean Value Theorem

A function $f : (a,b) \to \mathbb{R}$ is said to have a *local maximum* at $x_0 \in (a,b)$ if there is a $\delta > 0$ such that $f(x_0) \ge f(x)$ for all $x \in (x_0 - \delta, x_0 + \delta)$. A *local minimum* is defined similarly. The function f is said to have a *local extreme point* at x_0 if there is either a local maximum or minimum at x_0.

Lemma 5.3.8. *Suppose that $f : (a,b) \to \mathbb{R}$ has a local extreme point at $x_0 \in (a,b)$. If f has a derivative at x_0, then $f'(x_0) = 0$.*

Proof. Suppose that f has a local maximum at x_0; the case of a local minimum is similar.

There is a $\delta > 0$ such that $f(x) \le f(x_0)$ for all $x \in (x_0 - \delta, x_0 + \delta)$. This means that for $x_0 < x < x_0 + \delta$

$$\frac{f(x) - f(x_0)}{x - x_0} \le 0,$$

and for $x_0 - \delta < x < x_0$

$$\frac{f(x) - f(x_0)}{x - x_0} \ge 0.$$

Since the derivative is the limit of these difference quotients, it follows that $f'(x_0) \le 0$ and $f'(x_0) \ge 0$, which forces $f'(x_0) = 0$. □

Theorem 5.3.9. *(Rolle's Theorem) Suppose that $f : [a,b] \to \mathbb{R}$ is continuous on $[a,b]$, and differentiable on the open interval (a,b). If $f(a) = f(b) = 0$, then there is some point $x_0 \in (a,b)$ with $f'(x_0) = 0$.*

Proof. By Theorem 5.2.3 the function f has an extreme value at some point $x_0 \in [a,b]$. If the function f is zero at every point of $[a,b]$, then $f'(x_0) = 0$ for every $x_0 \in [a,b]$. Otherwise f must have a maximum or minimum at some point $x_0 \in (a,b)$, and $f'(x_0) = 0$ by Lemma 5.3.8

□

Rolle's theorem looks special because of the requirement that $f(a) = f(b) = 0$, but it is easy to use it to produce a more flexible result.

Theorem 5.3.10. *(Mean Value Theorem) Suppose that $g : [a,b] \to \mathbb{R}$ is continuous on $[a,b]$, and differentiable on the open interval (a,b). Then there is some point $x_0 \in (a,b)$ with*

$$g'(x_0) = \frac{g(b) - g(a)}{b - a}.$$

Proof. The idea is to apply Rolle's theorem to a modified function f. The new function is

$$f(x) = g(x) - g(a) - (x - a)\frac{g(b) - g(a)}{b - a}.$$

With this choice $f(a) = f(b) = 0$. By Rolle's theorem there is an $x_0 \in (a, b)$ such that $f'(x_0) = 0$, or

$$0 = g'(x_0) - \frac{g(b) - g(a)}{b - a},$$

as desired. □

The Mean Value Theorem may be used to show that functions with positive derivatives are increasing. Recall that $f : I \to \mathbb{R}$ is *increasing* if

$$f(x_1) \le f(x_2) \quad \text{whenever} \quad x_1 < x_2, \quad x_1, x_2 \in I.$$

The function f is *strictly increasing* if

$$f(x_1) < f(x_2) \quad \text{whenever} \quad x_1 < x_2, \quad x_1, x_2 \in I.$$

Obvious modifications provide the definitions of decreasing and strictly decreasing functions.

Theorem 5.3.11. *Suppose $f : [a, b] \to \mathbb{R}$ is continuous, and differentiable on (a, b), with $f'(x) > 0$ for $x \in (a, b)$. Then f is strictly increasing on $[a, b]$.*

Proof. If f is not strictly increasing, then there are points $x_1 < x_2$ in $[a, b]$ such that

$$\frac{f(x_2) - f(x_1)}{x_2 - x_1} \le 0.$$

By the Mean Value Theorem there is a point $x_0 \in (a, b)$ where $f'(x_0) \le 0$, contradicting the hypotheses. □

The Mean Value Theorem also implies that the amount a function $f(x)$ can change over the interval $[x_1, x_2]$ is controlled by the magnitude of the derivative $|f'(x)|$ on that interval. Notice in particular that the following result shows that functions with bounded derivatives are uniformly continuous.

Theorem 5.3.12. *Suppose that $f : (a, b) \to \mathbb{R}$ is differentiable and $m \le |f'(x)| \le M$ for $x \in (a, b)$. Then for all $x_1, x_2 \in (a, b)$*

$$m|x_2 - x_1| \le |f(x_2) - f(x_1)| \le M|x_2 - x_1|.$$

Proof. Beginning with the upper bound assumption $|f'(x)| \le M$, and arguing by contradiction, suppose there are two points x_1 and x_2 such that

$$|f(x_2) - f(x_1)| > M|x_2 - x_1|.$$

Without loss of generality, assume that $x_1 < x_2$ and $f(x_1) < f(x_2)$. Then

$$\frac{f(x_2) - f(x_1)}{x_2 - x_1} > M.$$

By the Mean Value Theorem there must be a point $x_0 \in (x_1, x_2)$ with

$$f'(x_0) = \frac{f(x_2) - f(x_1)}{x_2 - x_1} > M,$$

contradicting the assumed bound on $|f'(x)|$.

Similarly, suppose that $m \le |f'(x)|$, but there are two points x_1 and x_2 such that

$$m|x_2 - x_1| > |f(x_2) - f(x_1)|.$$

Again, it won't hurt to assume that $x_1 < x_2$ and $f(x_1) < f(x_2)$. Then

$$\frac{f(x_2) - f(x_1)}{x_2 - x_1} < m,$$

so by the Mean Value Theorem there must be a point $x_0 \in (x_1, x_2)$ with

$$f'(x_0) = \frac{f(x_2) - f(x_1)}{x_2 - x_1} < m,$$

again giving a contradiction.

□

Theorem 5.3.12 is helpful in studying inverse functions and their derivatives. Recall that a function f is *one-to-one* if $x_1 \ne x_2$ implies $f(x_1) \ne f(x_2)$. A one-to-one function $f(x)$ has an inverse function $f^{-1}(y)$ defined on the range of f by setting $f^{-1}(f(x)) = x$. The reader is invited to check that the identity $f(f^{-1}(y)) = y$ also holds.

Theorem 5.3.13. *(Inverse Function Theorem) Suppose that $f : (a, b) \to \mathbb{R}$ has a continuous derivative, and that $f'(x_1) > 0$ for some $x_1 \in (a, b)$. Let $f(x_1) = y_1$. Then there are numbers $x_0 < x_1 < x_2$, and $y_0 < y_1 < y_2$ such that $f : [x_0, x_2] \to \mathbb{R}$ is one-to-one, and the range of f with this domain is the interval $[y_0, y_2]$. The inverse function $f^{-1} : [y_0, y_2]$ is differentiable at y_1, with*

$$(f^{-1})'(y_1) = \frac{1}{f'(x_1)}.$$

Proof. Since $f'(x_0) > 0$ and $f'(x)$ is continuous on (a, b), there is a $\delta > 0$ such that $f'(x) > 0$ for x in the interval $I = (x_0 - \delta, x_0 + \delta)$. It follows that $f(x)$ is strictly increasing on I, hence one-to-one there, and so there is an inverse function $f^{-1}(y)$ on the range of $f : I \to \mathbb{R}$.

Pick points $x_0, x_2 \in I$ such that $x_0 < x_1 < x_2$, and define $y_0 = f(x_0)$, $y_2 = f(x_2)$. Since f is strictly increasing on $[x_0, x_2]$, it follows that $y_0 < y_1 < y_2$. The Intermediate Value Theorem Theorem 5.2.4 shows that the range of $f : [x_0, x_2] \to \mathbb{R}$ is the interval $[y_0, y_2]$.

To see that f^{-1} has a derivative at y_1, examine

$$\frac{f^{-1}(y) - f^{-1}(y_1)}{y - y_1} = \frac{x - x_1}{f(x) - f(x_1)}.$$

The right-hand side has the limit $1/f'(x_0)$ as $x \to x_0$. We want to show that

$$\lim_{y \to y_1} \frac{f^{-1}(y) - f^{-1}(y_1)}{y - y_1} = \lim_{x \to x_1} \frac{x - x_1}{f(x) - f(x_1)}.$$

To this end, suppose that $\epsilon > 0$, and find δ such that $0 < |x - x_1| < \delta$ implies

$$|\frac{x - x_1}{f(x) - f(x_1)} - \frac{1}{f'(x_1)}| < \epsilon.$$

On the compact interval $[x_0, x_2]$, the continuous function $f'(x)$ has a positive minimum m and maximum M. By Theorem 5.3.12 the inequality

$$m|x - x_1| \le |f(x) - f(x_1)|$$

holds for $x \in (x_0, x_2)$. Thus if

$$0 < |y - y_1| = |f(x) - f(x_1)| < m\delta = \delta_1,$$

then $|x - x_1| < \delta$, so that

$$|\frac{f^{-1}(y) - f^{-1}(y_1)}{y - y_1} - \frac{1}{f'(x_1)}| = |\frac{x - x_1}{f(x) - f(x_1)} - \frac{1}{f'(x_1)}| < \epsilon.$$

This establishes the desired limit equality, while also providing the value of the derivative. □

The assumption $f'(x_1) > 0$ in this theorem is for convenience in the proof. The hypothesis can be changed to $f'(x_1) \ne 0$.

5.3.3 Contractions

This section considers an application of the ideas in Theorem 5.3.12 to the root-finding algorithm known as Newton's method. We will be interested in functions f which map a compact interval $[a, b]$ back into itself, and for which all points $f(x_1)$ and $f(x_2)$ are closer to each other than x_1 and x_2 are. With this idea in mind, say that a function $f : [a, b] \to [a, b]$ is a *contraction* if there is a number α satisfying $0 \le \alpha < 1$ such that

$$|f(x_2) - f(x_1)| \le \alpha |x_2 - x_1|, \quad \text{for all} \quad x_1, x_2 \in [a, b].$$

By Theorem 5.3.12, a function $f : [a, b] \to [a, b]$ will be a contraction if f is differentiable and $|f'(x)| \le M < 1$ for all $x \in [a, b]$.

The first result is an easy exercise.

Lemma 5.3.14. *If $f : [a, b] \to [a, b]$ is a contraction, then f is uniformly continuous.*

The second observation is also straightforward. If $f : [a, b] \to [a, b]$ is continuous, then the graph of f must somewhere hit the line $y = x$.

Lemma 5.3.15. *If $f : [a,b] \to [a,b]$ is a continuous, then there is some $x_0 \in [a,b]$ such that $f(x_0) = x_0$.*

Proof. If $f(a) = a$ or $f(b) = b$ then there is nothing more to show, so we may assume that $f(a) > a$ and $f(b) < b$. This means that the function $f(x) - x$ is positive when $x = a$, and negative when $x = b$. By the Intermediate Value Theorem there is some $x_0 \in [a,b]$ such that $f(x_0) = x_0$. □

Solutions of the equation $f(x) = x$ are called *fixed points* of the function f. For functions $f : [a,b] \to [a,b]$ which are contractions there is a unique fixed point, and there is a constructive procedure for approximating the fixed point. The ideas in the next theorem generalize quite well, leading to many important applications.

Theorem 5.3.16. *(Contraction Mapping Theorem) Suppose that $f : [a,b] \to [a,b]$ is a contraction. Then there is a unique point $z_0 \in [a,b]$ such that $f(z_0) = z_0$. Moreover, if x_0 is any point in $[a,b]$, and x_n is defined by $x_{n+1} = f(x_n)$, then*

$$|x_n - z_0| \le \alpha^n |x_0 - z_0|,$$

so that $z_0 = \lim_{n\to\infty} x_n$.

Proof. By the previous lemma the function f has at least one fixed point. Let's first establish that there cannot be more than one. Suppose that $f(z_0) = z_0$ and $f(z_1) = z_1$. Using the definition of a contraction,

$$|z_1 - z_0| = |f(z_1) - f(z_0)| \le \alpha |z_1 - z_0|.$$

That is,

$$(1-\alpha)|z_1 - z_0| \le 0.$$

Since $0 \le \alpha < 1$ the factor $(1-\alpha)$ is positive, and the factor $|z_1 - z_0|$ is nonnegative. Since the product is less than or equal to 0, it must be that $|z_1 - z_0| = 0$, or $z_1 = z_0$.

The inequality

$$|x_n - z_0| \le \alpha^n |x_0 - z_0|$$

is proved by induction, with the first case $n = 0$ being trivial. Assuming the inequality holds in the n-th case, it follows that

$$|x_{n+1} - z_0| = |f(x_n) - f(z_0)| \le \alpha |x_n - z_0| \le \alpha\alpha^n |x_0 - z_0| = \alpha^{n+1}|x_0 - z_0|.$$

□

5.3.3.1 Rootfinding 2: Newton's Method

Newton's method is an old, popular, and powerful technique for obtaining numerical solutions of root finding problems $f(x) = 0$. Geometrically, the idea is to begin with an initial guess x_0. One then approximates f by the tangent line to the graph at x_0. If the slope $f'(x_0)$ is not 0, the point x_1 where the tangent line intercepts the x-axis is taken as the next estimate x_1, and the process is repeated. Thus the algorithm starts with the initial guess x_0 and defines a sequence of points

$$x_{n+1} = x_n - \frac{f(x_n)}{f'(x_n)}.$$

Theorem 5.3.17. *Suppose that $f : (a, b) \to \mathbb{R}$ has two continuous derivatives on the interval (a, b). Assume there is some $r \in (a, b)$ with $f(r) = 0$, but $f'(r) \neq 0$. Then if x_0 is chosen sufficiently close to r, the sequence*

$$x_{n+1} = x_n - \frac{f(x_n)}{f'(x_n)}$$

will converge to r.

Proof. To use the contraction idea, define

$$g(x) = x - \frac{f(x)}{f'(x)}.$$

A calculation gives

$$g(r) = r,$$

and

$$g'(x) = 1 - \frac{(f'(x))^2 - f(x)f''(x)}{(f'(x))^2} = \frac{f(x)f''(x)}{(f'(x))^2}.$$

Since $g'(x)$ is continuous for x near r, with $g'(r) = 0$, there is a $\delta > 0$ such that

$$|g'(x)| \le \frac{1}{2}, \quad r - \delta \le x \le r + \delta.$$

For $x \in [r - \delta, r + \delta]$ the combination of $g(r) = r$ and Theorem 5.3.12 implies

$$|g(x) - r| = |g(x) - g(r)| \le \frac{1}{2}|x - r| \le \frac{1}{2}\delta.$$

Since $g : [r - \delta, r + \delta] \to [r - \delta, r + \delta]$ is a contraction, and $x_{n+1} = g(x_n)$, Theorem 5.3.16 shows that

$$r = \lim_{n \to \infty} x_n$$

for any $x_0 \in [r - \delta, r + \delta]$. □

This proof showed that for x_0 close enough to r, the sequence x_n comes from iteration of a contraction with $\alpha \le 1/2$. This already guarantees rapid convergence of the sequence $\{x_n\}$ to r. Actually, as x_n gets close to r, the value of α will improve, further accelerating the rate of convergence. Additional information about Newton's method can be found in most basic numerical analysis texts.

5.3.4 Convexity

Calculus students spend a lot of time searching for extreme points of a function $f(x)$ by examining solutions of $f'(x) = 0$. Without additional information, such a critical point could be a local or global minimum or maximum, or none of these. This situation changes dramatically if f satisfies the additional condition $f'' > 0$. The positivity of the second derivative is closely related to a geometric condition called convexity. A simple model for convex functions is provided by the function $f(x) = x^2$, which is shown in Figure 7.b, along with the tangent line to this graph at $x = 2$, and the secant line joining $(1, f(1))$ and $(3, f(3))$. Notice that the graph of f lies below its secant line on the interval $[1, 3]$, and above the tangent line.

Given an interval I, a function $f : I \to \mathbb{R}$ is said to be convex if the graph of the function always lies beneath its secant lines. To make this precise, suppose that a and b are distinct points in I. The points on the line segment joining $(a, f(a))$ to $(b, f(b))$ may be written as

$$(tb + (1-t)a,\ tf(b) + (1-t)f(a)), \quad 0 \le t \le 1.$$

The function f is *convex* on I if for all distinct pairs $a, b \in I$

$$f(tb + (1-t)a)) \le tf(b) + (1-t)f(a), \quad 0 \le t \le 1. \tag{5.4}$$

The function f is *strictly convex* if the inequality is strict except at the end-points,

$$f(tb + (1-t)a) < tf(b) + (1-t)f(a), \quad 0 < t < 1.$$

The first result says that the graph of a convex function f lies above its tangent lines.

Theorem 5.3.18. *Suppose that $f : (c, d) \to \mathbb{R}$ is convex, and $f'(a)$ exists for some $a \in (c, d)$. Then*

$$f(a) + (b-a)f'(a) \le f(b) \quad \text{for all} \quad b \in (c, d). \tag{5.5}$$

Proof. After some algebraic manipulation (5.4) may be expressed as

$$f(t[b-a] + a) - f(a) \le t[f(b) - f(a)], \quad 0 \le t \le 1,$$

or, for $a \ne b$ and $t \ne 0$,

$$\frac{f(t[b-a] + a) - f(a)}{t[b-a]}(b-a) \le f(b) - f(a).$$

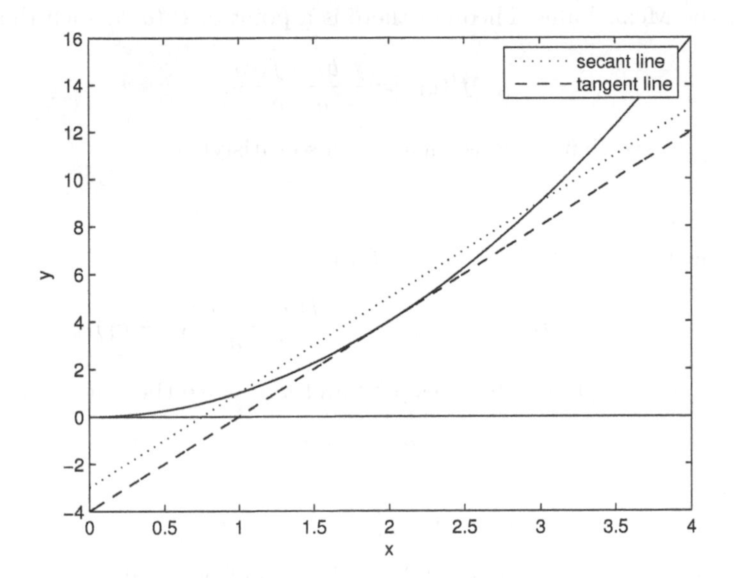

FIGURE 5.2
A convex function.

Take the limit as $h = t[b - a] \to 0$ to get

$$f'(a)(b-a) \le f(b) - f(a),$$

which is equivalent to (5.5). □

If f is differentiable on (c,d), then a converse to Theorem 5.3.18 holds.

Theorem 5.3.19. *Suppose that f is differentiable on (c,d). If*

$$f(a) + (b-a)f'(a) \le f(b) \quad \text{for all} \quad a, b \in (c,d),$$

then $f : (c,d) \to \mathbb{R}$ is convex.

Proof. Suppose that f is not convex on (c,d). Then there is a pair of points a, b with $c < a < b < d$, and some $t_1 \in (0,1)$, such that

$$f(t_1 b + (1-t_1)a) > t_1 f(b) + (1-t_1) f(a). \tag{5.6}$$

Define the function $g(x)$ whose graph is the line joining $(a, f(a))$ and $(b, f(b))$,

$$g(x) = f(a) + \frac{f(b) - f(a)}{b - a}(x - a), \quad a \le x \le b.$$

If $x_1 = t_1 b + (1-t_1)a$, then (5.6) says that

$$f(x_1) > g(x_1).$$

By the Mean Value Theorem there is a point $c_1 \in (a,b)$ such that

$$f'(c_1) = \frac{f(b) - f(a)}{b - a}.$$

It is easy to see that there is such a c_1 also satisfying

$$f(c_1) > g(c_1).$$

The tangent line to f at c_1 has the form

$$h(x) = f'(c_1)(x - c_1) + f(c_1) = \frac{f(b) - f(a)}{b - a}(x - c_1) + f(c_1).$$

Since $f(c_1) > g(c_1)$ and the lines $g(x)$ and $h(x)$ have the same slopes,

$$h(b) > g(b).$$

That is,

$$f'(c_1)(b - c_1) + f(c_1) > f(b),$$

so the inequality (5.5) is not valid for all pairs of points in (c,d).

□

There is a simple second derivative test that can be used to recognize convex functions.

Theorem 5.3.20. *Suppose that f is continuous on $[c,d]$, and has two derivatives on (c,d). If $f''(x) \ge 0$ for $x \in (c,d)$, then f is convex on $[c,d]$. If $f''(x) > 0$ for $x \in (c,d)$, then f is strictly convex on $[c,d]$.*

Proof. It is convenient to work with the contrapositive statement: if f is not convex on $[c,d]$, then there is some $x_1 \in (c,d)$ with $f''(x_1) < 0$.

Introduce the auxiliary function

$$g(t) = f(tb + (1-t)a)) - [tf(b) + (1-t)f(a))], \quad 0 \le t \le 1,$$

which satisfies $g(0) = 0 = g(1)$. If f is not convex, then there is some point $t_1 \in (0,1)$ such that $g(t_1) > 0$.

The continuous function $g : [0,1] \to \mathbb{R}$ has a positive maximum at some point $t_2 \in (0,1)$, with $g'(t_2) = 0$. In addition, an application of the Mean Value Theorem on the interval $[t_2, b]$ shows that there is a point $t_3 \in (t_2, b)$ with $g'(t_3) < 0$.

Now apply the Mean Value Theorem again, this time to the function g' on the interval $[t_2, t_3]$, obtaining

$$g''(t_4) = \frac{g'(t_2) - g'(t_3)}{t_2 - t_3} < 0, \quad \text{for some} \quad t_4 \in (t_2, t_3).$$

Finally, a chain rule calculation shows that

$$g''(t) = (b-a)^2 f''(tb + (1-t)a),$$

so

$$f''(t_4 b + (1 - t_4)a) = g''(t_4)/(b - a)^2 < 0.$$

The case $f'' > 0$ is handled in a similar fashion.

□

The previous theorem has a converse of sorts.

Theorem 5.3.21. *Suppose that f has two derivatives on (c, d), and $f''(x) < 0$ for all $x \in (c, d)$. Then f is not convex on (c, d).*

Proof. Picking distinct points $a, b \in (c, d)$, consider the function

$$g(x) = f(x) - f(a) - f'(a)(x - a).$$

By Theorem 5.3.11 the function $f'(x)$ is strictly decreasing on $[a, b]$. This implies $g'(x) < 0$ for $x > a$. Since $g(a) = 0$, it follows that $g(b) < 0$. This means

$$f(b) < f(a) + f'(a)(b - a),$$

so f cannot be convex by Theorem 5.3.18 □

Finally, here is the answer to a calculus student's prayers.

Theorem 5.3.22. *Suppose that $f : (c, d) \to \mathbb{R}$ is convex, and $f'(a_1) = 0$ for some $a_1 \in (c, d)$. Then a_1 is a global minimizer for f. If f is strictly convex, then f has at most one global minimizer.*

Proof. To see that a_1 is a global minimizer, simply apply Theorem 5.3.18 to conclude that

$$f(a_1) \le f(b), \quad \text{for all} \quad b \in (c, d).$$

Suppose that f is strictly convex, with global a minimizer a_1. If b is distinct from a_1, and $f(a_1) = f(b)$, the defining inequality for strict convexity gives

$$f(tb + (1 - t)a_1) < tf(b) + (1 - t)f(a_1) = f(a_1), \quad 0 < t < 1,$$

contradicting the assumption that a_1 is a global minimizer.

□

5.4 Problems

5.1. *Suppose we want to talk about the set of real valued rational functions of a real variable. For instance, we might say that the sum of any finite collection of rational functions is another rational function. Discuss the problem of defining a common domain for all rational functions. What is the appropriate domain for a fixed finite collection of rational functions?*

5.2. *Suppose that $\{a_n\}$ is a sequence of real numbers. Show that a function is defined by the rule $f(n) = a_n$. What is the domain?*

5.3. *Show that if*

$$\lim_{x \to \infty} f(x) = \infty$$

then

$$\lim_{x \to \infty} 1/f(x) = 0.$$

What can you say about the set of x where $f(x) = 0$?

5.4. *Suppose that $r(x) = p(x)/q(x)$ is a rational function, with*

$$p(x) = \sum_{k=0}^{m} a_k x^k, \quad a_m \neq 0,$$

and

$$q(x) = \sum_{k=0}^{n} b_k x^k, \quad b_n \neq 0.$$

Show that $\lim_{x \to \infty} r(x) = 0$ if $n > m$, and $\lim_{x \to \infty} r(x) = a_m/b_m$ if $m = n$.

5.5. *a) Suppose $\lim_{x \to x_0} f(x) = M$ and $M \neq 0$. Show there is a $\delta > 0$ such that $f(x) \neq 0$ for $0 < |x - x_0| < \delta$.*

b) State and prove an analogous result if $\lim_{x \to \infty} f(x) = M$ and $M \neq 0$.

5.6. *Complete the proofs of (iii) and (iv) in Theorem 5.2.1.*

5.7. *a) Suppose $\lim_{x \to x_0} f(x) = M$ and $M > 0$. Show there is a $\delta > 0$ such that*

$$M/2 \leq f(x) \leq 2M$$

for $0 < |x - x_0| < \delta$.

b) Take as a fact that

$$\lim_{x \to 0} \frac{\sin(x)}{x} = 1.$$

Show there is a $\delta > 0$ such that

$$x/2 \leq \sin(x) \leq 2x$$

for $0 < x < \delta$. What happens if $x < 0$?

5.8. *Show that the function* $f(x) = x$ *is continuous at every point* $x_0 \in \mathbb{R}$.

5.9. *a) Show that the function* $f(x) = |x|$ *is continuous at* $x_0 = 0$.
b) Show that $f(x) = |x|$ *is continuous at every point* $x_0 \in \mathbb{R}$.

5.10. *For each* $\epsilon > 0$, *find an explicit value of* $\delta > 0$ *such that* $|x^2 - 1| < \epsilon$ *if* $|x - 1| < \delta$.

5.11. *Functions* $f : [a, c] \to \mathbb{R}$ *are sometimes given piecewise definitions. That is, for* $a < b < c$ *the function* f *is defined by one formula on an interval* $[a, b]$ *and by a different formula on* $[b, c]$. *Suppose that* $f : [a, c] \to \mathbb{R}$ *is continuous when restricted to the intervals* $[a, b]$ *and* $[b, c]$. *Show that* f *is continuous on* $[a, c]$. *Is the conclusion still true if we only assume that* f *is continuous on the intervals* (a, b) *and* (b, c)*?*

5.12. *Show that the function*

$$f(x) = \begin{Bmatrix} |x|, & x < 1, \\ 1, & x \ge 1, \end{Bmatrix}$$

is continuous on $\mathbb{R}$.

5.13. *Suppose* $p(x)$ *is a polynomial. Where is the function* $f(x) = |p(x)|$ *continuous? Explain.*

5.14. *Produce two proofs that the function* $g(x) = 1/x$ *is continuous on the interval* $(0, \infty)$. *The first should use Theorem 5.2.1, while the second should be based on the definition of continuity.*

5.15. *Suppose that* $g : \mathbb{R} \to \mathbb{R}$ *is continuous. In addition, assume that the formula* $g(x) = x^2$ *holds for all rational values of* x. *Show that* $g(x) = x^2$ *for all* $x \in \mathbb{R}$.

5.16. *Let* $g : \mathbb{R} \to \mathbb{R}$ *be the function satisfying* $g(x) = 0$ *when* x *is irrational, while* $g(x) = x$ *when* x *is rational. Show that* $g(x)$ *is continuous at* $x_0 = 0$, *but at no other point.*

5.17. *Suppose* $f : [0, 1] \to \mathbb{R}$ *is a continuous function such that* $f(0) < 0$ *and* $f(1) > 1$. *Show that there is at least one point* $x_0 \in [0, 1]$ *such that* $f(x_0) = x_0$.

5.18. *Show that any polynomial*

$$p(x) = \sum_{k=0}^{n} a_k x^k, \quad a_n \ne 0,$$

with odd degree n *and real coefficients* a_k *has at least one real root.*

5.19. *Use the bisection method to approximate* $\sqrt{3}$ *by taking* $f(x) = x^2 - 3$, *with* $a = 0$ *and* $b = 2$. *Compute* a_n *and* b_n *for* $n \le 5$. *(Use a calculator.) How many iterations are required before* $|b_n - r| \le 10^{-10}$*?*

5.20. *Suppose* I_0 *and* I_1 *are open intervals, and* $f : I_0 \to \mathbb{R}$ *is continuous at* $x_0 \in I_0$. *Show that if* $x_k \in I_0$, $\lim_{k\to\infty} x_k = x_0$, *and* $f(x_0) \in I_1$, *then there is an* N *such that* $f(x_k) \in I_1$ *for* $k \ge N$. *How is this related to Theorem 5.2.5 ?*

5.21. *Suppose that* $f : I \to \mathbb{R}$ *and* f *satisfies the inequality*

$$|f(x) - f(y)| \le C|x - y|$$

for some constant C *and all* $x, y \in I$. *Show that* f *is uniformly continuous on the interval* I.

5.22. *Suppose that* f *is continuous on the interval* $(-\infty, \infty)$. *Assume in addition that* $f(x) \ge 0$ *for all* $x \in \mathbb{R}$, *and that*

$$\lim_{x\to\pm\infty} f(x) = 0.$$

Show that f *has a maximum at some* $x_0 \in \mathbb{R}$. *Find an example to show that* f *may not have a minimum.*

5.23. *Show that the function* $f(x) = |x|$ *does not have a derivative at* $x_0 = 0$.

5.24. *Show that the two definitions of* $f'(x_0)$,

$$f'(x_0) = \lim_{x\to x_0} \frac{f(x) - f(x_0)}{x - x_0},$$

and

$$f'(x_0) = \lim_{h\to 0} \frac{f(x_0 + h) - f(x_0)}{h},$$

are equivalent by showing that the existence of one limit implies the existence of the other, and that the two limits are the same.

5.25. *a) Prove directly from the definition that*

$$\frac{d}{dx}x^2 = 2x, \quad \frac{d}{dx}x^3 = 3x^2.$$

b) Prove that

$$\frac{d}{dx}x^n = nx^{n-1}, \quad n = 1, 2, 3, \dots.$$

5.26. *Given a function* f *defined on* $[a, b]$, *we sometimes wish to discuss the differentiability of* f *at* a *or* b *without looking at a larger interval. How would you define* $f'(a)$ *and* $f'(b)$ *using* $\lim_{x\to a^+}$ *and* $\lim_{x\to b^-}$ *?*

5.27. *a) Assume that* $f : [0, 1] \to \mathbb{R}$ *is differentiable at* $x_0 = 0$. *Suppose that there is a sequence* $x_n \in [0, 1]$ *such that* $f(x_n) = 0$ *and* $\lim_{n\to\infty} x_n = 0$. *Prove that* $f'(0) = 0$.

b) Define the function

$$g(x) = \begin{Bmatrix} x\sin(1/x), & x \ne 0 \\ 0, & x = 0 \end{Bmatrix}.$$

Is g *differentiable at* $x_0 = 0$*?*

5.28. *Show that the function*

$$f(x) = \begin{Bmatrix} x^2, & x \geq 0 \\ 0, & x < 0 \end{Bmatrix}$$

has a derivative at every real number x. Extend this example to a result about functions with piecewise definitions.

5.29. *Suppose $f'(x) = 0$ for all $x \in (a,b)$. Show that $f(x)$ is constant.*

5.30. *Here is another version of Theorem 5.3.11. Suppose $f : [a,b] \to \mathbb{R}$ is continuous, and f is differentiable on (a,b) with $f'(x) \geq 0$ for $x \in (a,b)$. Prove that f is increasing on $[a,b]$. Give an example to show that f may not be strictly increasing.*

5.31. *Show that the function $f(x) = x^5 + x^3 + x + 1$ has exactly one real root.*

5.32. *Suppose f and g are real valued functions defined on $[a,b)$. Show that if $f(a) = g(a)$ and if $f'(x) < g'(x)$ for $x \in (a,b)$, then $f(x) < g(x)$ for $x \in (a,b)$.*

5.33. *Assume that $f : \mathbb{R} \to \mathbb{R}$ is differentiable, that $|f(0)| \leq 1$, and that $|f'(x)| \leq 1$. What is the largest possible value for $|f(x)|$ if $x \geq 0$? Provide an example that achieves your bound.*

5.34. *Suppose that $f'(x) = g'(x)$ for all $x \in (a,b)$. Show that $f(x) = g(x)$ for all $x \in (a,b)$ if and only if there is some $x_0 \in (a,b)$ such that $f(x_0) = g(x_0)$.*

5.35. *Suppose that $f^{(n)}(x) = 0$ for all $x \in (a,b)$. Show that $f(x)$ is a polynomial of degree at most $n-1$.*

5.36. *Assume that $x_1 < x_2 < \cdots < x_N$, and define*

$$p(x) = (x - x_1) \cdots (x - x_N).$$

Show that $p'(x)$ has exactly $N-1$ real roots.

5.37. *Use Theorem 5.3.13 to show that the functions*

$$x^{1/n}, \quad n = 1, 2, 3, \ldots$$

are differentiable on $(0,\infty)$. Compute the derivative.

5.38. *Suppose that f is continuous on $[a,b]$, and has n derivatives on (a,b). Assume that there are points*

$$a \leq x_0 < x_1 < \cdots < x_n \leq b$$

such that $f(x_k) = 0$. Show that there is a point $\xi \in (a,b)$ such that $f^{(n)}(\xi) = 0$.

5.39. *a) Find an example of a function $f : [c,d] \to \mathbb{R}$ such that $|f'(x)| \leq \alpha < 1$ for $x \in [c,d]$, but f has no fixed point.*

b) Find an example of a function $f : (0,1) \to (0,1)$ such that $|f'(x)| \leq \alpha < 1$ for $x \in (0,1)$, but f has no fixed point in $(0,1)$.

5.40. *Prove Theorem 5.3.13 if the hypothesis $f'(x_1) > 0$ is replace by $f'(x_1) \neq 0$. Don't work too hard.*

5.41. *For $n \geq 0$, consider solving the equation*

$$x^{n+1} - \sum_{k=0}^{n} a_k x^k = 0, \quad a_k > 0,$$

by recasting it as the fixed point problem

$$x = f(x) = \sum_{k=0}^{n} a_k x^{k-n}.$$

a) Show that the problem has exactly one positive solution.
b) Show that $f : [a_n, f(a_n)] \to [a_n, f(a_n)]$.
c) Show that the sequence $x_0 = a_n$, $x_{m+1} = f(x_m)$ converges to the positive solution if

$$|\sum_{k=0}^{n-1} (k-n) \frac{a_k}{(a_n)^{n-k+1}}| < 1.$$

5.42. *Use Newton's method and a calculator to approximate $\sqrt{2}$. Use Theorem 5.3.16 to estimate the accuracy of the approximations.*

5.43. *a) Suppose*

$$f(x) = \sum_{k=1}^{n} c_k \exp(a_k x), \quad c_k > 0.$$

Show that if $\lim_{x \to \pm\infty} f(x) = \infty$, then f has a unique global minimum.
b) Find a strictly convex function $f : \mathbb{R} \to \mathbb{R}$ with no global minimizer.

5.44. *In addition to the hypotheses of Theorem 5.3.16, suppose that there is a constant C such that*

$$|f(x_2) - f(x_1)| \leq Cr|x_2 - x_1|, \quad x_1, x_2 \in [z_0 - r, z_0 + r].$$

Proceeding as in the proof of Theorem 5.3.16, one first has

$$|x_1 - z_0| = |f(x_0) - f(z_0)| \leq Cr|x_0 - z_0| \leq C|x_0 - z_0|^2.$$

Show that the convergence estimates improve to

$$|x_n - z_0| \leq C^{2^n - 1}|x_0 - z_0|^{2^n}.$$

5.45. *Assume that $f : \mathbb{R} \to \mathbb{R}$ is strictly convex. Show that there are no more than two distinct points x_i satisfying $f(x_i) = 0$.*

5.46. *Suppose f and g are convex functions defined on an interval I. Show that $f+g$ is also convex on I. Show that αf is convex if $\alpha \geq 0$. If $h : \mathbb{R} \to \mathbb{R}$ is convex and increasing, show that $h(f(x))$ is convex.*

5.47. *Suppose $f : [0,1] \to \mathbb{R}$ is an increasing function. Say that $f(x)$ has a jump at x_k if*

$$\lim_{x \to x_k^+} f(x) \neq \lim_{x \to x_k^-} f(x).$$

The size of the jump is

$$s(x_k) = \lim_{x \to x_k^+} f(x) - \lim_{x \to x_k^-} f(x).$$

(a) Show that f is continuous at x_k if and only if f does not have a jump at x_k.

(b) Show that for every $\epsilon > 0$ there are only finitely many points x_k with $s(x_k) > \epsilon$.

6

Integrals

One of the fundamental problems in calculus is the computation of the area between the graph of a function $f : [a,b] \to \mathbb{R}$ and the x-axis. The essential ideas are illustrated in Figure 6.1 and Figure 6.2. An interval $[a,b]$ is divided into n subintervals $[x_k, x_{k+1}]$, with $a = x_0 < x_1 < \cdots < x_n = b$. On each subinterval the area is approximated by the area of a rectangle, whose height is usually the value of the function $f(t_k)$ at some point $t_k \in [x_k, x_{k+1}]$. In the left figure, the heights of the rectangles are given by the values $f(x_k)$, while on the right the heights are $f(x_{k+1})$.

In elementary treatments it is often assumed that each subinterval has length $(b-a)/n$. One would then like to argue that the sum of the areas of the rectangles has a limit as $n \to \infty$. This limiting value will be taken as the area, which is denoted by the integral

$$\int_a^b f(x)\,dx.$$

As a beginning exercise, this idea will be carried out for the elementary functions x^m for $m = 0, 1, 2, \ldots$.

In general, there are both practical and theoretical problems that arise in trying to develop this idea for area computation. Although calculus texts emphasize algebraic techniques for integration, there are many important integrals which cannot be evaluated by such techniques. One then has the practical problem of selecting and using efficient algorithms to calculate integrals with high accuracy. This will require fairly explicit descriptions of the errors made by approximate integration techniques. On the theoretical side, problems arise because there are examples for which the area computation does not seem meaningful. A major problem is to describe a large class of functions for which the integral makes sense.

One example of a function whose integration is problematic is $f(x) = 1/x$. Consider an area computation on the interval $[0,1]$. Fix n and choose

$$x_k = \frac{k}{n}, \quad k = 0, \ldots, n.$$

Form rectangles with heights $f(x_{k+1})$, the value of the function at the right endpoint of each subinterval. Since $1/x$ is decreasing on the interval $(0,1]$, these rectangles will lie below the graph. The sum of the areas of the rectangles

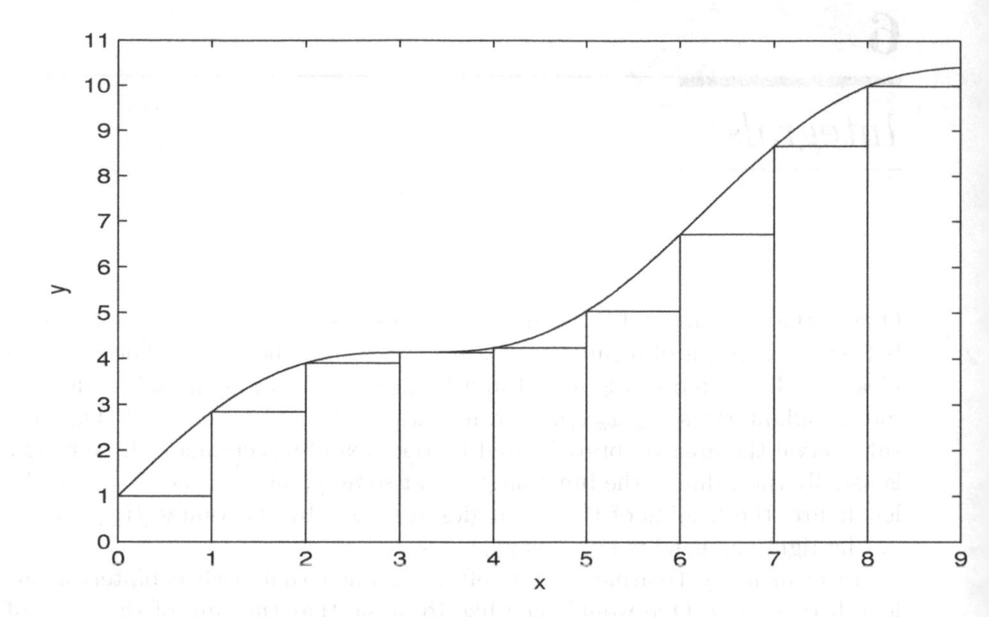

FIGURE 6.1
Graph of $1 + x + \sin(x)$ with lower Riemann sums.

is

$$s_n = \sum_{k=0}^{n-1} \frac{1}{n} f(x_{k+1}) = \frac{1}{n}[\frac{n}{1} + \frac{n}{2} + \frac{n}{3} + \cdots + \frac{n}{n}] = \sum_{k=1}^{n} \frac{1}{k}.$$

The areas s_n are the partial sums of the harmonic series, which diverges.

The problem becomes even worse if we consider the integral on $[-1, 1]$,

$$\int_{-1}^{1} \frac{1}{x}\, dx.$$

Recall that a signed area is intended when the function $f(x)$ is not positive; negative function values contribute negative area. While one is fairly safe in assigning the 'value' ∞ to $\int_0^1 1/x\, dx$, extreme caution is called for when trying to make sense of the expression

$$\int_{-1}^{1} \frac{1}{x}\, dx = \int_{-1}^{0} \frac{1}{x}\, dx + \int_{0}^{1} \frac{1}{x}\, dx = \infty - \infty.$$

A different sort of challenge is provided by the function

$$g(x) = \begin{Bmatrix} 1, & \text{x is rational,} \\ 0, & \text{x is irrational} \end{Bmatrix}. \tag{6.1}$$

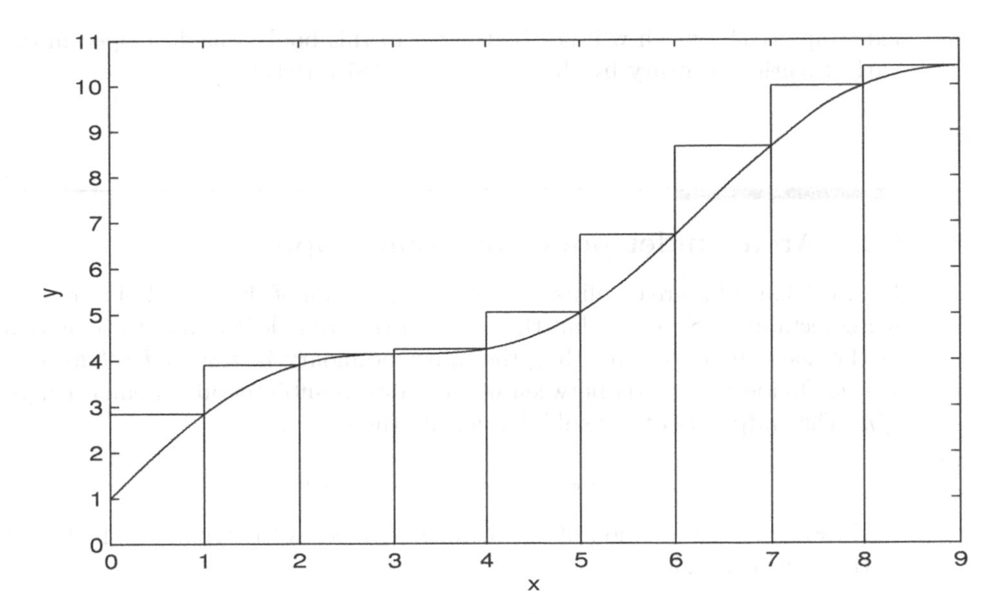

FIGURE 6.2
Graph of $1 + x + \sin(x)$ with upper Riemann sums.

Consider trying to compute $\int_0^1 g(x)\,dx$. Fix n, and take $x_k = k/n$ for $k = 0, \dots, n$. If the heights of the rectangles are given by $f(t_k)$, with $x_k \le t_k \le x_{k+1}$, the sum of the areas of the rectangles is

$$S_n = \sum_{k=0}^{n-1} \frac{1}{n} f(t_k).$$

If t_k is chosen to be the left endpoint x_k of the k-th subinterval, then since x_k is rational, $S_n = 1$. In contrast, if t_k is chosen to be an irrational number in the k-th subinterval, then $S_n = 0$. Regardless of how small the subintervals $[x_k, x_{k+1}]$ are, some of our computations result in an area estimate of 1, while others give an area estimate of 0.

These examples indicate that a certain amount of care is needed when trying to determine a class of functions which can be integrated. The approach we will follow, usually referred to as Riemann's theory of integration, was developed in the nineteenth century by Augustin-Louis Cauchy, Georg Friedrich Bernhard Riemann (1826–66) and Gaston Darboux (1842–1917) [5, pp. 956–961]. This development revived ideas of approximating areas under curves by sums of areas of simpler geometric figures that had antecedents in the work of ancient Greece, and then of Gottfried Wilhelm Leibniz. A still more sophisticated approach, which will not be treated in this book, was developed in the early twentieth century by Henri Lebesgue (1875–1941).

6.1 Areas under power function graphs

Figure 6.3 and Figure 6.4 illustrate the computation of the area A of a triangle using rectangles. Suppose that the height of the triangle is h, and the equation of the linear function providing the upper boundary is $f(x) = hx/b$ for $0 \le x \le b$. Divide the x-axis between 0 and b into n subintervals of equal length b/n. The endpoints of the subintervals are then

$$x_k = kb/n, \quad k = 0, \dots, n$$

In Figure 6.3 the union of the rectangles encloses the triangle. The height of the k-th rectangle is

$$f(x_k) = \frac{kb}{n}\frac{h}{b} = k\frac{h}{n}, \quad k = 1, \dots n.$$

The sum of the areas of the rectangles is

$$A_o = \sum_{k=1}^{n} \frac{b}{n} k \frac{h}{n} = \frac{bh}{n^2} \sum_{k=1}^{n} k.$$

Since

$$\sum_{k=1}^{n} k = \frac{n(n+1)}{2}$$

it follows that

$$A_o = \frac{bh}{n^2} \frac{n(n+1)}{2} = \frac{bh}{2} \frac{n^2+n}{n^2} = \frac{bh}{2}[1 + \frac{1}{n}].$$

In Figure 6.4 the triangle encloses the union of nonoverlapping rectangles. Starting the count now with $k = 0$ rather than $k = 1$, the height of the k-th rectangle is

$$f(x_k) = \frac{kb}{n}\frac{h}{b} = k\frac{h}{n}, \quad k = 0, \dots n-1.$$

The sum of the rectangular areas is

$$A_i = \sum_{k=0}^{n-1} \frac{b}{n} k \frac{h}{n} = \frac{bh}{n^2} \sum_{k=0}^{n-1} k = \frac{bh}{n^2} \frac{(n-1)n}{2}.$$

Thus

$$A_i = \frac{bh}{2}[1 - \frac{1}{n}].$$

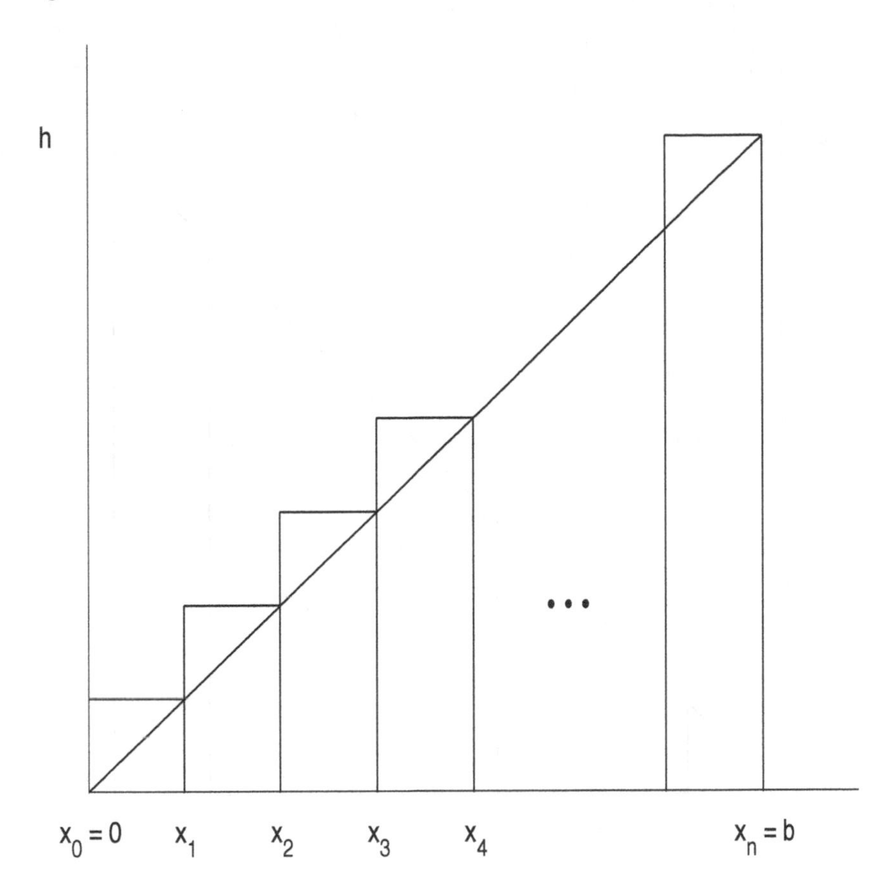

FIGURE 6.3
Triangle area 1.

Finally we have

$$A_i = \frac{bh}{2}[1 - \frac{1}{n}] < A < A_o = \frac{bh}{2}[1 + \frac{1}{n}].$$

Since this inequality is true for every positive integer n, the area A is neither smaller nor larger than $bh/2$, so that

$$A = bh/2.$$

The same ideas may be applied to the computation of the area lying under the graph of $f(x) = x^2$ for $0 \le x \le b$. Since the function is increasing for $x \ge 0$, the minimum and maximum values of the function x^2 on any subinterval $[x_k, x_{k+1}]$ are at x_k and x_{k+1} respectively. In this new case the function values

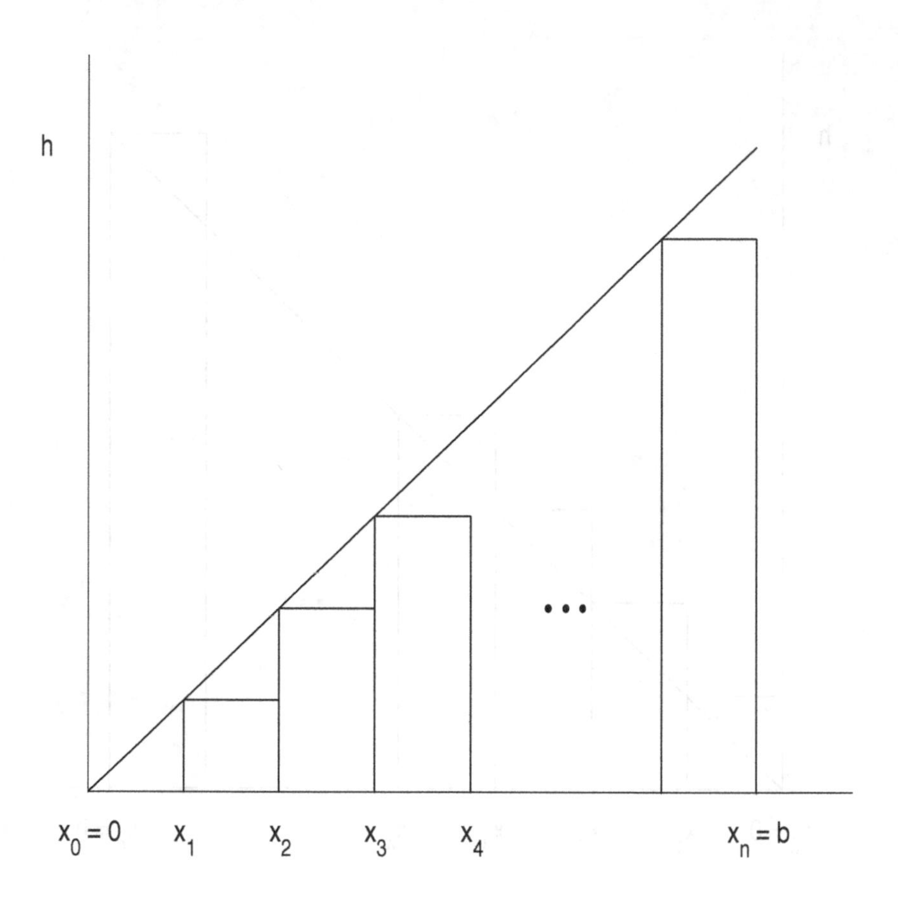

FIGURE 6.4
Triangle area 2.

are

$$f(x_k) = x_k^2 = \frac{k^2b^2}{n^2}$$

and

$$A_o = \sum_{k=1}^{n} \frac{b}{n}\frac{k^2b^2}{n^2} = \frac{b^3}{n^3}\sum_{k=1}^{n} k^2 = \frac{b^3}{n^3}\frac{n(n+1)(2n+1)}{6}$$

$$= \frac{b^3}{3}[1 + \frac{3}{2n} + \frac{1}{2n^2}].$$

Similarly

$$A_i = \sum_{k=0}^{n-1} \frac{b}{n}\frac{k^2b^2}{n^2} = \frac{b^3}{n^3}\frac{(n-1)n(2n-1)}{6},$$

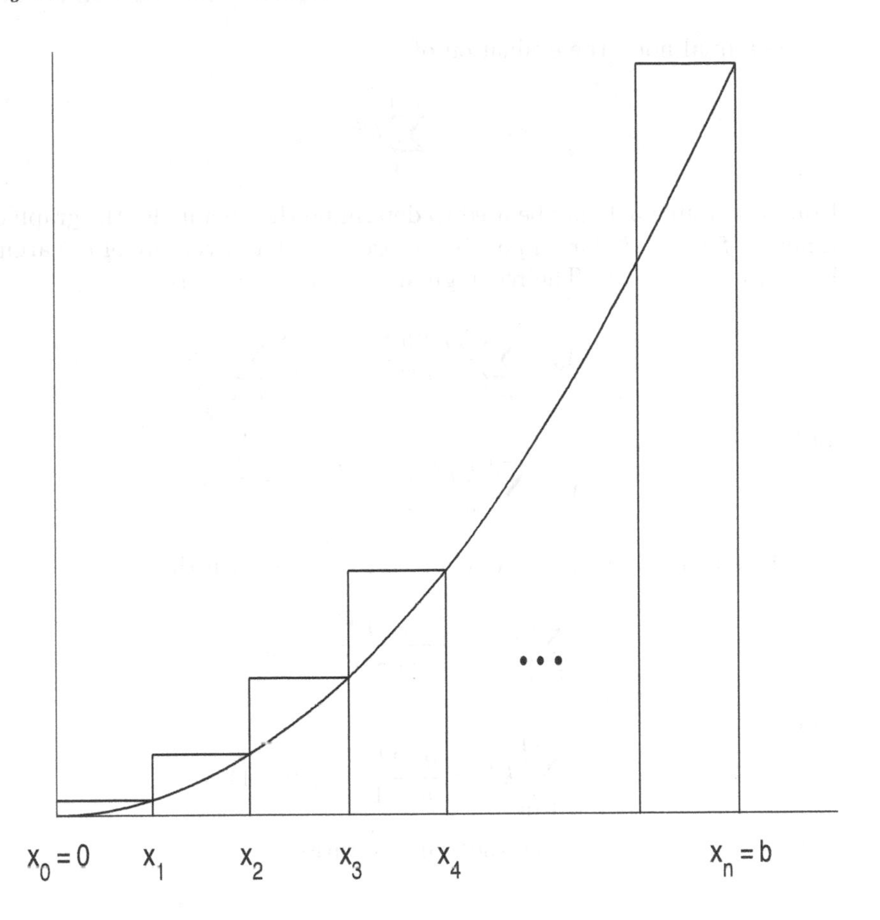

FIGURE 6.5
Parabolic area 1.

or

$$A_i = \frac{b^3}{3}\frac{2n^3 - 3n^2 + n}{2n^3} = \frac{b^3}{3}[1 - \frac{3}{2n} + \frac{1}{2n^2}].$$

This time we get

$$A_i = \frac{b^3}{3}[1 - \frac{3}{2n} + \frac{1}{2n^2}] < A < A_o = \frac{b^3}{3}[1 + \frac{3}{2n} + \frac{1}{2n^2}].$$

Since this inequality is true for every positive integer n, the area under the parabola is bigger than any number smaller than $b^3/3$, and smaller than any number bigger than $b^3/3$, or

$$A = b^3/3.$$

As a final note, the evaluation of

$$\sum_{k=0}^{n-1} k^m$$

from Theorem 4.3.4 may be used to determine the area under the graph of the function $f(x) = x^m$ for all positive integers m. The structure of the argument is the same as above. The rectangular areas have the form

$$A_o = \sum_{k=1}^{n} \frac{b}{n} \frac{k^m b^m}{n^m} = \frac{b^{m+1}}{n^{m+1}} \sum_{k=1}^{n} k^m$$

and

$$A_i = \sum_{k=0}^{n-1} \frac{b}{n} \frac{k^m b^m}{n^m} = \frac{b^{m+1}}{n^{m+1}} \sum_{k=0}^{n-1} k^m.$$

There is a polynomial q_0 of degree at most m such that

$$\sum_{k=1}^{n} k^m = \frac{(n+1)^{m+1}}{m+1} + q_0(n).$$

and

$$\sum_{k=0}^{n-1} k^m = \frac{n^{m+1}}{m+1} + q_0(n-1).$$

Notice that by the binomial theorem the ratio

$$\frac{(n+1)^{m+1}}{n^{m+1}}$$

may be written as

$$1 + \frac{q_1(n)}{n^{m+1}}$$

for some new polynomial $q_1(n)$ of degree at most m. This gives

$$A_i = \frac{b^{m+1}}{m+1}[1 + \frac{q_0(n-1)}{n^{m+1}}] < A < A_o = \frac{b^{m+1}}{m+1}[1 + \frac{q_2(n)}{n^{m+1}}]$$

for some polynomial $q_2(n)$ incorporating all the lower order terms in A_o. Thus the area under the graph of x^m from $x = 0$ to $x = b$ is

$$A = \frac{b^{m+1}}{m+1},$$

which is of course well known from calculus.

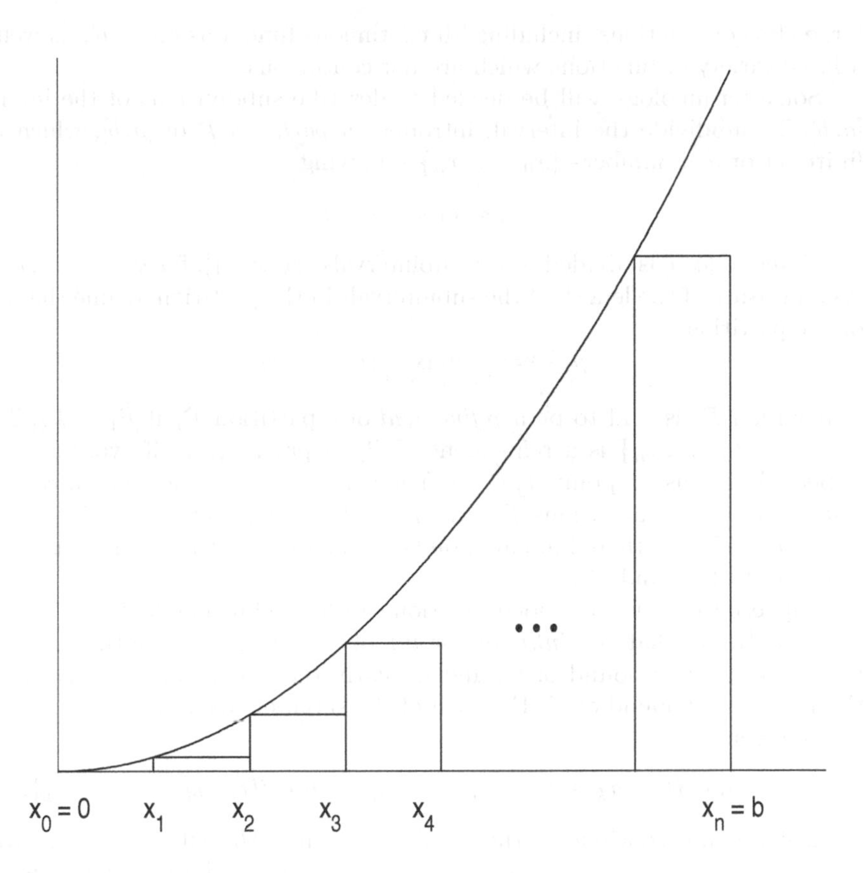

FIGURE 6.6
Parabolic area 2.

6.2 Integrable functions

Riemann's theory of integration treats the integral of a bounded function $f(x)$ defined on an interval $[a,b]$. Area computations are based on a process of estimation with two types of rectangles, as in Figure 6.1 and Figure 6.2. The strategy is easiest to describe for positive functions f, although the process works without any sign restrictions. When $f > 0$, upper rectangles are constructed with heights greater than the corresponding function values, and lower rectangles are constructed with heights less than the function values. Functions are considered integrable when the areas computed using upper and lower rectangles agree. This method permits integration of an extremely large class of functions, including all continuous functions on $[a, b]$, as well as a large variety of functions which are not continuous.

Some terminology will be needed to describe subdivisions of the interval $[a, b]$. To subdivide the interval, introduce a *partition* $\mathcal{P}$ of $[a, b]$, which is a finite set of real numbers $\{x_0, \dots, x_n\}$ satisfying

$$a = x_0 < x_1 < \dots < x_n = b.$$

The interval $[a, b]$ is divided into n subintervals $[x_k, x_{k+1}]$, for $k = 0, \dots, n-1$. As a measure of the length of the subintervals in the partition, define the *mesh* of the partition,

$$\mu(\mathcal{P}) = \max_{k=0,\dots,n-1} |x_{k+1} - x_k|.$$

A partition $\mathcal{P}_2$ is said to be a *refinement* of a partition $\mathcal{P}_1$ if $\mathcal{P}_1 \subset \mathcal{P}_2$. That is, $\mathcal{P}_2 = \{t_0, \dots, t_m\}$ is a refinement of $\mathcal{P}_1 = \{x_0, \dots, x_n\}$ if every $x_k \in \mathcal{P}_1$ appears in the list of points $t_j \in \mathcal{P}_2$. The partition $\mathcal{P}_3$ is said to be a *common refinement* of the partitions $\mathcal{P}_1$ and $\mathcal{P}_2$ if $\mathcal{P}_1 \subset \mathcal{P}_3$ and $\mathcal{P}_2 \subset \mathcal{P}_3$. The set $\mathcal{P}_3 = \mathcal{P}_1 \cup \mathcal{P}_2$ (with redundant points eliminated) is the smallest common refinement of $\mathcal{P}_1$ and $\mathcal{P}_2$.

Suppose that f is a bounded function which is defined on $[a, b]$ and satisfies $|f| \le N$. Recall that the *infimum*, or *inf*, of a set $U \subset \mathbb{R}$ is another name for the greatest lower bound of U, and similarly the *supremum*, or *sup*, of U is the least upper bound of U. For each of the subintervals $[x_k, x_{k+1}]$, introduce the numbers

$$m_k = \inf\{f(t),\ x_k \le t \le x_{k+1}\}, \quad M_k = \sup\{f(t),\ x_k \le t \le x_{k+1}\}.$$

Even if f is not continuous, the numbers m_k and M_k will exist, and satisfy $-N \le m_k \le M_k \le N$. No matter how pathological the function f is, our sense of area demands that

$$m_k|x_{k+1} - x_k| \le \int_{x_k}^{x_{k+1}} f(x)\, dx \le M_k|x_{k+1} - x_k|.$$

Adding up the contributions from the various subintervals, we obtain an *upper sum*

$$\mathcal{U}(f, \mathcal{P}) = \sum_{k=0}^{n-1} M_k|x_{k+1} - x_k|,$$

which will be larger than the integral, and a *lower sum*

$$\mathcal{L}(f, \mathcal{P}) = \sum_{k=0}^{n-1} m_k|x_{k+1} - x_k|,$$

which will be smaller than the integral. Since $|f| \le N$, and since the lower sums for a partition are always smaller than the upper sums for the same partition, the inequalities

$$-N[b - a] \le \sup_{\mathcal{P}} \mathcal{L}(f, \mathcal{P}) \le \inf_{\mathcal{P}} \mathcal{U}(f, \mathcal{P}) \le N[b - a]$$

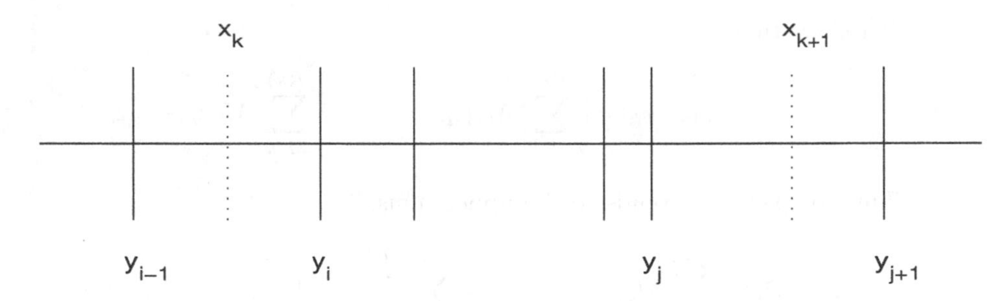

FIGURE 6.7
A common refinement of partitions.

are always valid.

The next lemma says that the lower sum for any partition is always smaller than the upper sum for any other partition.

Lemma 6.2.1. *Suppose that f is a bounded function defined on $[a,b]$, and that $\mathcal{P}_1$ and $\mathcal{P}_2$ are two partitions of $[a,b]$. Then*

$$\mathcal{L}(f,\mathcal{P}_1) \le \mathcal{U}(f,\mathcal{P}_2).$$

Proof. Given partitions $\mathcal{P}_1 = \{x_k | k = 0,\dots,n\}$ and $\mathcal{P}_2 = \{y_j | j = 0,\dots,m\}$, let

$$\mathcal{P}_3 = \{z_0,\dots,z_r\} = \mathcal{P}_1 \cup \mathcal{P}_2$$

be their smallest common refinement.

Let's compare the upper sums $\mathcal{U}(f,\mathcal{P}_1)$ and $\mathcal{U}(f,\mathcal{P}_3)$. Since $\mathcal{P}_3$ is a refinement of $\mathcal{P}_1$, each interval $[x_k,x_{k+1}]$ may be written as the union of one or more intervals $[z_l,z_{l+1}]$,

$$[x_k,x_{k+1}] = \cup_{l=I(k)}^{J(k)} [z_l,z_{l+1}], \quad x_k = z_{I(k)} < \cdots < z_{J(k)} = x_{k+1}.$$

Comparing

$$\widetilde{M}_l = \sup_{t\in[z_l,z_{l+1}]} f(t) \quad \text{and} \quad M_k = \sup_{t\in[x_k,x_{k+1}]} f(t),$$

we find that $M_k \ge \widetilde{M}_l$, since $[z_l,z_{l+1}] \subset [x_k,x_{k+1}]$.

Notice that the length of the interval $[x_k,x_{k+1}]$ is the sum of the lengths of the subintervals $[z_l,z_{l+1}]$ for $l = I(k),\dots,J(k)$,

$$|x_{k+1} - x_k| = \sum_{l=I(k)}^{J(k)} |z_{l+1} - z_l|.$$

It follows that

$$M_k|x_{k+1} - x_k| = \sum_{l=I(k)}^{J(k)} M_k|z_{l+1} - z_l| \ge \sum_{l=I(k)}^{J(k)} \widetilde{M}_l|z_{l+1} - z_l|.$$

This comparison extends to the upper sums,

$$\mathcal{U}(f, \mathcal{P}_1) = \sum_{k=0}^{j-1} M_k|x_{k+1} - x_k| \ge \sum_{k=0}^{n-1} \sum_{l=I(k)}^{J(k)} \widetilde{M}_l|z_{l+1} - z_l| = \mathcal{U}(f, \mathcal{P}_3).$$

Thus refinement of a partition reduces the upper sum. By a similar argument, refinement of a partition increases the lower sum. Since $\mathcal{P}_3$ is a common refinement of $\mathcal{P}_1$ and $\mathcal{P}_2$, and since the upper sum of a partition exceeds the lower sum of the same partition, it follows that

$$\mathcal{L}(f, \mathcal{P}_1) \le \mathcal{L}(f, \mathcal{P}_3) \le \mathcal{U}(f, \mathcal{P}_3) \le \mathcal{U}(f, \mathcal{P}_2).$$

□

If our expectations about area are correct, then the upper and lower sums should approach a common value as the mesh of the partition approaches 0. This expectation will be realized for "nice" functions, although pathological functions such as (6.1) will not fulfill our expectations. Say that a bounded function $f : [a, b] \to \mathbb{R}$ is *integrable* if the infimum of its upper sums, taken over all partitions $\mathcal{P}$, is equal to the supremum of the lower sums, or in abbreviated notation

$$\inf_{\mathcal{P}} \mathcal{U}(f, \mathcal{P}) = \sup_{\mathcal{P}} \mathcal{L}(f, \mathcal{P}). \tag{6.2}$$

If the function f is integrable, the *integral* is taken to be this common value

$$\int_a^b f(x)\, dx = \inf_{\mathcal{P}} \mathcal{U}(f, \mathcal{P}) = \sup_{\mathcal{P}} \mathcal{L}(f, \mathcal{P}).$$

It is often convenient to work with an alternative characterization of integrable functions. The straightforward proof of the next lemma is left as an exercise.

Lemma 6.2.2. *A bounded function f defined on the interval $[a, b]$ is integrable if and only if for every $\epsilon > 0$ there is a partition $\mathcal{P}$ such that*

$$\mathcal{U}(f, \mathcal{P}) - \mathcal{L}(f, \mathcal{P}) < \epsilon.$$

In some cases it is possible to show that f is integrable by explicitly bounding the difference of the upper and lower sums as a function of the mesh of the partition $\mathcal{P}$.

Theorem 6.2.3. *Suppose that $f : [a,b] \to \mathbb{R}$ is differentiable, and*

$$|f'(x)| \le C, \quad x \in [a,b].$$

Then f is integrable, and

$$\mathcal{U}(f,\mathcal{P}) - \mathcal{L}(f,\mathcal{P}) \le C\mu(\mathcal{P})[b-a].$$

Proof. Since f is differentiable, it is continuous on the compact interval $[a,b]$, and so bounded. In addition the continuity implies that for any partition $\mathcal{P} = \{x_0,\dots,x_n\}$, there are points u_k, v_k in $[x_k, x_{k+1}]$ such that

$$m_k = \inf_{t\in[x_k,x_{k+1}]} f(t) = f(u_k), \quad M_k = \sup_{t\in[x_k,x_{k+1}]} f(t) = f(v_k).$$

For this partition the difference of the upper and lower sums is

$$\mathcal{U}(f,\mathcal{P}) - \mathcal{L}(f,\mathcal{P}) = \sum_{k=0}^{n-1} f(v_k)[x_{k+1}-x_k] - \sum_{k=0}^{n-1} f(u_k)[x_{k+1}-x_k]$$

$$= \sum_{k=0}^{n-1} [f(v_k) - f(u_k)][x_{k+1}-x_k].$$

By the Mean Value Theorem

$$|f(v_k) - f(u_k)| \le C|v_k - u_k| \le C|x_{k+1} - x_k| \le C\mu(\mathcal{P}).$$

This gives the desired estimate of the difference of upper and lower sums,

$$\mathcal{U}(f,\mathcal{P}) - \mathcal{L}(f,\mathcal{P}) \le \sum_{k=0}^{n-1} C\mu(\mathcal{P})[x_{k+1}-x_k] = C\mu(\mathcal{P})[b-a].$$

By the previous lemma f is integrable, since the mesh $\mu(\mathcal{P})$ may be made arbitrarily small. □

When a computer is used to calculate integrals by geometric methods, it is important to relate the required number of arithmetic computations to the desired accuracy. Since

$$\mathcal{L}(f,\mathcal{P}) \le \int_a^b f(x)\,dx \le \mathcal{U}(f,\mathcal{P}),$$

Theorem 6.2.3 can be interpreted as a bound on the complexity of Riemann sum calculations. As a numerical technique the use of Riemann sums is rather inefficient.

The ideas used in the last proof may also be employed to show that continuous functions are integrable. In this case we lose the explicit connection between mesh size and the difference of the upper and lower sums.

Theorem 6.2.4. *If $f : [a, b] \to \mathbb{R}$ is continuous, then f is integrable. For any $\epsilon > 0$ there is a $\mu_0 > 0$ such that $\mu(\mathcal{P}) < \mu_0$ implies*

$$\mathcal{U}(f, \mathcal{P}) - \mathcal{L}(f, \mathcal{P}) < \epsilon.$$

Proof. As in the proof of Theorem 6.2.3, f is bounded and there are points u_k, v_k in $[x_k, x_{k+1}]$ such that

$$m_k = \inf_{t \in [x_k, x_{k+1}]} f(t) = f(u_k), \quad M_k = \sup_{t \in [x_k, x_{k+1}]} f(t) = f(v_k).$$

Since f is continuous on a compact interval, f is uniformly continuous. That is, for any $\eta > 0$ there is a δ such that

$$|f(x) - f(y)| < \eta \quad \text{whenever} \quad |x - y| < \delta.$$

Pick $\eta = \epsilon/(b - a)$, and let μ_0 be the corresponding δ. If $\mathcal{P}$ is any partition with $\mu(\mathcal{P}) < \mu_0$, then

$$\mathcal{U}(f, \mathcal{P}) - \mathcal{L}(f, \mathcal{P}) = \sum_{k=0}^{n-1} [f(v_k) - f(u_k)][x_{k+1} - x_k] < \sum_{k=0}^{n-1} \frac{\epsilon}{b - a}[x_{k+1} - x_k] = \epsilon.$$

□

Theorem 6.2.5. *Suppose that $f(x)$ is integrable on $[a, b]$. If $[c, d] \subset [a, b]$ then f is integrable on $[c, d]$.*

Proof. Any partition $\mathcal{P}$ of $[a, b]$ has a refinement $\mathcal{P}_1 = \{x_0, \dots, x_n\}$ which includes the points c, d. Let $\mathcal{P}_2 \subset \mathcal{P}_1$ be the corresponding partition of $[c, d]$.

Since

$$(M_k - m_k)[x_{k+1} - x_k] \ge 0,$$

it follows that

$$\mathcal{U}(f, \mathcal{P}_2) - \mathcal{L}(f, \mathcal{P}_2) \le \mathcal{U}(f, \mathcal{P}_1) - \mathcal{L}(f, \mathcal{P}_1) \le \mathcal{U}(f, \mathcal{P}) - \mathcal{L}(f, \mathcal{P}),$$

yielding the integrability of f on $[c, d]$. □

The next theorem allows us to construct examples of functions $g(x)$ which are integrable, but not continuous.

Theorem 6.2.6. *Suppose $a < b < c$, that $f_1(x)$ is integrable on $[a, b]$, and that $f_2(x)$ is integrable on $[b, c]$. If $g : [a, c] \to \mathbb{R}$ satisfies*

$$g(x) = \begin{Bmatrix} f_1(x), & a \le x < b, \\ f_2(x), & b < x \le c, \end{Bmatrix}$$

then g is integrable on $[a, c]$ and

$$\int_a^c g(x)\, dx = \int_a^b f_1(x)\, dx + \int_b^c f_2(x)\, dx.$$

Proof. For any $\epsilon > 0$ there are partitions $\mathcal{P}_1 = \{x_0, \dots, x_j\}$ of $[a, b]$ and $\mathcal{Q}_1 = \{y_0, \dots, y_n\}$ of $[b, c]$ such that

$$\mathcal{U}(f_1, \mathcal{P}_1) - \mathcal{L}(f_1, \mathcal{P}_1) < \frac{\epsilon}{3}, \quad \mathcal{U}(f_2, \mathcal{Q}_1) - \mathcal{L}(f_2, \mathcal{Q}_1) < \frac{\epsilon}{3}.$$

Since $g(b) \in \mathbb{R}$ there is a constant $C > 0$ such that

$$\max(|f_1(x)|, |f_2(x)|, |g(x)|) \le C, \quad x \in [a, c].$$

Thus there are points z_1, z_2 with $x_{j-1} < z_1 < b < z_2 < y_1$ such that $C(z_2 - z_1) < \epsilon/6$. Add z_1 to the partition $\mathcal{P}_1$ to get a new partition $\mathcal{P}_2$ of $[a, b]$, and similarly add z_2 to $\mathcal{Q}_1$ to get $\mathcal{Q}_2$.

Combine $\mathcal{P}_2$ and $\mathcal{Q}_2$ to get the partition $\mathcal{R} = \mathcal{P}_2 \cup \mathcal{Q}_2$ of $[a, c]$. With this partition, the upper and lower sums for g and f_1 or f_2 only differ on the intervals $[z_1, b]$ and $[b, z_2]$, so that

$$\mathcal{U}(g, \mathcal{R}) - \mathcal{L}(g, \mathcal{R})$$

$$\le [\mathcal{U}(f_1, \mathcal{P}_2) - \mathcal{L}(f_1, \mathcal{P}_2)] + [\mathcal{U}(f_2, \mathcal{Q}_2) - \mathcal{L}(f_2, \mathcal{Q}_2)] + 2C(z_2 - z_1) < \epsilon,$$

showing that $g(x)$ is integrable on $[a, c]$. Moreover

$$\mathcal{L}(f_1, \mathcal{P}_2) + \mathcal{L}(f_2, \mathcal{Q}_2) - 2C(z_2 - z_1) \le \mathcal{L}(g, \mathcal{R})$$

$$\le \mathcal{U}(g, \mathcal{R}) \le \mathcal{U}(f_1, \mathcal{P}_2) + \mathcal{U}(f_2, \mathcal{Q}_2) + 2C(z_2 - z_1),$$

so

$$\int_a^c g(x)\, dx = \int_a^b f_1(x)\, dx + \int_b^c f_2(x)\, dx.$$

□

When the functions f_k in the statement of Theorem 6.2.6 are continuous, the function $g : [a, c] \to \mathbb{R}$ is said to be *piecewise continuous.* Since continuous functions are integrable, so are piecewise continuous functions, which arise fairly often in applied mathematics.

Monotonic functions are a rich collection of functions which are integrable, but not necessarily continuous. Recall that a function $f : [a, b] \to \mathbb{R}$ is *increasing* if $f(x) \le f(y)$ whenever $x \le y$.

Theorem 6.2.7. *Every increasing (or decreasing) function $f : [a, b] \to \mathbb{R}$ is integrable.*

Proof. First note that $f(a) \le f(x) \le f(b)$, so f is bounded. If $\mathcal{P} = \{x_0, \dots, x_n\}$ is any partition of $[a, b]$, the fact that $f(x)$ is increasing means $f(x_{k+1}) - f(x_k) \ge 0$, and

$$\mathcal{U}(f, \mathcal{P}) - \mathcal{L}(f, \mathcal{P}) = \sum_{k=0}^{n-1} f(x_{k+1})(x_{k+1} - x_k) - \sum_{k=0}^{n-1} f(x_k)(x_{k+1} - x_k)$$

$$= \sum_{k=0}^{n-1} [f(x_{k+1}) - f(x_k)](x_{k+1} - x_k).$$

But

$$\sum_{k=0}^{n-1} [f(x_{k+1}) - f(x_k)] = [f(x_1) - f(x_0)] + [f(x_2) - f(x_1)] + \cdots + [f(x_n) - f(x_{n-1}]$$

$$= f(b) - f(a),$$

so

$$\mathcal{U}(f, \mathcal{P}) - \mathcal{L}(f, \mathcal{P}) \le [f(b) - f(a)]\mu(\mathcal{P}),$$

which goes to zero with the mesh of the partition.

□

It is often convenient, particularly for numerical calculations, to avoid the determination of the numbers m_k and M_k. Instead, given a partition $\mathcal{P}$, the values m_k and M_k will be replaced by $f(t_k)$ for an arbitrary $t_k \in [x_k, x_{k+1}]$. To simplify the notation define

$$\Delta x_k = x_{k+1} - x_k.$$

As an approximation to the integral one considers *Riemann sums*, which are sums of the form

$$\sum_{k=0}^{n-1} f(t_k)\Delta x_k, \quad t_k \in [x_k, x_{k+1}].$$

Theorem 6.2.8. *Suppose that $f : [a, b] \to \mathbb{R}$ is continuous, $\mathcal{P} = \{x_0, \dots, x_n\}$ is a partition, and*

$$\sum_{k=0}^{n-1} f(t_k)\Delta x_k, \quad t_k \in [x_k, x_{k+1}]$$

is a corresponding Riemann sum. For any $\epsilon > 0$ there is a $\mu_0 > 0$ such that $\mu(\mathcal{P}) < \mu_0$ implies

$$\Big| \int_a^b f(x)\, dx - \sum_{k=0}^{n-1} f(t_k)\Delta x_k \Big| < \epsilon.$$

Proof. On each interval $[x_k, x_{k+1}]$ we have

$$m_k \le f(t_k) \le M_k.$$

Multiplying by Δx_k and adding gives

$$\mathcal{L}(f, \mathcal{P}) = \sum_{k=0}^{n-1} m_k \Delta x_k \le \sum_{k=0}^{n-1} f(t_k)\Delta x_k \le \sum_{k=0}^{n-1} M_k \Delta x_k = \mathcal{U}(f, \mathcal{P}).$$

In addition,

$$\mathcal{L}(f,\mathcal{P}) \le \int_a^b f(x)\,dx \le \mathcal{U}(f,\mathcal{P}).$$

These inequalities imply

$$\Big|\int_a^b f(x)\,dx - \sum_{k=0}^{n-1} f(t_k)\Delta x_k\Big| \le \mathcal{U}(f,\mathcal{P}) - \mathcal{L}(f,\mathcal{P}).$$

Finally, Theorem 6.2.4 says that for any $\epsilon > 0$ there is a $\mu_0 > 0$ such that $\mu(\mathcal{P}) < \mu_0$ implies

$$\mathcal{U}(f,\mathcal{P}) - \mathcal{L}(f,\mathcal{P}) < \epsilon.$$

□

6.3 Properties of integrals

Theorem 6.3.1. *Suppose that $f(x)$ and $g(x)$ are integrable on $[a,b]$. For any constants c_1, c_2 the function $c_1 f(x) + c_2 g(x)$ is integrable, and*

$$\int_a^b c_1 f(x) + c_2 g(x)\,dx = c_1 \int_a^b f(x)\,dx + c_2 \int_a^b g(x)\,dx.$$

Proof. It suffices to prove that

$$\int_a^b c_1 f(x)\,dx = c_1 \int_a^b f(x)\,dx, \quad \int_a^b f(x)+g(x)\,dx = \int_a^b f(x)\,dx + \int_a^b g(x)\,dx.$$

Suppose that $\epsilon > 0$, and $\mathcal{P} = \{x_0, \dots, x_n\}$ is a partition such that

$$\mathcal{U}(f,\mathcal{P}) - \mathcal{L}(f,\mathcal{P}) < \epsilon.$$

If $c_1 \ge 0$ then

$$\inf_{x\in[x_k,x_{k+1}]} c_1 f(x) = c_1 \inf_{x\in[x_k,x_{k+1}]} f(x), \qquad \sup_{x\in[x_k,x_{k+1}]} c_1 f(x) = c_1 \sup_{x\in[x_k,x_{k+1}]} f(x),$$

while if $c_1 < 0$ then

$$\inf_{x\in[x_k,x_{k+1}]} c_1 f(x) = c_1 \sup_{x\in[x_k,x_{k+1}]} f(x), \qquad \sup_{x\in[x_k,x_{k+1}]} c_1 f(x) = c_1 \inf_{x\in[x_k,x_{k+1}]} f(x).$$

For $c_1 \ge 0$ it follows that

$$\mathcal{U}(c_1 f,\mathcal{P}) = c_1 \mathcal{U}(f,\mathcal{P}), \quad \mathcal{L}(c_1 f,\mathcal{P}) = c_1 \mathcal{L}(f,\mathcal{P}),$$

while for $c_1 < 0$

$$\mathcal{U}(c_1 f, \mathcal{P}) = c_1 \mathcal{L}(f, \mathcal{P}), \quad \mathcal{L}(c_1 f, \mathcal{P}) = c_1 \mathcal{U}(f, \mathcal{P}).$$

In either case

$$\mathcal{U}(c_1 f, \mathcal{P}) - \mathcal{L}(c_1 f, \mathcal{P}) < |c_1|\epsilon.$$

This is enough to show that $c_1 f(x)$ is integrable, with

$$\int_a^b c_1 f(x)\, dx = c_1 \int_a^b f(x)\, dx.$$

Suppose that $\mathcal{P}_1$ is a second partition such that

$$\mathcal{U}(g, \mathcal{P}_1) - \mathcal{L}(g, \mathcal{P}_1) < \epsilon.$$

By passing to a common refinement we may assume that $\mathcal{P} = \mathcal{P}_1$. For the function f let

$$m_k^f = \inf_{x \in [x_k, x_{k+1}]} f(x), \quad M_k^f = \sup_{x \in [x_k, x_{k+1}]} f(x).$$

Then for all $x \in [x_k, x_{k+1}]$,

$$f(x) + g(x) \le M_k^f + M_k^g,$$

so that

$$m_k^{f+g} \ge m_k^f + m_k^g, \quad M_k^{f+g} \le M_k^f + M_k^g,$$

and

$$\mathcal{U}(f+g, \mathcal{P}) \le \mathcal{U}(f, \mathcal{P}) + \mathcal{U}(g, \mathcal{P}), \quad \mathcal{L}(f+g, \mathcal{P}) \ge \mathcal{L}(f, \mathcal{P}) + \mathcal{L}(g, \mathcal{P}).$$

These inequalities imply

$$\mathcal{U}(f+g, \mathcal{P}) - \mathcal{L}(f+g, \mathcal{P}) < 2\epsilon,$$

so $f + g$ is integrable. In addition both numbers

$$\int_a^b f(x) + g(x)\, dx, \quad \text{and} \quad \int_a^b f(x)\, dx + \int_a^b g(x)\, dx$$

lie between $\mathcal{L}(f, \mathcal{P}) + \mathcal{L}(g, \mathcal{P})$ and $\mathcal{U}(f, \mathcal{P}) + \mathcal{U}(g, \mathcal{P})$, so that

$$\Big| \int_a^b f(x) + g(x)\, dx - \Big(\int_a^b f(x)\, dx + \int_a^b g(x)\, dx \Big) \Big| < 2\epsilon.$$

□

The next result will show that the product of integrable functions is integrable. In the proof it will be necessary to discuss the length of a set of the intervals $[x_k, x_{k+1}]$ from the partition $\mathcal{P} = \{x_0, \dots, x_n\}$ of $[a, b]$. The obvious notion of length is used; if B is a subset of the indices $\{0, \dots, n-1\}$, and

$$\mathcal{P}_B = \bigcup_{k \in B} [x_k, x_{k+1}],$$

then the length of $\mathcal{P}_B$ is

$$\text{length}(\mathcal{P}_B) = \sum_{k \in B} [x_{k+1} - x_k].$$

Theorem 6.3.2. *If $f(x)$ and $g(x)$ are integrable on $[a, b]$, then so is $f(x)g(x)$.*

Proof. The argument is somewhat simpler if f and g are assumed to be positive. It is a straightforward exercise to deduce the general case from this special case (see Problem 6.16).

Pick $\epsilon > 0$, and let $\mathcal{P}$ be a partition of $[a, b]$ such that

$$\mathcal{U}(f, \mathcal{P}) - \mathcal{L}(f, \mathcal{P}) < \epsilon^2, \quad \mathcal{U}(g, \mathcal{P}) - \mathcal{L}(g, \mathcal{P}) < \epsilon^2. \tag{6.3}$$

Let B_f be the set of indices k such that $M_k^f - m_k^f \ge \epsilon$. On one hand, the contributions from the intervals $[x_k, x_{k+1}]$ with $k \in B_f$ tend to make the difference between upper and lower sums big:

$$\mathcal{U}(f, \mathcal{P}) - \mathcal{L}(f, \mathcal{P}) = \sum_{k=0}^{n-1} (M_k^f - m_k^f)\Delta x_k \ge \sum_{k \in B_f} (M_k^f - m_k^f)\Delta x_k \ge \epsilon \sum_{k \in B_f} \Delta x_k.$$

On the other hand, (6.3) says the total difference between upper and lower sums is small. It follows that

$$\text{length}(\mathcal{P}_{B_f}) = \sum_{k \in B_f} \Delta x_k < \epsilon.$$

The analogous definition for B_g leads to the same conclusion.

Let J denote the set of indices k such that

$$M_k^f - m_k^f < \epsilon, \quad M_k^g - m_k^g < \epsilon, \quad k \in J.$$

The set of indices J is just the complement of the union of B_f and B_g.

The functions f and g are bounded. Assume that

$$0 \le f(x) \le L, \quad 0 \le g(x) \le L, \quad x \in [a, b].$$

To estimate $\mathcal{U}(fg, \mathcal{P}) - \mathcal{L}(fg, \mathcal{P})$, note that since $f(x) > 0$ and $g(x) > 0$,

$$M_k^{fg} - m_k^{fg} \le M_k^f M_k^g - m_k^f m_k^g$$

$$= M_k^f M_k^g - M_k^f m_k^g + M_k^f m_k^g - m_k^f m_k^g = M_k^f (M_k^g - m_k^g) + (M_k^f - m_k^f) m_k^g.$$

Thus for $k \in J$,

$$M_k^{fg} - m_k^{fg} < 2L\epsilon,$$

while for any k

$$M_k^{fg} - m_k^{fg} < L^2.$$

Putting these estimates together, the proof is completed by noting that the difference between upper and lower sums for the function fg is small for the partition $\mathcal{P}$.

$$\mathcal{U}(fg,\mathcal{P}) - \mathcal{L}(fg,\mathcal{P}) = \sum_{k=0}^{n-1} (M_k^{fg} - m_k^{fg}) \Delta x_k$$

$$\le \sum_{k \in J} (M_k^{fg} - m_k^{fg}) \Delta x_k + \sum_{k \in B_f} (M_k^{fg} - m_k^{fg}) \Delta x_k + \sum_{k \in B_g} (M_k^{fg} - m_k^{fg}) \Delta x_k$$

$$< 2L[b-a]\epsilon + L^2\epsilon + L^2\epsilon.$$

□

Theorem 6.3.3. *If f is integrable, so is $|f|$, and*

$$\Big| \int_a^b f(x)\, dx \Big| \le \int_a^b |f(x)|\, dx.$$

Proof. Again suppose that $\epsilon > 0$, and $\mathcal{P} = \{x_0, \dots, x_n\}$ is a partition such that

$$\mathcal{U}(f,\mathcal{P}) - \mathcal{L}(f,\mathcal{P}) < \epsilon.$$

For any numbers x, y

$$\big| |x| - |y| \big| \le |x - y|,$$

so that

$$M_k^{|f|} - m_k^{|f|} = \sup_{t_1,t_2} \big| |f(t_1)| - |f(t_2)| \big| \le \sup_{t_1,t_2} |f(t_1) - f(t_2)| = M_k^f - m_k^f.$$

This in turn implies

$$\mathcal{U}(|f|,\mathcal{P}) - \mathcal{L}(|f|,\mathcal{P}) \le \mathcal{U}(f,\mathcal{P}) - \mathcal{L}(f,\mathcal{P}) < \epsilon,$$

and $|f|$ is integrable.

One also checks easily that

$$|\mathcal{U}(f,\mathcal{P})| \le \mathcal{U}(|f|,\mathcal{P}),$$

leading to the desired inequality. □

We turn now to the Fundamental Theorem of Calculus. By relating integrals and derivatives, this theorem provides the basis for most of the familiar integration techniques of calculus. The theorem will be split into two parts: the first considers the differentiability of integrals, the second describes the integration of derivatives.

As part of the proof of the next theorem, the integral $\int_b^a f(x)\,dx$ with $a < b$. will be needed. This integral is defined by

$$\int_b^a f(x)\,dx = -\int_a^b f(x)\,dx, \quad a < b.$$

Theorem 6.3.4. *(Fundamental Theorem of Calculus: Part 1) Suppose that $f : (a,b) \to \mathbb{R}$ is continuous. If $x_0, x \in (a,b)$, then the function*

$$F(x) = \int_{x_0}^x f(t)\,dt$$

is differentiable, and

$$F'(x) = f(x).$$

Proof. For $h > 0$ the computation runs as follows.

$$F(x+h) - F(x) = \int_{x_0}^{x+h} f(t)\,dt - \int_{x_0}^x f(t)\,dt - \int_x^{x+h} f(t)\,dt.$$

Since f is continuous on $[x, x+h]$, there are u, v such that

$$f(u) = \min_{t\in[x,x+h]} f(t), \quad f(v) = \max_{t\in[x,x+h]} f(t).$$

The inequality

$$hf(u) \le \int_x^{x+h} f(t)\,dt \le hf(v),$$

then implies

$$f(u) \le \frac{F(x+h) - F(x)}{h} \le f(v).$$

The continuity of f at x means that for any $\epsilon > 0$ there is an h such that

$$|f(y) - f(x)| < \epsilon, \quad |y - x| < h.$$

Apply this inequality with u and v in place of y to obtain

$$f(x) - \epsilon < f(u), \quad f(v) < f(x) + \epsilon,$$

and so

$$f(x) - \epsilon \le \frac{F(x+h) - F(x)}{h} \le f(x) + \epsilon.$$

But this says that

$$\lim_{h \to 0^+} \frac{F(x+h) - F(x)}{h} = f(x).$$

For $h < 0$ the analogous computation begins with

$$F(x+h) - F(x) = -[F(x) - F(x+h)] = -\int_{x+h}^{x} f(t)\, dt = \int_{x}^{x+h} f(t)\, dt.$$

The rest of the computations, leading to

$$\lim_{h \to 0^-} \frac{F(x+h) - F(x)}{h} = f(x),$$

are left as an exercise.

□

The second part of the Fundamental Theorem of Calculus is now easy.

Theorem 6.3.5. *(Fundamental Theorem of Calculus: Part 2) Suppose that $F : (c,d) \to \mathbb{R}$ has a continuous derivative, $F'(x) = f(x)$, and that $[a,b] \subset (c,d)$. Then*

$$\int_a^b f(t)\, dt = F(b) - F(a).$$

Proof. For $x \in (c,d)$ the functions $F(x) - F(a)$ and $\int_a^x f(t)\, dt$ have the same derivative by the first part of the fundamental theorem of calculus. Consequently, these two functions differ by a constant C. Evaluation of the two functions at $x = a$ shows that $C = 0$. □

6.4 Arc length and trigonometric functions

The definition of arc length can be treated as part of the discussion of integration. Suppose $f : [a,b] \to \mathbb{R}$ has a continuous derivative. The graph G of f is a set of points in the $x - y$ plane,

$$G = \{(x, f(x)),\ a \le x \le b.\}$$

The length of the path G is approximated by using a piecewise linear function based on a partition $a = x_0 < x_1 < \dots < x_N = b$ of $[a,b]$. On each interval $[x_n, x_{n+1}]$ the approximating function $l(x)$ is the linear segment joining the endpoints $(x_n, f(x_n))$ and $(x_{n+1}, f(x_{n+1}))$.

If

$$\Delta x_n = x_{n+1} - x_n, \quad \Delta y_k = f(x_{n+1}) - f(x_n),$$

then

$$l(x) = f(x_n) + \frac{\Delta y_n}{\Delta x_n}(x - x_n).$$

The length of the linear segment is $\sqrt{(\Delta x_n)^2 + (\Delta y_n)^2}$. The length L of G is approximated by the sum of the segment lengths,

$$L \simeq \sum_{n=0}^{N-1} \sqrt{(\Delta x_n)^2 + (\Delta y_n)^2} = \sum_{n=0}^{N-1} \sqrt{1 + (\Delta y_n/\Delta x_n)^2}\Delta x_n.$$

By the Mean Value Theorem there is a point $t_n \in [x_n, x_{n+1}]$ with

$$f'(t_n) = \Delta y_n/\Delta x_n,$$

and the resulting expression

$$L \simeq \sum_{n=0}^{N-1} \sqrt{1 + (f'(t_n))^2}\Delta x_n$$

is a Riemann sum for the integral taken as the definition of arc length,

$$L = \int_a^b \sqrt{1 + (f'(x))^2}\, dx.$$

When the arc in question is part of a circle, the trigonometric functions $\sin(\theta)$, $\cos(\theta)$, and $\tan(\theta) = \sin(\theta)/\cos(\theta)$ appear. For $-1 < x < 1$ the upper half of the unit circle $\{(x, y)|x^2 + y^2 = 1\}$ may be described as the graph of the positive function

$$y(x) = \sqrt{1 - x^2}.$$

For $0 \le x_1 < 1$ define $\theta(x_1)$ to be the arc length of the graph of $y(x)$, starting the arc at $x = 0$ and extending to $x = x_1$. For $-1 < x < 0$ let $\theta(x) = -\theta(|x|)$. Then $\theta(x)$ is a strictly increasing differentiable function, and we define

$$\sin(\theta) = x(\theta), \quad \cos(\theta) = y(\theta).$$

(The arc length θ and the definition of the trigonometric functions are often based on the starting point $(1, 0)$ in the plane; the above choice seems convenient when working with the function $\sqrt{1 - x^2}$.)

Since

$$\frac{dy}{dx} = \frac{-x}{\sqrt{1 - x^2}},$$

the chain rule yields

$$1 = \frac{dx}{d\theta}\frac{d\theta}{dx} = \frac{d}{d\theta}\sin(\theta)\frac{d\theta}{dx} = \frac{d}{d\theta}\sin(\theta)\sqrt{1 + (dy/dx)^2},$$

or

$$\frac{d}{d\theta}\sin(\theta) = \frac{1}{\sqrt{1 + (dy/dx)^2}} = \sqrt{1 - x^2} = \sqrt{y^2} = \cos(\theta).$$

Similarly,

$$\frac{dy}{dx} = \frac{d}{d\theta}\cos(\theta)\sqrt{1 + (dy/dx)^2},$$

or

$$\frac{d}{d\theta}\cos(\theta) = \frac{-x}{\sqrt{1-x^2}}\sqrt{1-x^2} = -\sin(\theta).$$

The inverse trigonometric functions can be expressed as integrals of algebraic functions.

$$\sin^{-1}(x) = \int_0^x \frac{1}{\sqrt{1-t^2}}\,dt,$$

$$\cos^{-1}(x) = \frac{\pi}{2} - \int_0^x \frac{1}{\sqrt{1-t^2}}\,dt,$$

$$\tan^{-1}(x) = \int_0^x \frac{1}{1+t^2}\,dt,$$

6.5 Improper integrals

One often encounters integrals where the interval of integration is unbounded, or the integrand itself is unbounded. In an example like

$$\int_0^\infty \frac{e^{-x}}{\sqrt{x}}\,dx,$$

both the function and the interval of integration are unbounded. Integrals exhibiting such difficulties are often termed *improper.*

It is useful to draw analogies between the study of improper integrals and the study of infinite series. It is clear that some improper integrals such as

$$\int_0^\infty 1\,dx$$

represent an infinite area, and so will not have a real number value. Examples such as

$$\int_{-\infty}^\infty x\,dx$$

make even less sense. On the other hand, since an antiderivative of $1/(1+x^2)$ is $\tan^{-1}(x)$, and an antiderivative of $1/\sqrt{1-x^2}$ is $\sin^{-1}(x)$, it is reasonable to expect that

$$\int_0^1 \frac{1}{\sqrt{1-x^2}}\,dx = \sin^{-1}(1) - \sin^{-1}(0) = \pi/2, \tag{6.4}$$

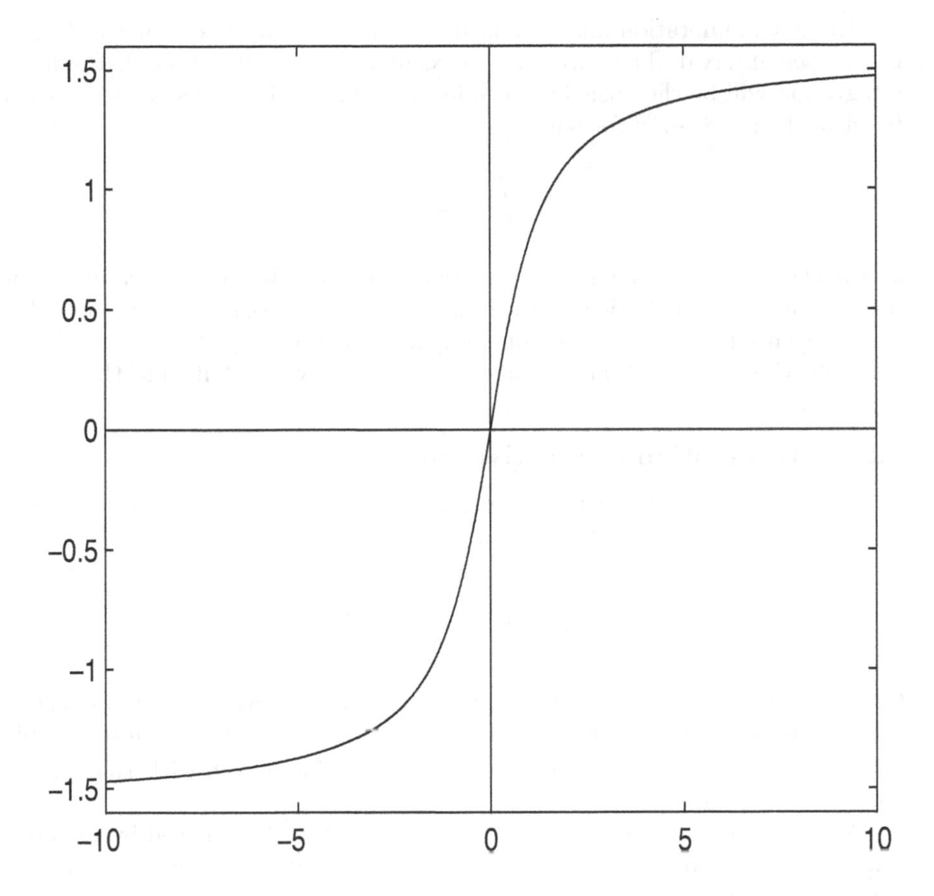

FIGURE 6.8
Graph of $\tan^{-1}(x)$.

and that (see Figure 6.8)

$$\int_{-\infty}^{\infty} \frac{1}{1+x^2}\,dx = \lim_{N\to\infty} \int_{-N}^{N} \frac{1}{1+x^2}\,dx \tag{6.5}$$

$$= \lim_{N\to\infty} \tan^{-1}(N) - \tan^{-1}(-N) = \pi/2 - (-\pi/2) = \pi.$$

As in the case of infinite series, where sums of positive terms provide the foundation of the theory, integration of positive functions has the central role for the study of improper integrals. After extending the theory of integration to handle improper integrals of positive functions, the ideas are extended to functions whose absolute values are well behaved. Finally, more general cases of conditionally convergent integrals are considered.

To fix some notation and standard assumptions, suppose that $(\alpha, \beta) \subset \mathbb{R}$ is an open interval. The cases $\alpha = -\infty$ and $\beta = \infty$ are allowed. Riemann integration will be the basis for considering improper integrals. Thus for each function $f : (\alpha, \beta) \to \mathbb{R}$ for which

$$\int_\alpha^\beta f(x)\ dx$$

is considered, it is assumed that f is Riemann integrable on each compact interval $[a, b] \subset (\alpha, \beta)$. Notice that the function f is not assumed to be bounded on the open interval (α, β), just on compact subintervals $[a, b] \subset (\alpha, \beta)$. As an example, the function $f(x) = x$ on the interval $(-\infty, \infty)$ falls into this class.

6.5.1 Integration of positive functions

In addition to the standard assumptions above, suppose that $f(x) \ge 0$. Say that the integral $\int_\alpha^\beta f(x)\ dx$ *converges* to the number I if

$$\sup_{[a,b] \subset (\alpha,\beta)} \int_a^b f(x)\ dx = I < \infty.$$

Otherwise the integral *diverges*. If the integral converges, the number I is taken to be the value of the integral. Of course if $f(x)$ is Riemann integrable on $[\alpha, \beta]$, then $\int_\alpha^\beta f(x)\ dx$ converges, and the value agrees with that of the Riemann integral.

An important role in the study of improper integrals is played by the least upper bound axiom. Recall that this axiom says that any set of real numbers with an upper bound has a least upper bound, or supremum. This axiom plays a role analogous to that of the bounded monotone sequence theorem in our study of positive series, in that we are able to establish the existence of limits without finding them explicitly.

The first lemma restates the definition of convergence of an integral in a more convenient form.

Lemma 6.5.1. *Suppose that $f(x) \ge 0$ for $x \in (\alpha, \beta)$, and f is integrable for every $[a, b] \subset (\alpha, \beta)$. Then the integral*

$$\int_\alpha^\beta f(x)\ dx$$

converges to the value I if and only if the following two conditions hold:

i) for every $[a, b] \subset (\alpha, \beta)$,

$$\int_a^b f(x)\ dx \le I,$$

ii) for every $\epsilon > 0$ *there is an interval* $[a_1, b_1] \subset (\alpha, \beta)$ *such that*

$$I - \int_{a_1}^{b_1} f(x)\, dx < \epsilon.$$

As with the comparison test for infinite series of positive terms, the convergence of the improper integral of a positive function can be established by showing that the integral of a larger function converges.

Theorem 6.5.2. *(Comparison test) Assume that* $0 \le f(x) \le g(x)$ *for all* $x \in (\alpha, \beta)$*. If* $\int_\alpha^\beta g(x)\, dx$ *converges to* $I_g \in \mathbb{R}$*, then* $\int_\alpha^\beta f(x)\, dx$ *converges to a number* I_f*, and* $I_f \le I_g$*. If* $\int_\alpha^\beta f(x)\, dx$ *diverges, so does* $\int_\alpha^\beta g(x)\, dx$*.*

Proof. For every interval $[a, b] \subset (\alpha, \beta)$ the inequality

$$\int_a^b f(x)\, dx \le \int_a^b g(x)\, dx,$$

holds for the Riemann integrals. Thus

$$\sup_{[a,b]\subset(\alpha,\beta)} \int_a^b f(x)\, dx \le \sup_{[a,b]\subset(\alpha,\beta)} \int_a^b g(x)\, dx = I_g.$$

Since the set of values $\int_a^b f(x)\, dx$ for $[a, b] \subset (\alpha, \beta)$ is bounded above, it has a supremum $I_f \le I_g$, which by definition is the integral $\int_a^b f(x)\, dx$.

In the other direction, if $\int_\alpha^\beta f(x)\, dx$ diverges, then for any $M > 0$ there is an interval $[a, b]$ such that

$$M \le \int_a^b f(x)\, dx \le \int_a^b g(x)\, dx.$$

It follows that

$$\sup_{[a,b]\subset(\alpha,\beta)} \int_a^b g(x)\, dx = \infty.$$

□

In calculus, improper integrals are analyzed with limit computations; the next result connects limits with our definition.

Theorem 6.5.3. *Suppose that* $f(x)$ *is positive and integrable on every subinterval* $[a, b] \subset (\alpha, \beta)$*.*

If the integral $\int_\alpha^\beta f(x)\, dx$ *converges, then for any point* $c \in (\alpha, \beta)$ *there are real numbers* I_1 *and* I_2 *such that*

$$\lim_{a\to\alpha^+} \int_a^c f(x)\, dx = I_1, \quad \text{and} \quad \lim_{b\to\beta^-} \int_c^b f(x)\, dx = I_2.$$

In the opposite direction, if there is any point $c \in (\alpha, \beta)$, and real numbers I_1 and I_2 such that

$$\lim_{a \to \alpha^+} \int_a^c f(x)\, dx = I_1, \quad \text{and} \quad \lim_{b \to \beta^-} \int_c^b f(x)\, dx = I_2,$$

then the integral $\int_\alpha^\beta f(x)\, dx$ converges to $I_1 + I_2$.

Proof. Suppose that the integral $\int_\alpha^\beta f(x)\, dx$ converges to I. Take any point $c \in (\alpha, \beta)$. Suppose that $a \le c \le b$. Since $f(x) \ge 0$ for all $x \in (\alpha, \beta)$, the number $g(b) = \int_c^b f(x)\, dx$ increases as b increases, and $h(a) = \int_a^c f(x)\, dx$ increases as a decreases. In addition

$$\int_a^c f(x)\, dx \le I, \quad \text{and} \quad \int_c^b f(x)\, dx \le I.$$

Since the numbers $\int_c^b f(x)\, dx$ are bounded above, they have a least upper bound I_2. For $\epsilon > 0$ there is a number d with $c \le d < \beta$ such that

$$I_2 - \epsilon < g(d) = \int_c^d f(x)\, dx \le I_2.$$

Since $g(b)$ increases as b increases, we conclude that

$$\lim_{b \to \beta^-} \int_c^b f(x)\, dx = I_2.$$

A similar argument applies to the integrals $\int_a^c f(x)\, dx$.

Now suppose that there is a point $c \in (\alpha, \beta)$, and real numbers I_1 and I_2 such that

$$\lim_{a \to \alpha^+} \int_a^c f(x)\, dx = I_1, \quad \text{and} \quad \lim_{b \to \beta^-} \int_c^b f(x)\, dx = I_2.$$

Let $[a, b] \subset (\alpha, \beta)$. Since $\int_a^b f(x)\, dx$ increases as b increases, or as a decreases, there no loss of generality in assuming that $a \le c \le b$, and we have

$$\int_a^b f(x)\, dx = \int_a^c f(x)\, dx + \int_c^b f(x)\, dx \le I_1 + I_2. \tag{6.6}$$

Thus the integral $\int_\alpha^\beta f(x)\, dx$ converges.

Finally, for any $\epsilon > 0$ there are points $a_1 \le c$ and $b_1 \ge c$ such that

$$(I_1 + I_2) - \int_{a_1}^{b_1} f(x)\, dx = (I_1 - \int_{a_1}^c f(x)\, dx) + (I_2 - \int_c^{b_1} f(x)\, dx) < \epsilon.$$

Together with (6.6), this shows that

$$\sup_{[a,b] \subset (\alpha,\beta)} \int_a^b f(x)\, dx = I_1 + I_2.$$

□

To illustrate the last theorem, consider the integrals of x^{-n}, $n = 1, 2, 3, \dots$, which are convenient for comparisons. First look at

$$\int_1^\infty \frac{1}{x^n}.$$

For $n \neq 1$,

$$\int_1^N \frac{1}{x^n} = \frac{x^{1-n}}{1-n}\Big|_1^N = \frac{N^{1-n}}{1-n} - \frac{1}{1-n}.$$

Thus the integral

$$\int_1^\infty \frac{1}{x^n}.$$

converges if $n > 1$. If $n = 1$, then

$$\int_1^N \frac{1}{x} \geq \sum_{k=1}^N \frac{1}{k} \to \infty,$$

since the harmonic series is divergent.

Mathematicians spend a considerable effort showing that various integrals converge. One of the most useful tools is the following inequality.

Theorem 6.5.4. *(Cauchy–Schwarz inequality) Suppose that $f(x)$ and $g(x)$ are positive and integrable on every subinterval $[a, b] \subset (\alpha, \beta)$. If $\int_\alpha^\beta f^2(x)\,dx$ and $\int_\alpha^\beta g^2(x)\,dx$ both converge, then so does $\int_\alpha^\beta f(x)g(x)\,dx$, and*

$$\Big[\int_\alpha^\beta f(x)g(x)\,dx\Big]^2 \leq \Big[\int_\alpha^\beta f^2(x)\,dx\Big]\Big[\int_\alpha^\beta g^2(x)\,dx\Big]. \tag{6.7}$$

Proof. For notational convenience, let A, B, and C be the positive numbers defined by

$$A^2 = \int_\alpha^\beta f^2(x)\,dx, \quad B^2 = \int_\alpha^\beta g^2(x)\,dx, \quad C = \int_\alpha^\beta f(x)g(x)\,dx.$$

If $B = 0$ then both sides of (6.7) are 0. Suppose that $B \neq 0$.

Define the quadratic polynomial

$$p(t) = \int_\alpha^\beta [f(x) - tg(x)]^2\,dx = A^2 - 2tC + t^2B^2.$$

The minimum, which is at least 0 since we are integrating the square of a function, occurs when

$$t = \frac{C}{B^2}.$$

Thus

$$0 \leq A^2 - 2\frac{C^2}{B^2} + \frac{C^2}{B^2} = A^2 - \frac{C^2}{B^2},$$

or

$$C^2 \le A^2 B^2$$

as desired. □

As an example illustrating the use of the Cauchy-Schwarz inequality, consider

$$I = \int_1^\infty \frac{1}{x^p e^x} \, dx.$$

For $p > 1/2$ this integral may be estimated as follows.

$$I^2 \le \int_1^\infty x^{-2p} \int_1^\infty e^{-2x} = (\frac{1}{2p-1})(\frac{1}{2e^2}).$$

6.5.2 Integrals and sums

Theorem 6.5.5. *(Integral Test) Suppose $f(x)$ is a decreasing positive continuous function defined for $x \ge 0$, and that $c_k = f(k)$. Then the series $\sum_{k=0}^\infty c_k$ converges if and only if*

$$\int_0^\infty f(x) \, dx < \infty.$$

Proof. Consider left and right endpoint Riemann sums approximations to the integral, with subintervals $[x_k, x_{k+1}] = [k, k+1]$. Since the function f is decreasing, the left endpoint sums give

$$\int_0^n f(x) \, dx \le s_n = \sum_{k=0}^{n-1} f(k) = \sum_{k=0}^{n-1} c_k.$$

The right endpoint sums yield the estimate

$$c_0 + \sum_{k=1}^n c_k = c_0 + \sum_{k=1}^n f(k) \le c_0 + \int_0^n f(x) \, dx.$$

Thus the sequence of partial sums is bounded if and only if

$$\int_0^\infty f(x) \, dx < \infty.$$

□

Apply this theorem to the function

$$g(x) = \frac{1}{(x+1)^2}.$$

Since

$$\int_0^n \frac{1}{(x+1)^2} \, dx = -(x+1)^{-1}\Big|_0^n = 1 - \frac{1}{n+1},$$

and

$$\lim_{n\to\infty} 1 - \frac{1}{n+1} = 1,$$

the series

$$\sum_{k=0}^{\infty} \frac{1}{(k+1)^2} = \sum_{m=1}^{\infty} \frac{1}{m^2}$$

converges.

6.5.3 Absolutely convergent integrals

Infinite series containing both positive and negative terms behave well if the series converges absolutely. Integrals behave in a similar manner. Suppose that $f : (\alpha, \beta) \to \mathbb{R}$ is integrable on every compact subinterval $[a, b] \subset (\alpha, \beta)$. We no longer assume that f is positive. Say that $\int_\alpha^\beta f(x)\, dx$ *converges absolutely* if $\int_\alpha^\beta |f(x)|\, dx$ converges.

Define the positive and negative parts of a real valued function f as follows:

$$f^+(x) = \begin{Bmatrix} f(x), & f(x) > 0 \\ 0, & f(x) \le 0 \end{Bmatrix},$$

$$f^-(x) = \begin{Bmatrix} f(x), & f(x) < 0 \\ 0, & f(x) \ge 0 \end{Bmatrix}.$$

Lemma 6.5.6. *If $f : [a, b] \to \mathbb{R}$ is Riemann integrable, then so are f^+ and f^-.*

The integral $\int_\alpha^\beta f(x)\, dx$ converges absolutely if and only if the integrals

$$\int_\alpha^\beta f^+(x)\, dx, \quad \text{and} \quad \int_\alpha^\beta -f^-(x)\, dx$$

converge.

Proof. Observe that

$$f^+(x) = \frac{f(x) + |f(x)|}{2}, \quad f^-(x) = \frac{f(x) - |f(x)|}{2}. \tag{6.8}$$

The function $|f(x)|$ is integrable by Theorem 6.3.3. Since f^+ and f^- can be written as the sum of two integrable functions, they are integrable by Theorem 6.3.1.

The remaining conclusions are straightforward consequences of (6.8) and the associated formula

$$|f(x)| = f^+(x) - f^-(x).$$

□

If the integral $\int_\alpha^\beta f(x)\,dx$ converges absolutely, define

$$\int_\alpha^\beta f(x)\,dx = \int_\alpha^\beta f^+(x)\,dx - \int_\alpha^\beta -f^-(x)\,dx.$$

Theorem 6.5.7. *Suppose that the integral $\int_\alpha^\beta f(x)\,dx$ converges absolutely. For any $c \in (\alpha,\beta)$ the limits*

$$I_1 = \lim_{a\to\alpha^+} \int_a^c f(x)\,dx \quad \text{and} \quad I_2 = \lim_{b\to\beta^-} \int_c^b f(x)\,dx$$

exist, and $\int_\alpha^\beta f = I_1 + I_2$.

Proof. Suppose that $c \le b < \beta$. Then by Lemma 6.5.6 we have

$$\int_c^b f(x)\,dx = \int_c^b f^+(x)\,dx - \int_c^b -f^-(x)\,dx$$

and similarly for $\int_a^c f$ if $a < c$. Now use Theorem 6.5.3 to conclude that

$$\int_\alpha^\beta f = \int_\alpha^\beta f^+ - \int_\alpha^\beta -f^- = \lim_{a\to\alpha^+} \int_a^c f^+(x)\,dx + \lim_{b\to\beta^-} \int_c^b f^+(x)\,dx$$

$$- \lim_{a\to\alpha^+} \int_a^c -f^-(x)\,dx - \lim_{b\to\beta^-} \int_c^b -f^-(x)\,dx = I_1 + I_2.$$

□

6.5.4 Conditionally convergent integrals

Recall that there are infinite series such as $\sum_{n=1}^\infty (-1)^n 1/n$ which are convergent (by the alternating series test), but not absolutely convergent. A similar situation arises in the study of improper integrals. For example, the integral

$$\int_1^\infty \frac{\sin(x)}{x}\,dx \tag{6.9}$$

is not absolutely convergent, but there is a number

$$L = \lim_{N\to\infty} \int_1^N \frac{\sin(x)}{x}\,dx.$$

Assume, as before, that f is Riemann integrable on each compact subinterval $[a,b]$ of the open interval (α,β). Say that the integral $\int_\alpha^\beta f(x)\,dx$ *converges* if there is some $c \in (\alpha,\beta)$ and numbers I_1 and I_2 such that the limits

$$I_1 = \lim_{a\to\alpha^+} \int_a^c f(x)\,dx, \quad \text{and} \quad I_2 = \lim_{b\to\beta^-} \int_c^b f(x)\,dx$$

exist. We then define

$$\int_{\alpha}^{\beta} f(x)\, dx = I_1 + I_2.$$

By Lemma 6.5.6 an absolutely convergent integral is convergent. An integral which is convergent, but not absolutely convergent, is said to be *conditionally convergent.*

It appears that the value of a conditionally convergent integral might depend on the choice of the point c. The first result will show that this is not the case.

Theorem 6.5.8. *Suppose the integral $\int_{\alpha}^{\beta} f(x)\, dx$ converges. Let $d \in (\alpha, \beta)$, and let I_1 and I_2 be as above. There are real numbers J_1 and J_2 such that*

$$J_1 = \lim_{a \to \alpha^+} \int_a^d f(x)\, dx, \quad J_2 = \lim_{b \to \beta^-} \int_d^b f(x)\, dx,$$

and $J_1 + J_2 = I_1 + I_2$.

Proof. For ease of notation assume that $c < d$. Then

$$J_2 = \lim_{b \to \beta^-} \int_d^b f(x)\, dx = \lim_{b \to \beta^-} [\int_c^b f(x)\, dx - \int_c^d f(x)\, dx] = I_2 - \int_c^d f(x)\, dx,$$

so J_2 exists. A similar argument gives

$$J_1 = I_1 + \int_c^d f(x)\, dx,$$

which in turn leads to $J_1 + J_2 = I_1 + I_2$. □

The next results display a strong link between convergent integrals and convergent series.

Theorem 6.5.9. *Assume that $f : [c, \beta) \to \mathbb{R}$ is integrable on every compact subinterval $[a, b] \subset [c, \beta)$. Suppose there is a sequence of points $\{x_k\}$ such that $x_1 = c$, $x_k < x_{k+1}$, $\lim_{k \to \infty} x_k = \beta$, and for $k = 1, 2, 3, \dots$ we have $f(x) \ge 0$ when $x \in [x_{2k-1}, x_{2k}]$, while $f(x) \le 0$ when $x \in [x_{2k}, x_{2k+1}]$. Then the integral $\int_c^{\beta} f(x)\, dx$ converges if and only if the series*

$$\sum_{k=1}^{\infty} \int_{x_k}^{x_{k+1}} f(x)\, dx \tag{6.10}$$

converges.

Proof. If the integral $\int_c^{\beta} f(x)\, dx$ converges, then the limit $\lim_{b \to \beta^-} \int_c^b f(x)\, dx$ exists. This implies that the sequence of partial sums

$$s_n = \sum_{k=1}^{n} \int_{x_k}^{x_{k+1}} f(x)\, dx = \int_c^{x_{n+1}} f(x)\, dx$$

has a limit as $n \to \infty$, so the series (6.10) converges.

Now assume that the sequence of partial sums

$$s_n = \sum_{k=1}^{n} \int_{x_k}^{x_{k+1}} f(x)\ dx = \int_c^{x_{n+1}} f(x)\ dx$$

converges to a limit L. Since the terms of the series are $a_k = \int_{x_k}^{x_{k+1}} f(x)\ dx$, and the terms of a convergent series have limit 0, it follows that

$$\lim_{k\to\infty} \int_{x_k}^{x_{k+1}} f(x)\ dx = 0.$$

For any $\epsilon > 0$ there is an N such that

$$|s_n - L| < \epsilon/2, \quad n \ge N,$$

and

$$|\int_{x_k}^{x_{k+1}} f(x)\ dx| < \epsilon/2, \quad k \ge N.$$

Since f does not change sign in the interval $[x_k, x_{k+1}]$, it follows that

$$|\int_{x_k}^{b} f(x)\ dx| < \epsilon/2, \quad k \ge N, \quad x_k \le b \le x_{k+1}.$$

Suppose now that $x_m \le b \le x_{m+1}$, with $m > N$. Then by the triangle inequality

$$|\int_c^b f(x)\ dx - L| = |\int_c^{x_m} f(x)\ dx + \int_{x_m}^b f(x)\ dx - L|$$

$$\le |\int_c^{x_m} f(x)\ dx - L| + |\int_{x_m}^b f(x)\ dx| < \epsilon.$$

Thus

$$\lim_{b\to\beta^-} \int_c^b f(x)\ dx = L.$$

□

This theorem, in conjunction with the alternating series test for infinite series, can be applied to integrals such as (6.9).

Theorem 6.5.10. *Suppose that $f : [0,\infty) \to \mathbb{R}$ is integrable on every compact subinterval $[a,b] \subset [0,\infty)$. Assume that f is positive, decreasing, and $\lim_{x\to\infty} f(x) = 0$. Then the integrals*

$$\int_0^\infty f(x)\sin(x)\ dx \quad \text{and} \quad \int_0^\infty f(x)\cos(x)\ dx \tag{6.11}$$

converge.

Proof. We will treat the first case; the second is similar. The sequence $x_k = (k-1)\pi$ will satisfy the hypotheses of Theorem 6.5.9 since the sign of $f(x)\sin(x)$ is the same as that of $\sin(x)$, which changes at the points $k\pi$.

The numbers

$$a_k = \int_{(k-1)\pi}^{k\pi} f(x)\sin(x)\,dx, \quad k = 1, 2, 3, \ldots$$

have alternating signs and decreasing magnitudes. The second claim is verified with the following computation, which makes use of the identity $\sin(x+\pi) = -\sin(x)$:

$$|a_k| - |a_{k+1}| = \int_{(k-1)\pi}^{k\pi} |f(x)\sin(x)|\,dx - \int_{k\pi}^{(k+1)\pi} |f(x)\sin(x)|\,dx$$

$$= \int_{(k-1)\pi}^{k\pi} |f(x)\sin(x)| - |f(x+\pi)\sin(x+\pi)|\,dx$$

$$= \int_{(k-1)\pi}^{k\pi} [f(x) - f(x+\pi)]\,|\sin(x)|\,dx.$$

Since $f(x)$ is decreasing, $|a_k| - |a_{k+1}| \ge 0$, and the sequence $\{|a_k|\}$ is decreasing.

The estimate

$$|a_k| = \int_{k\pi}^{(k+1)\pi} |f(x)\sin(x)|\,dx \le \pi f(k\pi),$$

together with $\lim_{x\to\infty} f(x) = 0$, imply that $\lim_{k\to\infty} a_k = 0$. The series $\sum a_k = \sum(-1)^{k+1}|a_k|$ converges by the alternating series test. By Theorem 6.5.9 the integral $\int_0^\infty f(x)\sin(x)$ converges. □

6.6 Problems

6.1. *Following the example of $f(x) = x^2$ in the text, use left and right endpoint Riemann sums and the formula $\sum_{k=0}^n k^3 = n^2(n+1)^2/4$ to show that*

$$\int_0^b x^3\,dx = b^4/4.$$

6.2. *Calculate $\mathcal{U}(f,\mathcal{P}) - \mathcal{L}(f,\mathcal{P})$ if*

$$x_k = a + k\frac{b-a}{n}, \quad k = 0, \ldots, n$$

and $f(x) = Cx$ for some constant C. How big should n be if you want

$$\mathcal{U}(f,\mathcal{P}) - \mathcal{L}(f,\mathcal{P}) < 10^{-6}?$$

6.3. *Denote by L_n and R_n respectively the left and right endpoint Riemann sums for the integral*

$$\int_a^b f(x)\ dx.$$

Assume that the interval $[a,b]$ is divided into n subintervals of equal length. If the function $f(x)$ is decreasing, then

$$R_n \le \int_a^b f(x)\ dx \le L_n.$$

a) Determine which rectangular areas appear in both left and right endpoint sums, and use this observation to show that

$$L_n - R_n = \frac{b-a}{n}[f(a) - f(b)].$$

b) Now show that

$$|\int_a^b f(x)\ dx - L_n| \le \frac{b-a}{n}[f(a) - f(b)],$$

and similarly for R_n.

c) Show that either

$$|\int_a^b f(x)\ dx - L_n| \ge \frac{b-a}{2n}[f(a) - f(b)],$$

or

$$|\int_a^b f(x)\ dx - R_n| \ge \frac{b-a}{2n}[f(a) - f(b)].$$

d) If you use left or right endpoint Riemann sums to compute

$$\log(10) = \int_1^{10} 1/x\ dx,$$

how big should n be to insure that the error in your computation is less than 10^{-6}?

6.4. *Assume that $g : [a,b] \to \mathbb{R}$ is integrable. For $c \in \mathbb{R}$, show that $f(x) = g(x-c)$ is integrable on $[a+c, b+c]$ and*

$$\int_a^b g(x)\ dx = \int_{a+c}^{b+c} f(x)\ dx.$$

(Hint: draw a picture.)

6.5. *Suppose that* $f : [a, b] \to \mathbb{R}$ *is continuous, and* $f(x) \geq 0$ *for all* $x \in [a, b]$. *Show that if*

$$\int_a^b f(x)\ dx = 0,$$

then $f(x) = 0$ *for all* $x \in [a, b]$. *Is the same conclusion true if* f *is merely integrable?*

6.6. *Prove Lemma 6.2.2.*

6.7. *Show that for any* $n \geq 1$ *there is a partition of* $[a, b]$ *with* $n + 1$ *points* $x_0, \ldots, x_n$ *such that*

$$\int_{x_j}^{x_{j+1}} t^2\ dt = \int_{x_k}^{x_{k+1}} t^2\ dt$$

for any j *and* k *between* 0 *and* $n - 1$. *You may use calculus to evaluate the integrals.*

6.8. *Show that if* f *and* g *are integrable, and* $f(x) \leq g(x)$ *for all* $x \in [a, b]$, *then*

$$\int_a^b f(x)\ dx \leq \int_a^b g(x)\ dx.$$

6.9. *Suppose that* $f : [a, b] \to \mathbb{R}$ *is continuous and*

$$\int_c^d f(x)\ dx = 0$$

for all $a \leq c < d \leq b$. *Show that* $f(x) = 0$ *for all* $x \in [a, b]$.

6.10. *Suppose that* $f : [a, b] \to \mathbb{R}$ *is integrable, and that* $f(x) = 0$ *for all rational numbers* x. *Show that*

$$\int_a^b f(x)\ dx = 0.$$

6.11. *Suppose that* $f : [a, b] \to \mathbb{R}$ *is integrable, that* $g(x)$ *is real valued, and that* $g(x) = f(x)$ *except at finitely many points* $t_1, \ldots, t_m \in [a, b]$. *Show that* g *is integrable.*

6.12. *Fill in the details of the proof for Theorem 6.3.4 when* $h < 0$.

6.13. *Prove that if* $w(x) \geq 0$ *is integrable, and* $f(x)$ *is continuous, then for some* $\xi \in [a, b]$

$$\int_a^b f(x)w(x)\ dx = f(\xi) \int_a^b w(x)\ dx.$$

Hint: Start by showing that for some $x_1, x_2 \in [a, b]$

$$f(x_1) \int_a^b w(x)\ dx \leq \int_a^b f(x)w(x)\ dx \leq f(x_2) \int_a^b w(x)\ dx.$$

Now use the Intermediate Value Theorem.

6.14. *Is* $f : [0,1] \to \mathbb{R}$ *integrable if*

$$f(x) = \begin{Bmatrix} \sin(1/x) & 0 < x < 1, \\ 0 & x = 0 \end{Bmatrix}?$$

6.15. *Assume that* $f : [a,b] \to \mathbb{R}$ *is bounded, and that* f *is continuous on* (a,b). *Show that* f *is integrable on* $[a,b]$

6.16. *Complete the proof of Theorem 6.3.2 for functions* f *and* g *which are not necessarily positive. (Hint: Find* c_f *and* c_g *such that* $f + c_f \ge 0$ *and* $g + c_g \ge 0$. *Now compute.)*

6.17. *Find functions* $f : [0,1] \to \mathbb{R}$ *and* $g : [0,1] \to \mathbb{R}$ *which are not integrable, but whose product* fg *is integrable.*

6.18. *Rephrase Theorem 6.3.4, and give a proof, if the function* $f(x)$ *is merely assumed to be integrable on* $[a,b]$, *and continuous at the point* x.

6.19. *Suppose* $f : [a,b] \to \mathbb{R}$ *is integrable. Show that*

$$F(x) = \int_a^x f(t)\, dt$$

is a continuous function on $[a,b]$.

6.20. *Show that the conclusions of Theorem 6.3.5 are still valid if* $F'(x) = f(x)$ *for* $x \in (a,b)$, *and if* $f(x), F(x)$ *are continuous on* $[a,b]$.

6.21. *(The substitution method.) Suppose the functions* $F(x)$ *and* $g(x)$ *have continuous derivatives, with* $F'(x) = f(x)$. *Show that*

$$\int_a^b f(g(t))g'(t)\, dt = \int_{g(a)}^{g(b)} f(x)\, dx.$$

6.22. *Consider two curves in the upper half plane. The first is the semicircle* $r = r_0$ *for* $0 \le \theta \le \pi$. *The second satisfies*

$$r(\theta) = r_0 \frac{2\pi - \theta}{\pi}$$

for $0 \le \theta \le \pi$.

a) Divide the outer curve into n *equal angle subarcs with endpoints at angles* θ_k *so that* $\theta_{k+1} - \theta_k = \pi/n$. *Find expressions for* θ_k *and* $r(\theta_k)$.

b) Consider the area A_k *of the* $k-th$ *angular sector bounded by the outer curve and lines from the origin at angles* θ_k *and* θ_{k+1}. *Compare these areas to the areas of sectors of disks with radii* $r(\theta_k)$ *and* $r(\theta_{k+1})$.

c) Using a Riemann sum like calculation, calculate the area of the region between the two curves for $0 \le \theta \le \pi$. *(This type of result was known to Archimedes (287–212 B.C.) [5, pp. 114–115].)*

d) Replace the formula $r(\theta) = r_0[2\pi - \theta]$ *by others for which the analogous calculation can be made.*

6.23. *Suppose that $f'(x)$ is continuous and nonnegative for $x \in [a,b]$. Use the following approach to estimate the error made when using left endpoint Riemann sums to approximate the integral of f.*

a) Show that

$$\int_a^b f(x)\ dx - \sum_{k=0}^{n-1} f(x_k)\Delta x_k = \sum_{k=0}^{n-1} \int_{x_k}^{x_{k+1}} f(x) - f(x_k)\ dx.$$

b) If $C_k = \min_{t \in [x_k, x_{k+1}]} f'(t)$, show that

$$\int_{x_k}^{x_{k+1}} f(x) - f(x_k)\ dx \geq C_k (\Delta x_k)^2/2.$$

c) What is the analogous result if $f' \leq 0$? Use this to obtain a lower bound on the error of approximating $\int_{1/2}^1 e^{-x^2}\ dx$ using left endpoint Riemann sums.

6.24. *a) Show that*

$$\lim_{n\to\infty} \sum_{k=0}^{n-1} \frac{1}{n}\left(\frac{k}{n}\right)^m = \frac{1}{m+1}, \qquad m = 0, 1, 2, \ldots.$$

(Hint: Use calculus and Riemann sums.)

b) Show that

$$\lim_{n\to\infty} \sum_{k=0}^{n-1} \frac{1}{n} \frac{1}{(1+k/n)^m} = \frac{1-2^{1-m}}{m-1}, \qquad m = 2, 3, \ldots.$$

6.25. *a) Show that*

$$\lim_{n\to\infty} \sum_{k=0}^{n-1} \pi \frac{1}{n} \sin(k\pi/n) = 2.$$

b) Show that

$$\lim_{n\to\infty} \sum_{k=0}^{n-1} \frac{n}{n^2+k^2} = \pi/4.$$

6.26. *a) Show that the series*

$$\sum_{k=1}^{\infty} \sin\left(\frac{1}{k^2}\right)$$

converges, but that the series

$$\sum_{k=1}^{\infty} \sin\left(\frac{1}{k}\right)$$

diverges.

b) Does the series

$$\sum_{k=1}^{\infty}[1-\cos(\frac{1}{k})]$$

converge ?

c) For what values of p does the series

$$\sum_{k=1}^{\infty}[\log(k^p+1)-\log(k^p)]$$

converge ?

6.27. *Suppose $f(x)$ has a continuous derivative on $[0,\infty)$ and*

$$f(0)=0, \quad \int_0^{\infty}(f'(x))^2\,dx<\infty.$$

Show there is a constant C such that

$$|f(x)|\le C\sqrt{x}, \quad 0\le x<\infty.$$

7

The Natural logarithm

7.1 Introduction

Logarithms are usually introduced with a base of 10, with subsequent extensions to other (usually integer) bases. Suppose numbers α, β are written as powers of ten,

$$\alpha = 10^x, \quad \beta = 10^y.$$

The addition rule for exponents relates the product of α and β to the sum of x, y,

$$\alpha \cdot \beta = 10^x \cdot 10^y = 10^{x+y}.$$

This suggests that it might be possible to reduce problems of multiplication to problems of addition. Similarly, a computation like

$$\sqrt{\alpha} = (10^x)^{1/2} = 10^{x/2},$$

is simplified when the representation of numbers as powers of ten is known.

The implementation of this idea to simplify calculations is credited to John Napier (1550–1617), who published a table of logarithms in 1614. The use of 10 as the base was introduced shortly afterward by Henry Briggs (1561–1631). Equipped with a table of logarithms, calculations of products, quotients, and roots could be reduced to simpler sums, differences, or simple quotients. These logarithm tables were soon mechanized into a popular computing machine, the slide rule.

From the point of view of calculus the most useful logarithm function is the natural logarithm, which arises from geometric considerations. The context is the problem of finding the area under the graph of the function $f(t) = 1/t$, say between two positive numbers $t = a$ and $t = b$. In the language of calculus the problem is to calculate

$$A_1 = \int_a^b \frac{1}{t}\, dt.$$

Consider attacking this problem by constructing Riemann sum approximations of the area A_1 (Figure 7.1). For instance we can use left and right endpoint Riemann sums, with n subintervals of equal length,

$$L_n = \sum_{k=0}^{n-1} f(t_k) \frac{b-a}{n} = \frac{b-a}{n} \sum_{k=0}^{n-1} \frac{1}{t_k}, \quad t_k = a + k\frac{b-a}{n}$$

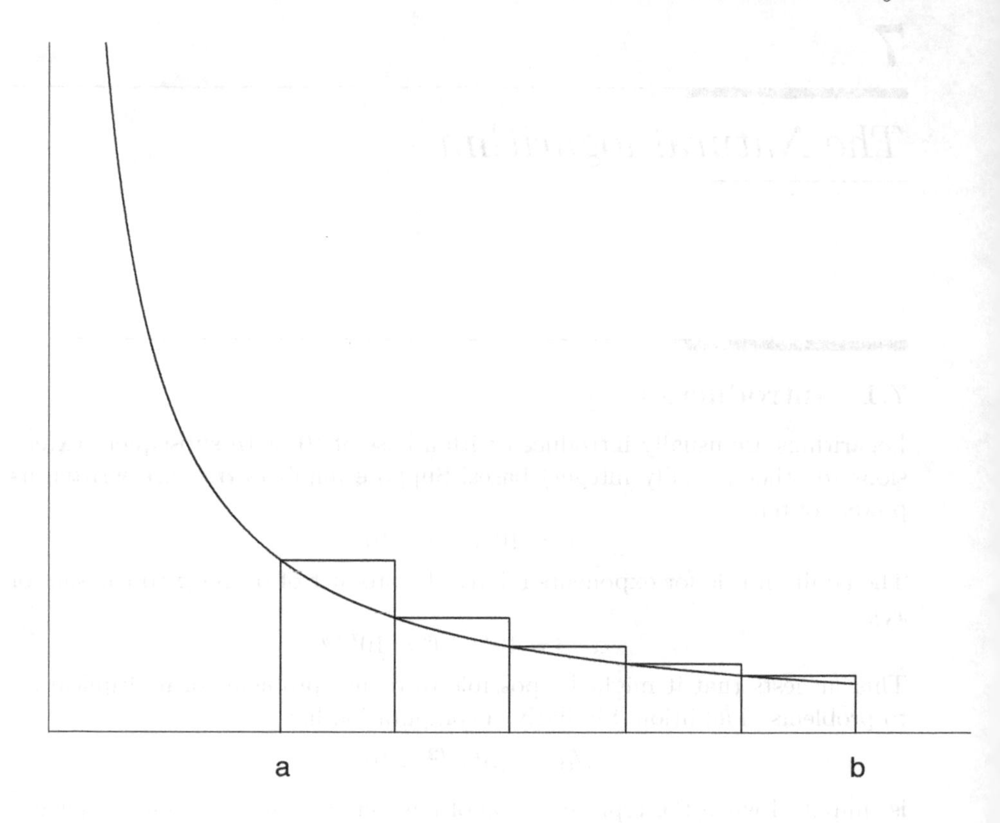

FIGURE 7.1
Area under $f(x) = 1/x$.

and

$$R_n = \sum_{k=0}^{n-1} f(t_{k+1}) \frac{b-a}{n} = \frac{b-a}{n} \sum_{k=0}^{n-1} \frac{1}{t_{k+1}}.$$

Since the function $f(t) = 1/t$ is decreasing for $t > 0$,

$$R_n < A_1 < L_n,$$

and

$$\lim_{n \to \infty} R_n = A_1 = \lim_{n \to \infty} L_n.$$

A key observation establishes a relationship among similar areas. Let m be a positive number, and consider the new area (Figure 7.2)

$$A_2 = \int_{ma}^{mb} \frac{1}{t} \, dt, \quad m > 0.$$

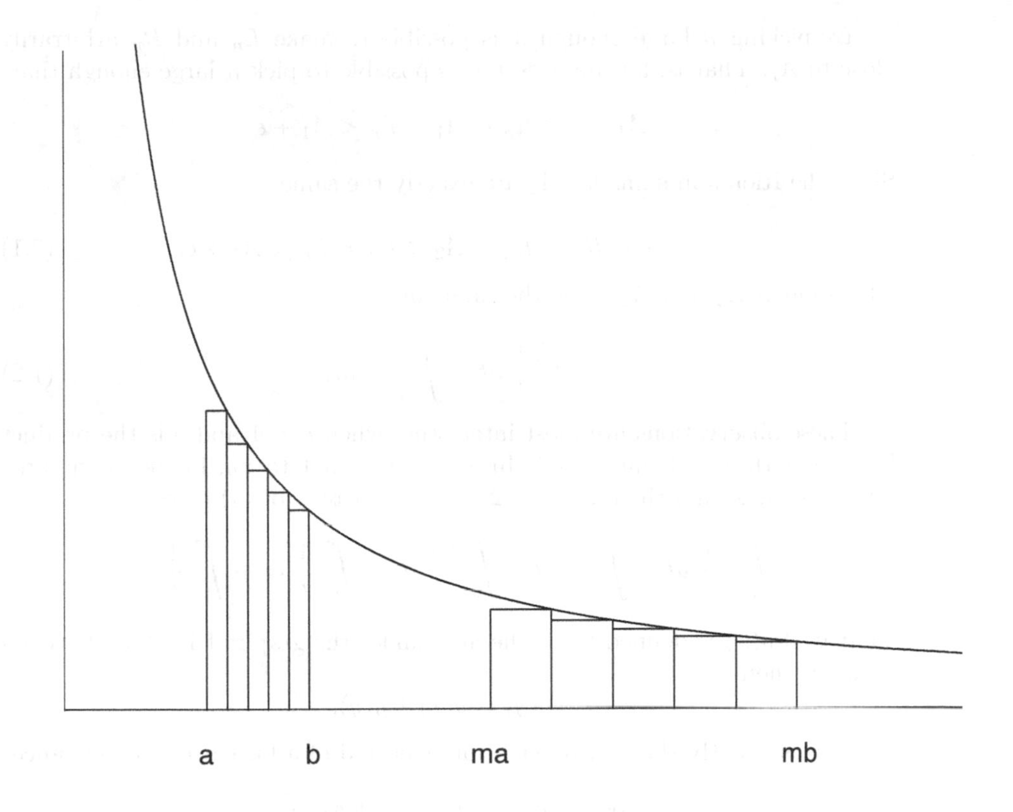

FIGURE 7.2
Two areas under $f(x) = 1/x$.

Since the interval $[a, b]$ has been scaled by the number m, the same will be true of the n subintervals of equal length. The new sample points are

$$T_k = ma + k\frac{mb - ma}{n} = m[a + k\frac{b-a}{n}] = mt_k.$$

The corresponding left endpoint Riemann sum is

$$\tilde{L}_n = \sum_{k=0}^{n-1} f(T_k)\frac{mb - ma}{n} = m\frac{b-a}{n}\sum_{k=0}^{n-1}\frac{1}{mt_k}.$$

Notice that the factors m and $1/m$ cancel, so that

$$\tilde{L}_n = \sum_{k=0}^{n-1} f(T_k)\frac{mb - ma}{n} = \sum_{k=0}^{n-1} f(t_k)\frac{b-a}{n} = L_n.$$

The right endpoint result $\tilde{R}_n = R_n$ holds as well.

By picking n large enough it is possible to make L_n and R_n arbitrarily close to A_1. That is, for any $\epsilon > 0$, it is possible to pick n large enough that

$$A_1 - \epsilon < R_n < A_1 < L_n < A_1 + \epsilon.$$

Since the Riemann sums for A_2 are exactly the same,

$$A_1 - \epsilon < R_n = \tilde{R}_n < A_2 < L_n = \tilde{L}_n < A_1 + \epsilon, \tag{7.1}$$

which forces A_1 and A_2 to be the same, or

$$\int_a^b \frac{1}{t}\, dt = \int_{ma}^{mb} \frac{1}{t}\, dt. \tag{7.2}$$

These observations are most interesting when $a = 1$ and b is the product $b = xy$, with $x > 1$ and $y > 1$. Integrating from 1 to xy, breaking the area into two parts, and then using (7.2) (with x instead of m) gives

$$\int_1^{xy} \frac{1}{t}\, dt = \int_1^x \frac{1}{t}\, dt + \int_x^{xy} \frac{1}{t}\, dt = \int_1^x \frac{1}{t}\, dt + \int_1^y \frac{1}{t}\, dt.$$

That is, if $g(x)$ is defined to be the area under the graph of $1/t$ for t between 1 and x, then

$$g(xy) = g(x) + g(y).$$

This is exactly the sort of equation expected of a logarithm; for instance

$$\log_{10}(xy) = \log_{10}(x) + \log_{10}(y).$$

Also, since there is zero area if $x = 1$,

$$g(1) = 0.$$

The function

$$g(x) = \int_1^x \frac{1}{t}\, dt$$

is then defined to be our *natural logarithm*. So far the values of $\log(x)$ for numbers $0 < x < 1$ haven't been considered. A geometric treatment of these cases is left for the problem section.

Before the completed development of calculus, a result similar to the following was published in 1649 by Alfons A. de Sarasa (1618–67).

Theorem 7.1.1. *The function*

$$\log(x) = \int_1^x \frac{1}{t}\, dt, \quad x > 0,$$

satisfies $\log(1) = 0$ *and*

$$\log(xy) = \log(x) + \log(y), \quad x, y > 0. \tag{7.3}$$

Proof. Suppose $0 < x \le y$. Then

$$\log(x,y) = \int_1^{xy} \frac{1}{t}\, dt = \int_1^{x} \frac{1}{t}\, dt + \int_x^{xy} \frac{1}{t}\, dt.$$

Using Problem 6.21, the substitution $t = xu$ gives

$$\int_x^{xy} \frac{1}{t}\, dt = \int_1^{y} \frac{1}{xu} x\, du = \int_1^{y} \frac{1}{u}\, du = \log(y). \tag{7.4}$$

□

Since the natural logarithm has the same kind of behavior as $\log_{10}(x)$, it is natural to ask if there is a base for the natural logarithms. In other words, is there a number b such that if $x = \log(y)$, then $y = b^x$?

7.2 The natural exponential function

Having established the existence of a natural logarithm, there are two paths to explore. To see the options, let's return to the setting of base 10 logarithms and exponentiation. Suppose $y = 10^x$. Since solving for x gives $x = \log_{10}(y)$, the exponential function 10^x is the inverse of the logarithm $\log_{10}(y)$. One option is to define the natural exponential function to be the inverse of the natural logarithm. The second option is to look for a number playing the role of 10, the base of the natural logarithm. Both options will be considered.

The function

$$\log(x) = \int_1^x \frac{1}{t}\, dt, \quad x > 0,$$

is strictly increasing. By the Fundamental Theorem of Calculus it is differentiable, with

$$\frac{d}{dx} \log(x) = \frac{1}{x}.$$

A comparison with the harmonic series shows that $\lim_{x\to\infty} \log(x) = \infty$. For $x > 0$,

$$0 = \log(1) = \log(x\frac{1}{x}) = \log(x) + \log(\frac{1}{x}),$$

so

$$\log(1/x) = -\log(x), \tag{7.5}$$

and $\lim_{x\to 0^+} \log(x) = -\infty$.

Define the natural exponential function $\exp(x)$ to be the inverse of the natural logarithm $\log(x)$. Since the domain of $\log(x)$ is $(0,\infty)$ while the range is $(-\infty,\infty)$, the function $\exp(x)$ has domain $\mathbb{R}$ and range $(0,\infty)$.

Theorem 7.2.1. *For all $x \in \mathbb{R}$ the function* $\exp(x)$ *is differentiable with*

$$\frac{d}{dx}\exp(x) = \exp(x).$$

In addition exp *satisfies the identity*

$$\exp(\alpha + \beta) = \exp(\alpha)\exp(\beta), \quad \alpha, \beta \in \mathbb{R}.$$

Proof. The Inverse Function Theorem (Theorem 5.3.13) applies to $\log(x)$, with

$$\frac{d}{dx}\exp(x) = \frac{d}{dx}\log^{-1}(x) = \frac{1}{1/\log^{-1}(x)} = \exp(x).$$

For any $\alpha, \beta \in \mathbb{R}$ there are positive numbers x, y such that $\alpha = \log(x)$ and $\beta = \log(y)$. Applying exp to the formula $\log(xy) = \log(x) + \log(y)$ yields

$$\exp(\alpha + \beta) = \exp(\log(xy)) = xy = \exp(\alpha)\exp(\beta).$$

□

Turning to the question of a base for the natural logarithm, let e denote the number with property that $\log(e) = 1$. There must be a unique choice of e since $\log(x)$ is continuous and strictly increasing with range $(-\infty, \infty)$. Riemann sums help provide some crude information about the value of e. Using left endpoint Riemann sums with subintervals of length $1/2$ we find that

$$\int_1^2 1/t \; dt < \frac{1}{2}[1 + \frac{2}{3}] = 5/6 < 1.$$

A bit more work with right endpoint Riemann sums and subintervals of length $1/2$ leads to

$$\int_1^4 1/t \; dt > 1.$$

That is, the number e satisfies $2 < e < 4$. (Of course your calculator, which has more sophisticated knowledge embedded inside, will report $e = 2.71828\ldots$.)

Having convinced ourselves that there is some number e satisfying

$$\int_1^e \frac{1}{t} \; dt = 1,$$

it is reasonable to ask if this number behaves as the base of the natural logarithm.

Theorem 7.2.2. *For every rational number r,*

$$\log(e^r) = r = \log_e(e^r), \quad e^r = \exp(r).$$

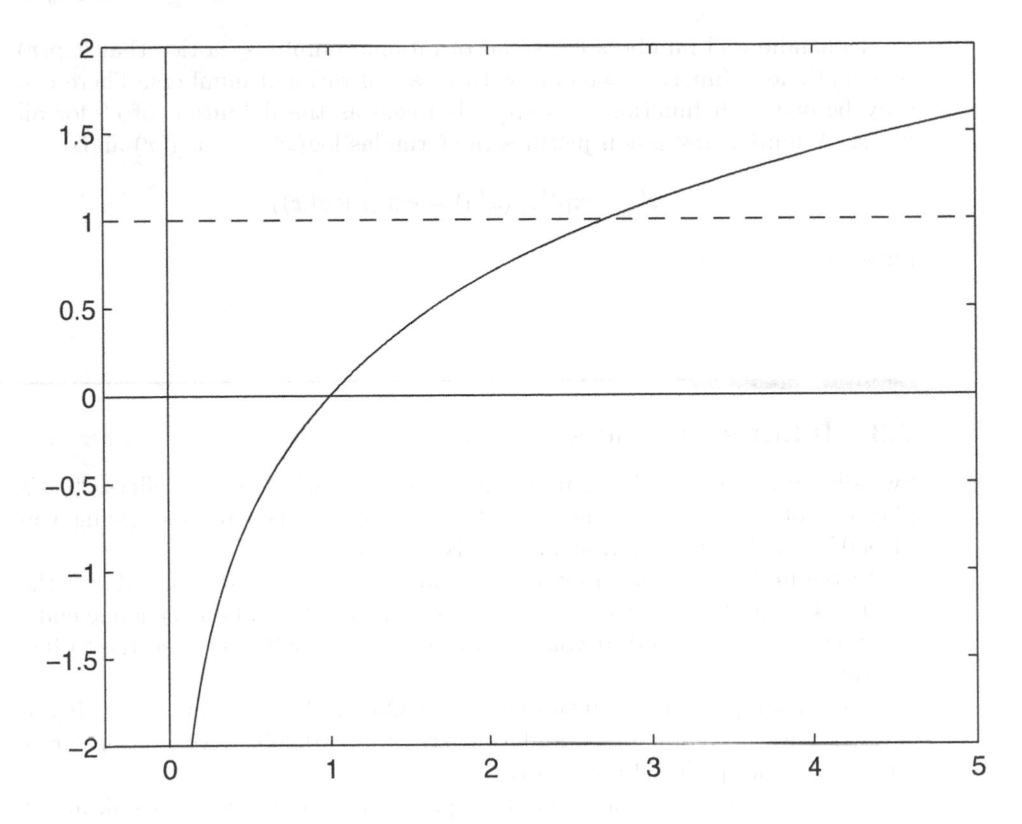

FIGURE 7.3
Graph of log(x).

Proof. Write r as the quotient of integers, $r = m/n$, with $n > 0$. Noting that $e > 1$, use (7.4) for the following calculation

$$1 = \log(e) = \log([e^{1/n}]^n)$$

$$= \int_1^{e^{1/n}} \frac{1}{t}\,dt + \int_{e^{1/n}}^{e^{2/n}} \frac{1}{t}\,dt + + \cdots + \int_{e^{(n-1)/n}}^{e^{n/n}} \frac{1}{t}\,dt = n \int_1^{e^{1/n}} \frac{1}{t}\,dt.$$

That is,

$$\log(e^{1/n}) = \int_1^{e^{1/n}} \frac{1}{t}\,dt = 1/n.$$

For $m > 0$ the formula (7.3) extends to give

$$\log(e^{m/n}) = m/n,$$

while (7.5) handles the case when $m < 0$. Since exp is the inverse of log, $e^r = \exp(r)$ follows. □

To handle real numbers r instead of rational numbers, notice that $\exp(r)$ is a continuous function agreeing with e^r at all rational numbers. There can only be one such function, so $\exp(r)$ is taken as the definition of e^r for all $r \in \mathbb{R}$. A similar discussion justifies the formulas $\log(x^r) = r\log(x)$ and

$$x^r = \exp(\log(x^r)) = \exp(r\log(x))$$

for all $x > 0$, $r \in \mathbb{R}$.

7.3 Infinite products

Swindler Stan, the gambling man, comes to you with a pair of offers. "Let's play one of these games," he says. "Each game starts with me giving you $1,000." Needless to say, your interest is aroused.

In the first game you draw a ball from an urn once a day. On day k the urn has k white balls and 1 black ball. If you draw a black ball the game ends, and you pay Stan $2,000. If you never draw a black ball, you keep the $1,000 forever.

The second game has you throwing dice. On day k you throw k dice. If you throw k ones the game ends, and you pay Stan $10,000. If you never throw all ones, you keep the $1,000 forever.

You remember a bit of elementary probability, and start to think about game two. Suppose you play for n days. You keep the $1,000 at the end of n days if you manage to avoid throwing all ones on each day. Since the probability of throwing k ones is $(1/6)^k$, the probability of avoiding all ones on day k is

$$p_k = 1 - (1/6)^k.$$

The throws on different days are independent, so the probability of hanging on to the money at the end of n days is

$$P_n = p_1 \cdot p_2 \cdot \dots \cdot p_n = \prod_{k=1}^{n} (1 - 6^{-k}).$$

Turning back to game one, the probability of avoiding the black ball on day k is $q_k = 1 - 1/(k+1)$. Reasoning in a similar fashion, you conclude that the probability of retaining the money after n days in game one is

$$Q_n = q_1 \cdot q_2 \cdot \dots \cdot q_n = \prod_{k=1}^{n} (1 - \frac{1}{k+1}).$$

You reasonably conclude that the probability of retaining your money forever in game one is

$$Q = \lim_{n\to\infty} \prod_{k=1}^{n} (1 - \frac{1}{k+1}),$$

while for game two the probability is

$$P = \lim_{n\to\infty} \prod_{k=1}^{n} (1 - 6^{-k}).$$

Just as an infinite series is defined by

$$\sum_{k=1}^{\infty} a_k = \lim_{n\to\infty} \sum_{k=1}^{n} a_k,$$

infinite products can be defined by

$$\prod_{k=1}^{\infty} p_k = \lim_{n\to\infty} \prod_{k=1}^{n} p_k,$$

if the limit exists. The first challenge is to develop some understanding of when such limits exist. Afterward (see Problem 7.6 and Problem 7.7) we can consider the attractiveness of Stan's offer.

By making use of the logarithm function, it is straightforward to understand the limit process for products. Suppose the numbers c_k are positive, so that $\log(c_k)$ is defined. Since the log of a product is the sum of the logs,

$$\log(\prod_{k=1}^{n} c_k) = \sum_{k=1}^{n} \log(c_k).$$

If the last series converges, the continuity of the exponential function leads to

$$\lim_{n\to\infty} \prod_{k=1}^{n} c_k = \lim_{n\to\infty} \exp(\sum_{k=1}^{n} \log(c_k)) = \exp(\lim_{n\to\infty} \sum_{k=1}^{n} \log(c_k)) = \exp(\sum_{k=1}^{\infty} \log(c_k)).$$

Recall (Lemma 3.3.1) that in order for the series to converge it is necessary (but not sufficient) to have

$$\lim_{k\to\infty} \log(c_k) = 0.$$

Since the exponential function is continuous this means that

$$\lim_{k\to\infty} c_k = \lim_{k\to\infty} e^{\log(c_k)} = e^{\lim_{k\to\infty} \log(c_k)} = e^0 = 1.$$

This being the case, we will write

$$c_k = 1 + a_k$$

and look for conditions on the sequence $a_k > -1$ which insure the convergence of the series

$$\sum_{k=1}^{\infty} \log(1 + a_k).$$

The essential observation is that for $|a_k|$ small,

$$|\log(1 + a_k)| \simeq |a_k|.$$

To make this more precise, start with the definition

$$\log(y) = \int_1^y \frac{1}{t}\, dt,$$

which implies

$$\log(1 + x) = \int_1^{1+x} \frac{1}{t}\, dt.$$

Some simple estimates coming from the interpretation of the integral as an area will establish the following result.

Lemma 7.3.1. *For* $|x| \le 1/2$,

$$|x|/2 \le |\log(1 + x)| \le 2|x|.$$

Proof. For $x \ge 0$

$$\frac{x}{1 + x} \le \log(1 + x) \le x,$$

and the desired inequality holds if $0 \le x \le 1$. For $-1 < x < 0$ write

$$1 + x = 1 - |x|.$$

Then

$$|x| \le |\log(1 - |x|)| \le \frac{|x|}{1 - |x|},$$

and the desired inequality holds if $-1/2 \le x < 0$. Pasting these two cases together gives the final result. □

This prepares us for the main result on infinite products.

Theorem 7.3.2. *Suppose that* $1 + a_k > 0$. *If the series*

$$\sum_{k=1}^{\infty} |a_k| \tag{7.6}$$

converges, then the sequence

$$p_n = \prod_{k=1}^{n} (1 + a_k)$$

has a limit $0 < p < \infty$, *where*

$$p = \exp(\sum_{k=1}^{\infty} \log(1 + a_k)),$$

Proof. If the series in (7.6) converges, then

$$\lim_{k\to\infty} |a_k| = 0;$$

in particular there is an integer N such that $|a_k| < 1/2$ for $k \ge N$. For $n \ge N$ the estimate of Lemma 7.3.1 gives

$$\sum_{k=N}^{n} |\log(1+a_k)| \le 2 \sum_{k=N}^{n} |a_k|.$$

Consequently the series

$$\sum_{k=1}^{\infty} \log(1+a_k)$$

converges absolutely, with sum $S \in \mathbb{R}$.

As noted above, the continuity of the exponential function gives

$$\lim_{n\to\infty} \prod_{k=1}^{n} (1+a_k) = \lim_{n\to\infty} \exp(\sum_{k=1}^{n} \log(1+a_k)) = \exp(\lim_{n\to\infty} \sum_{k=1}^{n} \log(1+a_k)) = e^S.$$

□

There are a few points to address before considering whether the converse of Theorem 7.3.2 is valid. If a single factor $1+a_j$ is equal to 0, then all of the partial products $\prod_{k=1}^{n}(1+a_k)$ for $n \ge j$ will be 0 and the sequence of partial products will converge regardless of the values of the other terms $1+a_k$. Also, if

$$\lim_{n\to\infty} \sum_{k=1}^{n} \log(1+a_k) = -\infty,$$

then the sequence of partial products will have limit 0. It would seem desirable to avoid these cases. Problems also arise if $1+a_k < 0$ for an infinite set of indices k. In this case the sequence of partial products must have an infinite subsequence of positive terms and an infinite subsequence of negative terms. If the sequence of partial products has a limit in this setting, the limit must be 0.

Despite the difficulty that arises if even a single factor is 0, there are good reasons to want to allow at least a finite number of factors which are negative or 0. We therefore say that an infinite product $\prod_{k=1}^{\infty}(1+a_k)$ converges if there is an integer N such that $1+a_k > 0$ for $k \ge N$, and if the modified sequence of partial products $\prod_{k=N}^{n}(1+a_k)$ has a positive limit.

The proof of the next result is left as an exercise.

Theorem 7.3.3. *Suppose that the sequence*

$$p_n = \prod_{k=1}^{n}(1+a_k)$$

has a limit p, with $0 < p < \infty$, *and* $1 + a_k > 0$ *for* $k \ge N$. *The series*

$$\sum_{k=N}^{\infty} \log(1 + a_k)$$

converges. If the terms a_k *are all positive or all negative, then the series* $\sum_{k=1}^{\infty} |a_k|$ *converges.*

7.4 Stirling's formula

There are many problems in counting and probability where one needs to understand the size of $n!$. The basic estimate was discovered in 1730 thanks to a collaboration between Abraham DeMoivre (1667–1754) and James Stirling (1692–1770). The result is

$$n! \sim \sqrt{2\pi} e^{-n} n^{n+1/2} = \sqrt{2\pi n} (\frac{n}{e})^n. \tag{7.7}$$

Here the symbol $\sim$ means that

$$\lim_{n \to \infty} \frac{\sqrt{2\pi} e^{-n} n^{n+1/2}}{n!} = 1.$$

It is possible to use very elementary methods to get a slightly inferior result. The first idea is to consider $\log(n!)$ rather than $n!$. Since the logarithm of a product is the sum of the logarithms,

$$\log(n!) = \sum_{k=1}^{n} \log(k). \tag{7.8}$$

The plan is to compare this sum to a convenient integral.

Thinking of (7.8) as a left endpoint Riemann sum (Figure 7.4) leads to the inequality

$$\sum_{k=1}^{n} \log(k) \le \int_1^{n+1} \log(x) \, dx. \tag{7.9}$$

The inequality follows from the fact that $\log(x)$ is an increasing function, so the left endpoint Riemann sums are smaller than the corresponding integral.

Recognizing that $\log(1) = 0$ allows us to rewrite (7.8) as

$$\log(n!) = \sum_{k=2}^{n} \log(k). \tag{7.10}$$

Interpreting (7.10) as a right endpoint Riemann sum shows that

$$\log(n!) = \sum_{k=2}^{n} \log(k) \geq \int_1^n \log(x)\, dx. \tag{7.11}$$

Recall that $\log(x)$ has an elementary antiderivative,

$$\int_1^n \log(x)\, dx = (x\log(x) - x)\Big|_1^n = n\log(n) - (n-1).$$

Using this formula with the inequalities (7.9) and (7.11) leads to

$$n\log(n) - (n-1) \leq \log(n!) \leq (n+1)\log(n+1) - n.$$

The exponential function is increasing, so exponentiation doesn't alter the order of the terms. After noting that $n\log(n) = \log(n^n)$, the resulting inequality is

$$n^n e^{1-n} \leq n! \leq (n+1)^{n+1} e^{-n}. \tag{7.12}$$

By refining these ideas it is possible to get much closer to Stirling's formula (7.7).

The first modification is to replace the left and right endpoint Riemann sums with a midpoint Riemann sum (Figure 7.5). The following observation about midpoint Riemann sums will be important. On each subinterval the midpoint approximation

$$\int_{x_i}^{x_{i+1}} f(x)\, dx \simeq f(\frac{x_i + x_{i+1}}{2})[x_{i+1} - x_i]$$

gives the same area as if we used the tangent line to the graph of f at the midpoint,

$$f(\frac{x_i + x_{i+1}}{2})[x_{i+1} - x_i] = \int_{x_i}^{x_{i+1}} f(\frac{x_i + x_{i+1}}{2}) + f'(\frac{x_i + x_{i+1}}{2})[x - \frac{x_i + x_{i+1}}{2}]\, dx.$$

Since the function $\log(x)$ is concave down (the first derivative is decreasing), the tangent line at the midpoint lies above the graph of the function, and the midpoint approximation is greater than the integral. (See Figure 7.6)

Interpreting (7.8) as a midpoint Riemann sum also requires shifting the integral by 1/2. That is,

$$\sum_{k=1}^{n} \log(k) \geq \int_{1/2}^{n+1/2} \log(x)\, dx = x\log(x) - x\Big|_{1/2}^{n+1/2} \tag{7.13}$$

$$= (n + 1/2)\log(n + 1/2) - \frac{1}{2}\log(1/2) - n.$$

Another estimate for (7.8) results from using the trapezoidal rule for integrals (Figure 7.7), which is just averaging of the left and right endpoint

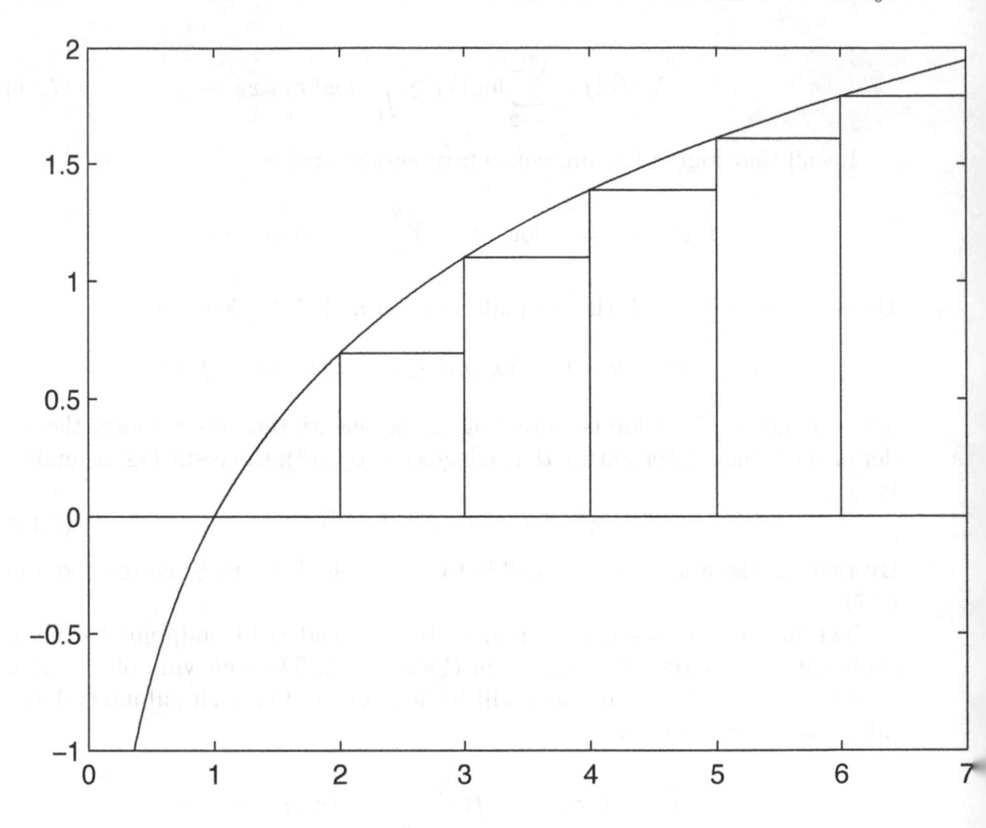

FIGURE 7.4
Riemann sum for $\int_1^x \log(t)\, dt$.

Riemann sums. This time the fact that $\log(x)$ is concave down means that the trapezoidal sums underestimate the integral. (See Problem 7.14) Thus

$$\frac{1}{2}\sum_{k=1}^{n}[\log(k) + \log(k+1)] \leq \int_1^{n+1} \log(x)\, dx,$$

or, rewriting the left-hand side,

$$\sum_{k=1}^{n} \log(k) + \frac{1}{2}\log(n+1) \leq \int_1^{n+1} \log(x)\, dx.$$

This is the same as

$$\sum_{k=1}^{n} \log(k) \leq \int_1^{n+1} \log(x)\, dx - \frac{1}{2}\log(n+1) = (n+1/2)\log(n+1) - n. \quad (7.14)$$

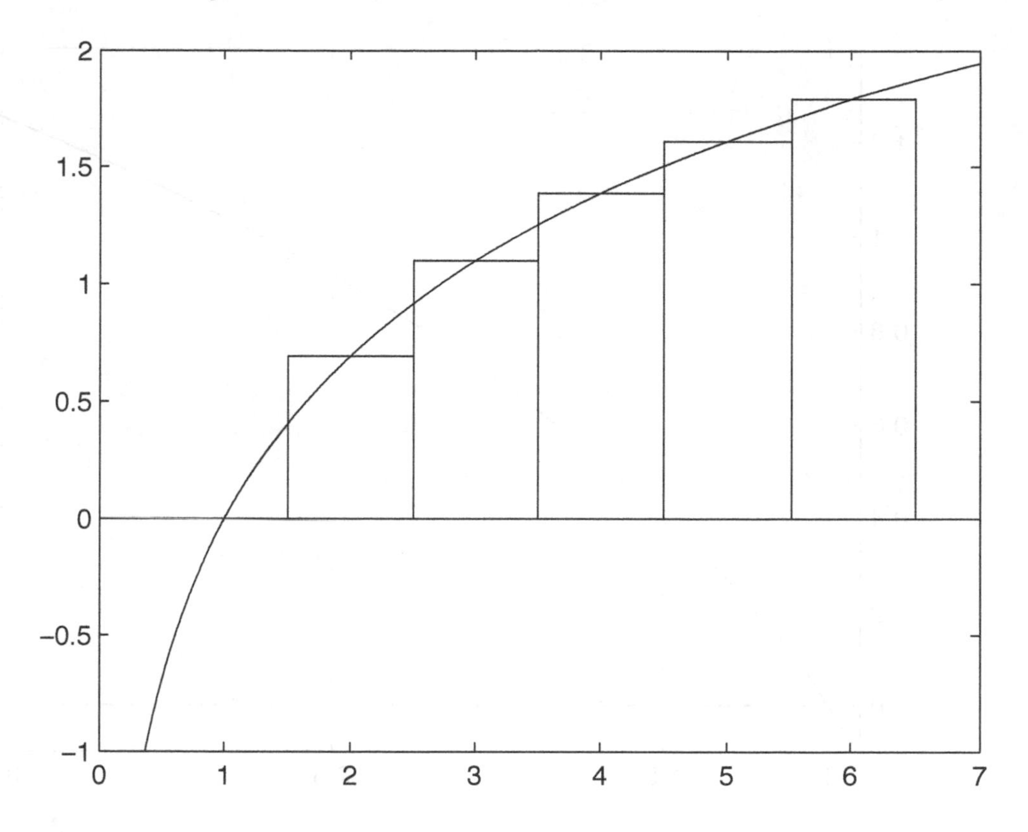

FIGURE 7.5
Midpoint sum for $\int_1^x \log(t)\ dt$.

Together the estimates (7.13) and (7.14) are

$$(n+1/2)\log(n+1/2) - \frac{1}{2}\log(1/2) - n \le \sum_{k=1}^{n} \log(k) \tag{7.15}$$

$$\le (n+1/2)\log(n+1) - n.$$

Exponentiation gives

$$\sqrt{2}(n+1/2)^{n+1/2}e^{-n} \le n! \le (n+1)^{n+1/2}e^{-n},$$

which may also be written as

$$\sqrt{2}n^{n+1/2}(1+\frac{1}{2n})^{n+1/2}e^{-n} \le n! \le n^{n+1/2}(1+1/n)^{n+1/2}e^{-n}. \tag{7.16}$$

The reader may recall from calculus that

$$\lim_{n\to\infty}(1+\frac{1}{n})^n = e, \quad \lim_{n\to\infty}(1+\frac{1}{2n})^n = e^{1/2},$$

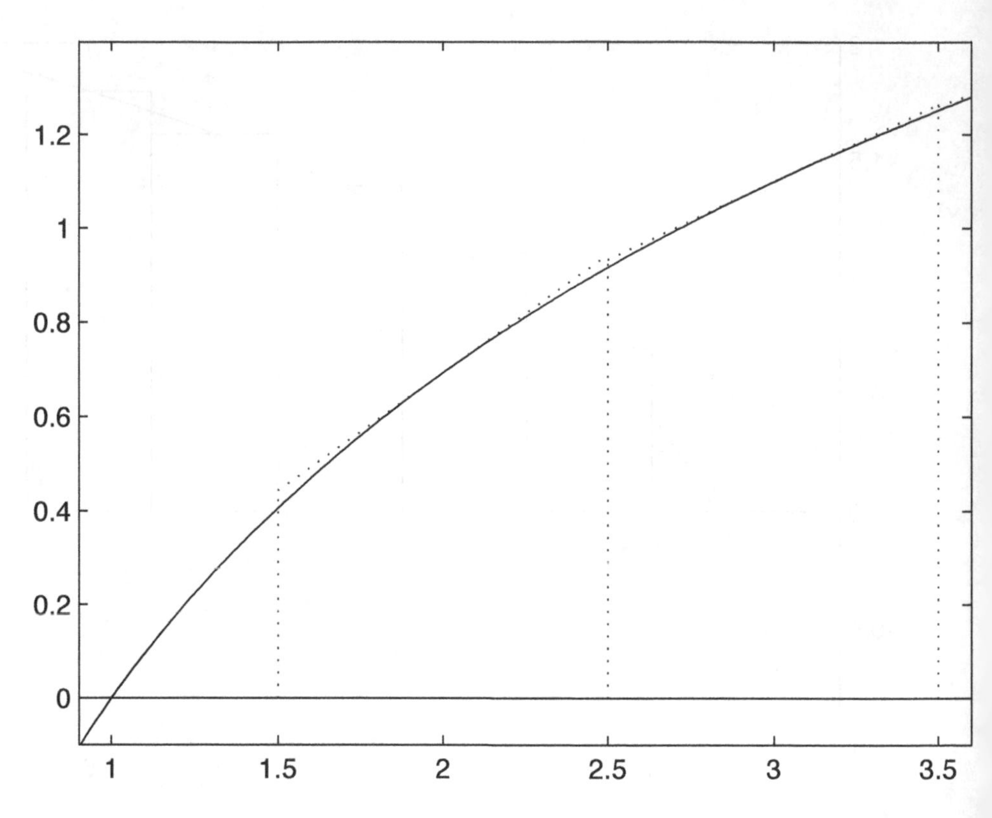

FIGURE 7.6
Tangent Riemann sum for $\int_1^x \log(t)\ dt$.

which suggests that our expressions can be simpified. That simplification is our next order of business.

Notice that for $x \geq 0$

$$\log(1+x) \leq x.$$

This follows from $\log(1) = 0$ and

$$\frac{d}{dx}\log(1+x) = \frac{1}{1+x}$$

so that

$$\frac{d}{dx}\log(1+x) \leq 1 = \frac{d}{dx}x, \quad x \geq 0.$$

A simple logarithmic calculation now gives

$$(n+1/2)\log(1+\frac{1}{n}) \leq (n+1/2)\frac{1}{n} = 1 + \frac{1}{2n},$$

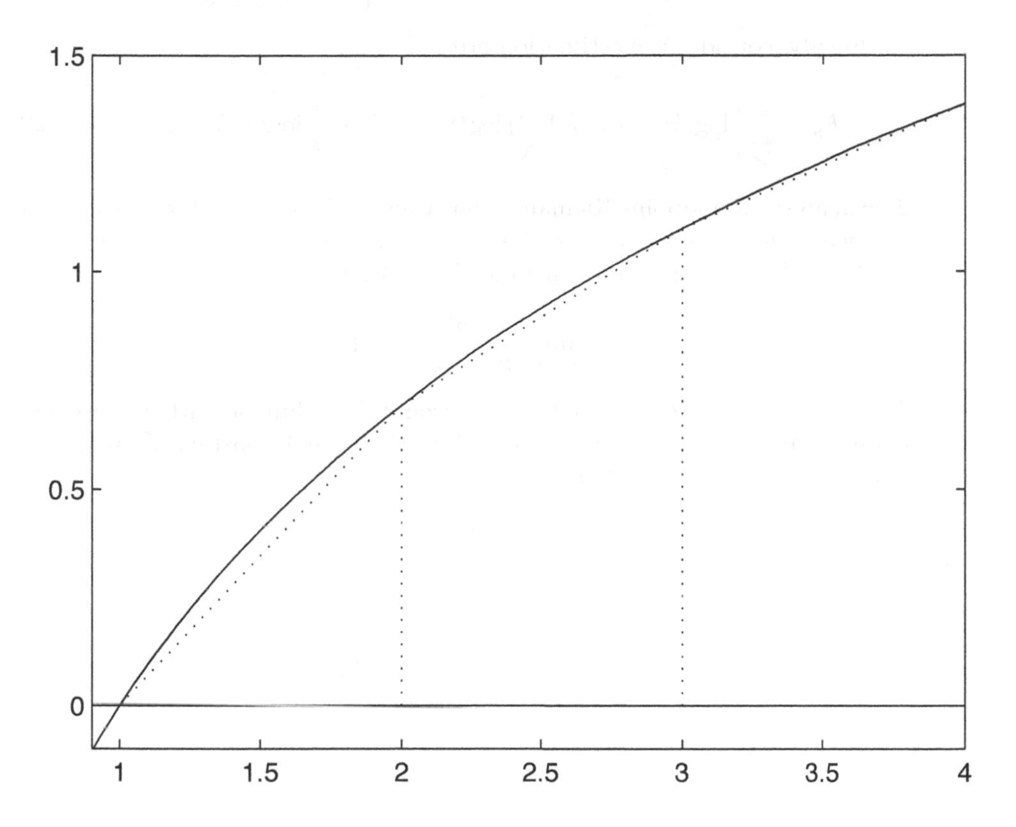

FIGURE 7.7
Trapezoidal sum for $\int_1^x \log(t)\ dt$.

or

$$(1+\frac{1}{n})^{n+1/2} \le e^{1+1/(2n)}. \tag{7.17}$$

Similarly, the calculation

$$\log(1+x) = \int_1^{1+x} \frac{1}{t}\ dt \ge \frac{x}{1+x}, \quad x \ge 0$$

gives

$$(n+1/2)\log(1+\frac{1}{2n}) \ge \frac{2n+1}{2}\frac{1}{2n}\frac{2n}{2n+1} = 1/2,$$

or

$$e^{1/2} \le (1+\frac{1}{2n})^{n+1/2}. \tag{7.18}$$

Using (7.17) and (7.18) in (7.16) produces the inequality

$$\sqrt{2e}n^{n+1/2}e^{-n} \le n! \le e^{1+(2n)^{-1}}n^{n+1/2}e^{-n}. \tag{7.19}$$

Finally, consider the estimation error

$$E_n = \sum_{k=1}^{n} \log(k) - [(n + 1/2)\log(n + 1/2) - \frac{1}{2}\log(1/2) - n]. \qquad (7.20)$$

The analysis of midpoint Riemann sums which led to (7.13) shows that this is an increasing sequence. It's not hard to check that the sequence is bounded, implying the existence of a constant C such that

$$\lim_{n\to\infty} \frac{n!}{n^{n+1/2}e^{-n}} = C.$$

The actual value of C does not emerge from this technique, although we have good bounds. Our uncertainty regarding the actual constant $C = \sqrt{2\pi}$ is expressed by the inequalities

$$\sqrt{2e} \le C \le e.$$

7.5 Problems

7.1. *Suppose a table of base 10 logarithms has the values*

$$\log_{10}(1.5) = 0.1761, \quad \log_{10}(1.6) = 0.2041.$$

Estimate the value of $\log_{10}(1.53)$ *by assuming that the function is approximately linear between the given values.*

7.2. *What modifications to the argument in the text are needed to show geometrically that*

$$\log(xy) = \log(x) + \log(y), \quad 0 < x, y < 1?$$

7.3. *Argue that*

$$\log(N) = \int_1^N \frac{1}{t}\ dt \geq \sum_{n=2}^N \frac{1}{n}.$$

As a consequence, conclude that $\lim_{x\to\infty} \log(x) = \infty$.

7.4. *a) Suppose that the numbers* a, b, m *are all positive, and* $a < b$. *Using a geometric calculation as illustrated in Figure 7.2, show that*

$$m^{r+1} \int_a^b t^r\ dt = \int_{ma}^{mb} t^r\ dt$$

Verify the geometric argument by explicitly calculating the integrals.

b) Show that the function

$$g(x) = \int_1^x t^r\ dt, \quad x \geq 1$$

satisfies

$$g(xy) = g(x) + x^{r+1} g(y), \quad x, y \geq 1.$$

7.5. *Show that if* p/q *is a rational number, and* $x > 0$, *then*

$$\log(x^{p/q}) = \frac{p}{q} \log(x).$$

7.6. *If*

$$Q = \lim_{N\to\infty} \prod_{n=1}^N (1 - \frac{1}{n+1}),$$

is $Q > 0$? *Should you play game one with Swindler Stan?*

7.7. *If*

$$P = \lim_{N\to\infty} \prod_{n=1}^{N} (1 - 6^{-n}),$$

is $P > 0$. Should you play game two with Swindler Stan?

7.8. *Estimate the sum*

$$\sum_{k=1}^{n} \frac{1}{k^{1/3}} \tag{7.21}$$

by the following process.

(a) Sketch the graph of $x^{-1/3}$ for $1 \le x \le \infty$. Include in your sketch rectangular boxes with base $[k, k+1]$ on the real axis, and height $k^{-1/3}$.

(b) Argue that

$$\sum_{k=1}^{n} \frac{1}{k^{1/3}} \ge \int_1^{n+1} x^{-1/3}\, dx.$$

(c) By a similar argument obtain an estimate of the form

$$\sum_{k=2}^{n} \frac{1}{k^{1/3}} \le \int_1^{n} x^{-1/3}\, dx.$$

(d) Evaluate the integrals and obtain upper and lower bounds for the sum (7.21). Is the sum bounded as $n \to \infty$?

7.9. *a) Guided by the treatment of Stirling's formula, show that*

$$\int_1^{n+1} \frac{1}{x}\, dx < \sum_{k=1}^{n} \frac{1}{k} < 1 + \int_1^{n} \frac{1}{x}\, dx,$$

or equivalently,

$$\log(n+1) < \sum_{k=1}^{n} \frac{1}{k} < 1 + \log(n). \tag{7.22}$$

b) Observe that

$$\log(k+1) - \log(k) = \int_k^{k+1} \frac{1}{x}\, dx < \frac{1}{k},$$

and conclude that the function

$$f(n) = (\sum_{k=1}^{n} \frac{1}{k}) - \log(n+1), \quad n \ge 1$$

increases with n.

c) Show that $f(n) \le 1$ for all $n \ge 1$.

The number

$$\gamma = \lim_{n\to\infty}\Big(\sum_{k=1}^{n}\frac{1}{k} - \log(n+1)\Big)$$

is called the Euler constant. The actual value is approximately $\gamma = .577\ldots$. *It is not known if* γ *is a rational number.*

d) Use midpoint and trapezoidal sums to show that

$$.5 \le \gamma \le \log(2) = .69\ldots.$$

7.10. *Based on our treatment of Stirling's formula*

$$\sqrt{2e}n^{n+1/2}e^{-n} \le n! \le e^2 n^{n+1/2}e^{-n}, \quad n \ge 1.$$

Use these inequalities to obtain good upper and lower bounds for

$$\binom{n}{k} = \frac{n!}{k!(n-k)!}.$$

7.11. *For fixed* n *we may consider* $\binom{n}{k}$ *as a function of* k, *with* $0 \le k \le n$. *This function increases until* k *reaches the middle of its range. Use Stirling's formula to estimate the size of* $\binom{n}{n/2}$ *when* n *is even and* $\binom{n}{(n-1)/2}$ *when* n *is odd.*

7.12. *Show that the sequence* E_n *in* (7.20) *is increasing and bounded. (Hint: Consider* (7.15).*)*

7.13. *Prove Theorem 7.3.3.*

7.14. *Suppose* $f : [a,b] \to \mathbb{R}$ *is convex and differentiable. Let* $x_k = a + (b-a)k/n$, *and let* M_n *and* T_n *respectively denote the midpoint and trapezoidal rule approximations to*

$$I = \int_a^b f(x)\, dx.$$

Show that $M_n \le I \le T_n$.

8

Taylor polynomials and series

Infinite sequences and infinite series are very useful ways to represent numbers and functions. To pick one application, many differential equations can be solved by essentially algebraic methods when solutions are represented using infinite (power) series. Infinite series can also be used to represent the antiderivatives of functions when more elementary methods are inadequate.

However, the use of infinite processes such as infinite sums requires extra diligence. This point is illustrated by the important geometric series,

$$S(x) = \sum_{k=0}^{\infty} x^k = 1 + x + x^2 + x^3 + \dots.$$

Formal manipulations suggest that

$$xS(x) = x + x^2 + x^3 + \dots,$$

and by subtraction

$$(1 - x)S(x) = 1, \quad S(x) = \frac{1}{1-x}.$$

This claim seems reasonable if $x = 0.1$, where it yields

$$\frac{1}{0.9} = \frac{10}{9} = 1.1111\dots,$$

or if $x = 1/2$, where it gives

$$2 = 1 + \frac{1}{2} + \frac{1}{4} + \frac{1}{8} + \dots.$$

The claim is more doubtful if $x = -1$, where it suggests that

$$1 + (-1) + 1 + (-1) + \dots = \sum_{k=0}^{\infty} (-1)^k = \frac{1}{2}.$$

Even more troubling is the case $x = 2$, when the formula produces

$$1 + 2 + 4 + \dots = \sum_{k=0}^{\infty} 2^k = \frac{1}{-1} = -1.$$

The pitfalls associated with the use of infinite processes led the ancient Greeks to avoid them [1, p. 13–14], [5, p. 176]. As calculus was developed [5, p. 436–467] the many successful calculations achieved with infinite processes, including infinite series, apparently reduced the influence of the cautious critics. A reconciliation between the success of cavalier calculations and the careful treatment of foundational issues was not achieved until the nineteenth century.

For this particular case of the geometric series, the issues are clarified by noting the formula for the partial sums,

$$S_n(x) = 1 + x + x^2 + \cdots + x^{n-1} = \sum_{k=0}^{n-1} x^k = \frac{1 - x^n}{1 - x} = \frac{1}{1 - x} - \frac{x^n}{1 - x}, \quad x \neq 1.$$

If $|x| < 1$ then the numbers x^n shrink to 0 as n gets larger. In these cases $\lim_{n\to\infty} S_n(x) = S(x) = 1/(1 - x)$. The limits of the partial sums fail to exist in the problematic cases $x = -1$ or $x = 2$; the finite sums

$$\sum_{k=0}^{n-1} (-1)^k = \frac{1}{2} - \frac{(-1)^n}{2},$$

and

$$\sum_{k=0}^{n-1} 2^k = \frac{1}{-1} - \frac{2^n}{-1},$$

are not approaching the numbers $1/2$ or -1 respectively.

The example of the geometric series has been used to slip in another idea, that a function might be represented by an infinite series. In the case of the geometric series, the function is

$$S(x) = \frac{1}{1 - x},$$

with the corresponding series

$$\sum_{k=0}^{\infty} x^k = 1 + x + x^2 + \ldots.$$

Notice that the function $1/(1 - x)$ is defined for all values of x except $x = 1$, but the infinite series only converges for $|x| < 1$.

The geometric series looks like a polynomial with infinitely many terms. More generally, an infinite series of the form

$$\sum_{k=0}^{\infty} a_k (x - x_0)^k.$$

is called a *power series*. The number x_0 is called the *center* of the series. In the case of the geometric series the center is 0.

A second example of a power series can be constructed by a simple modification of the geometric series. Replacing x by $-x$ gives

$$\frac{1}{1+x} = \frac{1}{1-(-x)} = \sum_{k=0}^{\infty} (-1)^k x^k, \quad |x| < 1,$$

where the coefficients are $a_k = (-1)^k$.

Since power series look so much like polynomials, it is tempting to treat them in the same way. Yielding to this temptation, consider using a term by term integration of the last series to obtain the conjectured formula

$$\log(1+x) = \int_0^x \frac{1}{1+t}\,dt = \sum_{j=0}^{\infty} (-1)^j \int_0^x t^j\,dt$$

$$= \sum_{j=0}^{\infty} (-1)^j \frac{x^{j+1}}{j+1} = \sum_{k=1}^{\infty} (-1)^{k-1} \frac{x^k}{k}, \quad |x| < 1,$$

In this case $a_0 = 0$ and $a_k = (-1)^{k-1}/k$ for $k \ge 1$.

To the extent that a power series may be manipulated like a polynomial, such a representation for a function is extremely convenient. Differentiation is no challenge, and more importantly the term by term integration of power series is trivial. This is quite different from the problem of trying to find elementary antiderivatives, where examples such as

$$\int_0^x e^{-t^2}\,dt$$

prove to be an impossible challenge.

So far the discussion of power series has emphasized formal algebraic manipulations, with some analysis to help with the justification. In the remainder of this chapter, questions about power series will be considered more generally, and with more analytical precision. There are some basic questions to consider. Which functions may be represented by a power series, what are the coefficients of the power series, and how much error is made when a partial sum of the power series is used instead of the entire infinite series?

8.1 Taylor polynomials

Suppose that the function $f(x)$ has a power series representation

$$f(x) = \sum_{k=0}^{\infty} a_k x^k = a_0 + a_1 x + a_2 x^2 + a_3 x^3 + \dots.$$

The first problem is to decide what the coefficients a_k are. Notice that evaluation of the function $f(x)$ at $x = 0$ gives the first coefficient,

$$f(0) = a_0.$$

To get a_1 we formally differentiate the power series,

$$f'(x) = \sum_{k=1}^{\infty} k a_k x^{k-1} = a_1 + 2a_2 x + 3a_3 x^2 + \dots,$$

and evaluate the derivative at $x = 0$, obtaining the second coefficient,

$$f'(0) = a_1.$$

Continuing in this manner leads to

$$f^{(2)}(x) = \sum_{k=2}^{\infty} k(k-1) a_k x^{k-2} = 2a_2 + 6a_3 x + \dots, \qquad f^{(2)}(0) = 2a_2.$$

Differentiating term by term n times gives

$$f^{(n)}(x) = \sum_{k=n}^{\infty} k(k-1)\cdots(k-(n-1)) a_k x^{k-n}.$$

Now evaluate both sides at $x = 0$ to get

$$f^{(n)}(0) = n! a_n, \quad a_n = f^{(n)}(0)/n!.$$

Similar computations may be carried out for more general power series with center at x_0,

$$f(x) = \sum_{k=0}^{\infty} c_k (x - x_0)^k = c_0 + c_1(x - x_0) + c_2(x - x_0)^2 + c_3(x - x_0)^3 + \dots.$$

In this case $f(x_0) = c_0$, $f'(x_0) = c_1$, and in general (see Problem 8.4)

$$f^{(n)}(x_0) = n! c_n, \quad c_n = f^{(n)}(x_0)/n!.$$

The infinite series

$$\sum_{k=0}^{\infty} \frac{f^{(k)}(x_0)}{k!} (x - x_0)^k$$

is called the Taylor series (named for Brook Taylor (1685–1731)) for the function $f(x)$ centered at x_0. This series is defined as long as $f(x)$ has derivatives of all orders at x_0, but the series may not converge except at x_0. If it does converge, it is possible that the series will not converge to the function $f(x)$.

Taylor polynomials are truncated versions of these series. Given a function

$f(x)$ with at least n derivatives at x_0, its Taylor polynomial of degree n based at $x = x_0$ is the polynomial

$$P_n(x) = \sum_{k=0}^{n} c_k(x - x_0)^k = c_0 + c_1(x - x_0) + c_2(x - x_0)^2 + \cdots + c_n(x - x_0)^n,$$

with

$$c_k = f^{(k)}(x_0)/k!,$$

or

$$P_n(x) = \sum_{k=0}^{n} \frac{f^{(k)}(x_0)}{k!}(x - x_0)^k.$$

This is the unique polynomial of degree n whose 0 through n-th derivatives at x_0 agree with those of f.

It is worth considering a few examples to see in what sense the polynomials $P_n(x)$ 'look like' a function $f(x)$. First take $x_0 = 0$ and $f(x) = e^x$. Since

$$\frac{d^k}{dx^k}e^x = e^x,$$

the coefficients of the Taylor series for e^x based at $x = 0$ are

$$a_k = \frac{1}{k!},$$

and the Taylor polynomial of degree n based at $x_0 = 0$ is

$$P_n(x) = \sum_{k=0}^{n} \frac{x^k}{k!}.$$

That is,

$$P_0(x) = 1, \quad P_1(x) = 1 + x, \quad P_2(x) = 1 + x + \frac{x^2}{2},$$

$$P_3(x) = 1 + x + \frac{x^2}{2} + \frac{x^3}{3 \cdot 2}, \quad \ldots.$$

Figure 8.1 shows the graph of e^x and the first order Taylor polynomial $P_1(x)$ with center $x_0 = 0$. Figure 8.2 is similar with the third order Taylor polynomial $P_3(x)$.

As a second example take $x_0 = \pi/2$ and $f(x) = \cos(x)$. In this case

$$\cos'(x) = -\sin(x), \quad \cos^{(2)}(x) = -\cos(x),$$

$$\cos^{(3)}(x) = \sin(x), \quad \cos^{(4)}(x) = \cos(x),$$

and in general

$$\cos^{(2m)}(x) = (-1)^m \cos(x),$$

$$\cos^{(2m+1)}(x) = (-1)^{m+1} \sin(x), \quad m = 0, 1, 2, \ldots.$$

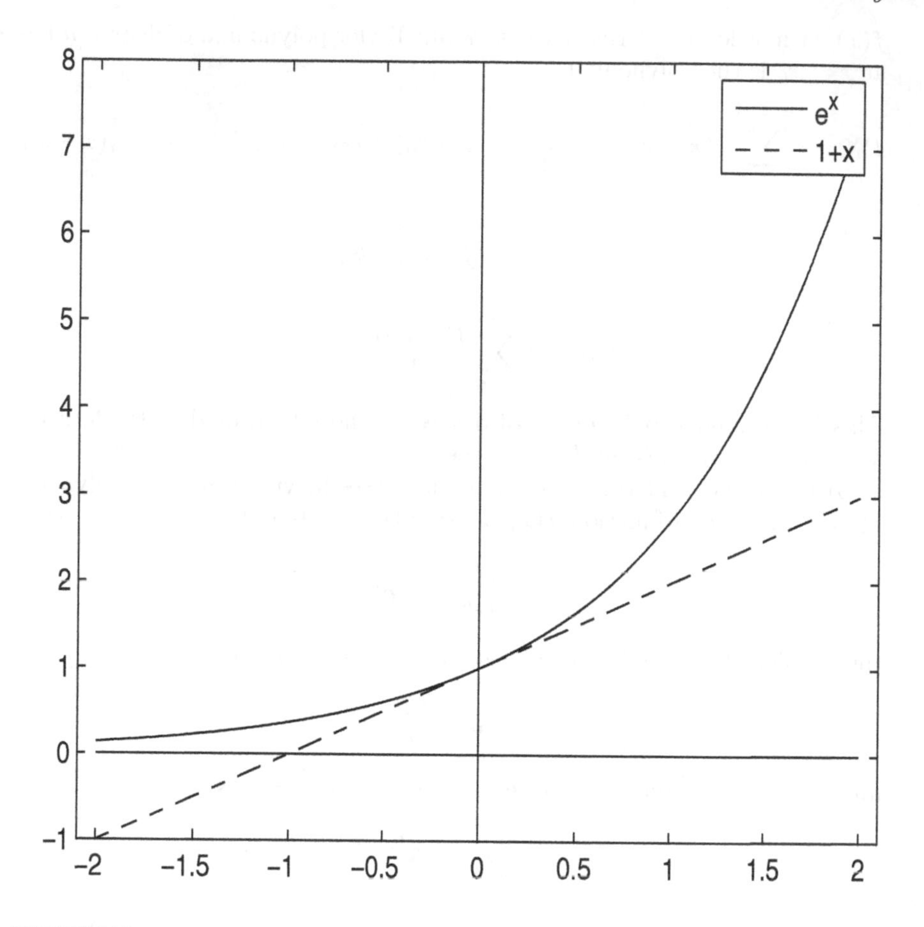

FIGURE 8.1
First order Taylor polynomial for e^x.

Evaluation of the derivatives at $x_0 = \pi/2$ gives

$$\cos^{(2m)}(\pi/2) = (-1)^m \cos(\pi/2) = 0,$$

$$\cos^{(2m+1)}(\pi/2) = (-1)^{m+1} \sin(\pi/2) = (-1)^{m+1}, \quad m = 0, 1, 2, \dots .$$

Since the coefficients with even index k vanish we find

$$P_0(x) = 0, \quad P_1(x) = -(x - \pi/2),$$

$$P_2(x) = -(x - \pi/2), \quad P_3(x) = -(x - \pi/2) + \frac{(x - \pi/2)^3}{6}, \quad \dots .$$

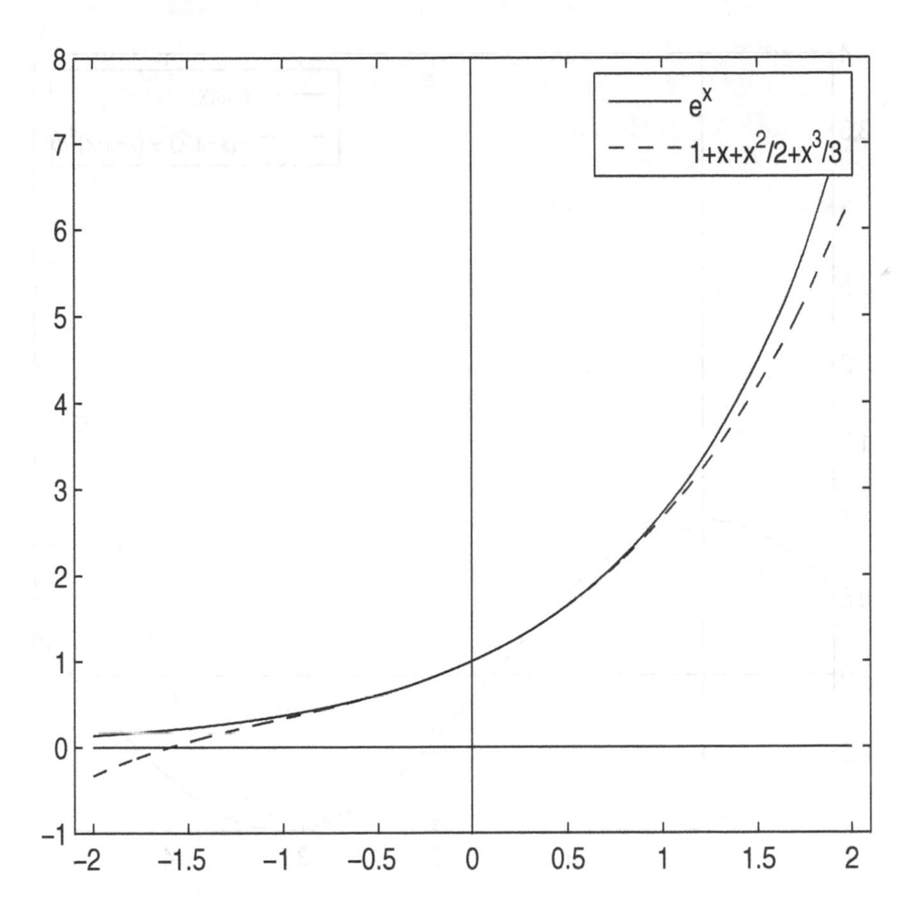

FIGURE 8.2
Third order Taylor polynomial for e^x.

The general form of this Taylor polynomial of order n based at $x_0 = \pi/2$ is

$$P_n(x) = \sum_{k=0}^{\lfloor (n-1)/2 \rfloor} (-1)^{k+1} \frac{(x - \pi/2)^{2k+1}}{(2k+1)!}.$$

Here $\lfloor x \rfloor$ is the largest integer less than or equal to x, and there is no summation (the sum is 0) if the upper limit is negative.

Figure 8.3 shows the graph of $\cos(x)$ and its third order Taylor polynomial with center $x_0 = \pi/2$.

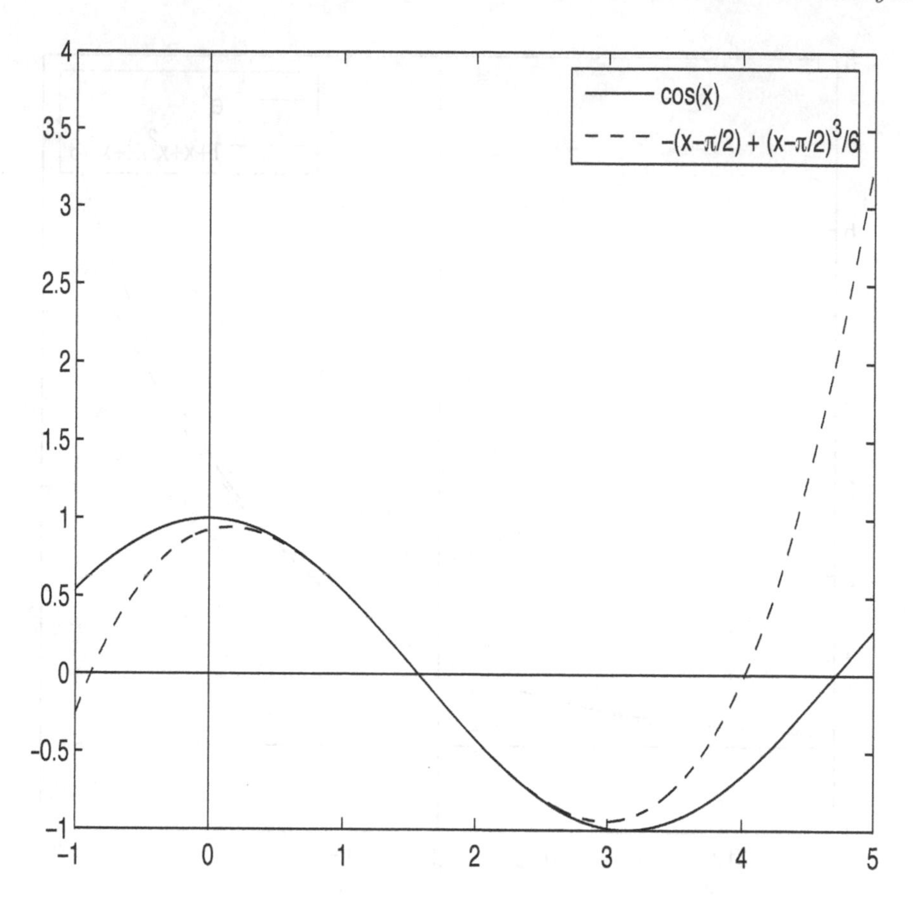

FIGURE 8.3
Taylor polynomial for $\cos(x)$ with center $x_0 = \pi/2$.

8.2 Taylor's Theorem

Taylor's Theorem provides an exact description of the difference between a function $f(x)$ and its Taylor polynomials. Let's start with a simple motivating example,

$$f(x) = e^x, \quad x_0 = 0.$$

By the Fundamental Theorem of Calculus:

$$f(x) - f(x_0) = \int_{x_0}^{x} f'(t)\, dt.$$

For the example this gives

$$e^x - e^0 = \int_0^x e^t \, dt.$$

Now recall the integration by parts formula

$$\int_{x_0}^x h'(t)g(t) \, dt = h(x)g(x) - h(x_0)g(x_0) - \int_{x_0}^x h(t)g'(t) \, dt.$$

For our example, take $g(t) = e^t$ and $h'(t) = 1$. A convenient choice of $h(t)$ with $h'(t) = 1$ is $h(t) = t - x$. Here we are thinking of x as fixed for the moment.

Since $e^0 = 1$,

$$e^x = 1 + \int_0^x e^t \, dt = 1 + (t - x)e^t \Big|_{t=0}^{t=x} - \int_0^x (t - x)e^t \, dt$$

$$= 1 + (x - x)e^x - (0 - x)e^0 + \int_0^x (x - t)e^t \, dt$$

That is

$$e^x = 1 + x + \int_0^x (x - t)e^t \, dt.$$

Now use the same idea again with $g(t) = e^t$ and $h'(t) = (x - t)$. Take

$$h(t) = -\frac{(x - t)^2}{2}$$

to get

$$e^x = 1 + x - \frac{(t - x)^2}{2} e^t \Big|_0^x + \int_0^x \frac{(x - t)^2}{2} e^t \, dt$$

$$= 1 + x + \frac{x^2}{2} + \int_0^x \frac{(x - t)^2}{2} e^t \, dt$$

Such formulas can give us information about the function. For instance, notice that in the last formula the integrand is nonnegative, so that the integral is positive (if $x \geq 0$). Dropping the last term makes the remaining expression smaller, which implies that for $x \geq 0$

$$e^x \geq 1 + x + \frac{x^2}{2}.$$

The techniques that worked in the example also work in the general case. Start with the Fundamental Theorem of Calculus to write

$$f(x) - f(x_0) = \int_{x_0}^x f'(t) \, dt$$

Since $\frac{d}{dt}(t-x) = 1$,

$$f(x) = f(x_0) + \int_{x_0}^{x} f'(t)\, dt = f(x_0) + \int_{x_0}^{x} f'(t)\frac{d}{dt}(t-x)\, dt.$$

Now use integration by parts to get

$$f(x) = f(x_0) + \int_{x_0}^{x} f'(t)\frac{d}{dt}(t-x)\, dt = f(x_0) + (t-x)f'(t)\Big|_{x_0}^{x} - \int_{x_0}^{x} (t-x)f^{(2)}(t)\, dt$$

$$f(x) = f(x_0) + f'(x_0)(x-x_0) + \int_{x_0}^{x} (x-t)f^{(2)}(t)\, dt$$

Of course we are assuming here that $f(x)$ can be differentiated as many times as indicated, always with continuous derivatives. By repeatedly using the same integration by parts idea the following theorem is obtained.

Theorem 8.2.1. *(Taylor's Theorem) Suppose that the function f(x) has $n+1$ continuous derivatives on the open interval (a, b), and that x_0 and x are in this interval. Then*

$$f(x) = \sum_{k=0}^{n} f^{(k)}(x_0)\frac{(x-x_0)^k}{k!} + \int_{x_0}^{x} \frac{(x-t)^n}{n!} f^{(n+1)}(t)\, dt$$

or, rewriting in alternate notation,

$$f(x) = f(x_0) + f'(x_0)(x-x_0) + f^{(2)}(x_0)\frac{(x-x_0)^2}{2!} + \dots$$

$$+ f^{(n)}(x_0)\frac{(x-x_0)^n}{n!} + \int_{x_0}^{x} \frac{(x-t)^n}{n!} f^{(n+1)}(t)\, dt.$$

Proof. The formal proof is by induction. Notice that the Fundamental Theorem of Calculus shows that the formula is true when $n = 0$. Suppose that the formula is true for $n \le K$. Then using integration by parts we find that

$$f(x) = f(x_0) + f'(x_0)(x-x_0) + f^{(2)}(x_0)\frac{(x-x_0)^2}{2!} + \dots +$$

$$f^{(K)}(x_0)\frac{(x-x_0)^K}{K!} + \int_{x_0}^{x} \frac{(x-t)^K}{K!} f^{(K+1)}(t)\, dt$$

$$= f(x_0) + f'(x_0)(x-x_0) + f^{(2)}(x_0)\frac{(x-x_0)^2}{2!} + \dots + f^{(K)}(x_0)\frac{(x-x_0)^K}{K!}$$

$$+ \frac{-(x-t)^{K+1}}{(K+1)!} f^{(K+1)}(t)\Big|_{x_0}^{x} - \int_{x_0}^{x} \frac{-(x-t)^{K+1}}{(K+1)!} f^{(K+2)}(t)\, dt$$

$$= f(x_0) + f'(x_0)(x-x_0) + f^{(2)}(x_0)\frac{(x-x_0)^2}{2!} + \dots$$

$$+ \frac{(x-x_0)^{K+1}}{(K+1)!} f^{(K+1)}(x_0) + \int_{x_0}^{x} \frac{(x-t)^{K+1}}{(K+1)!} f^{(K+2)}(t)\, dt.$$

When the formula is correct for $n \le K$ it is also true for $n = K+1$, so the proof is finished. ☐

8.3 The remainder

The last term in the formula of Taylor's Theorem is called the *remainder*,

$$R_n(x) = \int_{x_0}^{x} \frac{(x-t)^n}{n!} f^{(n+1)}(t)\, dt.$$

A description of the remainder is attributed to Joseph-Louis Lagrange (1736–1813). The size of this remainder is interesting since the error made in replacing $f(x)$ with the polynomial approximation

$$f(x_0) + f'(x_0)(x - x_0) + f^{(2)}(x_0)\frac{(x-x_0)^2}{2!} + \cdots + f^{(n)}(x_0)\frac{(x-x_0)^n}{n!}$$

is just

$$|R_n(x)| = |f(x) - \sum_{k=0}^{n} f^{(k)}(x_0)\frac{(x-x_0)^k}{k!}|.$$

The definition of convergence of an infinite series immediately leads to the next result.

Theorem 8.3.1. *Suppose that $f(x)$ has derivatives of all orders at x_0. The Taylor series*

$$\sum_{k=0}^{\infty} \frac{f^{(k)}(x_0)}{k!}(x-x_0)^k$$

converges to the value $f(x)$ if and only if

$$\lim_{n\to\infty} R_n(x) = 0.$$

To determine how accurately $f(x)$ is approximated by its Taylor polynomials an estimate for $|R_n(x)|$ is usually employed. It is common to use the fact (Theorem 6.3.3) that for a continuous function $f(x)$ on the interval $[a, b]$,

$$\left|\int_a^b f(x)\, dx\right| \le \int_a^b |f(x)|\, dx. \tag{8.1}$$

Theorem 8.3.2. *Under the hypotheses of Taylor's Theorem, if*

$$M = \max_{x_0 \le t \le x} |f^{(n+1)}(t)|,$$

then

$$|R_n(x)| \le M\frac{|x-x_0|^{n+1}}{(n+1)!}.$$

Proof. Suppose first that $x \ge x_0$. Based on (8.1) the remainder satisfies the inequality

$$|R_n(x)| = |\int_{x_0}^{x} \frac{(x-t)^n}{n!} f^{(n+1)}(t)\,dt| \le \int_{x_0}^{x} \frac{|x-t|^n}{n!} |f^{(n+1)}(t)|\,dt.$$

The integrand will be even larger if $|f^{(n+1)}(t)|$ is replaced by

$$M = \max_{x_0 \le t \le x} |f^{(n+1)}(t)|.$$

Since M is a constant,

$$|R_n(x)| \le M \int_{x_0}^{x} \frac{|x-t|^n}{n!}\,dt.$$

Because $x \ge x_0$ the term $x - t$ is nonnegative, and

$$\int_{x_0}^{x} \frac{|x-t|^n}{n!}\,dt = \int_{x_0}^{x} \frac{(x-t)^n}{n!}\,dt = \frac{(x-x_0)^{n+1}}{(n+1)!}.$$

This gives

$$|R_n(x)| \le M \frac{(x-x_0)^{n+1}}{(n+1)!} = M \frac{|x-x_0|^{n+1}}{(n+1)!},$$

as desired.

If $x < x_0$ the intermediate computations are the same except for a possible factor -1, and the final result is the same. □

The next result uses a different argument to analyze the remainder.

Theorem 8.3.3. *(Lagrange) Under the hypotheses of Taylor's Theorem, there is some c between x_0 and x such that*

$$R_n(x) = \int_{x_0}^{x} \frac{(x-t)^n}{n!} f^{(n+1)}(t)\,dt = f^{(n+1)}(c) \frac{(x-x_0)^{n+1}}{(n+1)!}.$$

Proof. Suppose for simplicity that $x_0 \le x$; the other case may be handled similarly. Since $f^{(n+1)}(t)$ is continuous on the closed interval from x_0 to x, it has a minimum $f^{(n+1)}(x_1)$ and a maximum $f^{(n+1)}(x_2)$, with $x_0 \le x_1 \le x$ and $x_0 \le x_2 \le x$. Since the values of the function $f^{(n+1)}$ at x_1 and x_2 are just constants, and

$$\frac{(x-t)^n}{n!} \ge 0, \quad x_0 \le t \le x,$$

we have

$$R_n(x) = \int_{x_0}^{x} \frac{(x-t)^n}{n!} f^{(n+1)}(t)\,dt \le \int_{x_0}^{x} \frac{(x-t)^n}{n!} f^{(n+1)}(x_2)\,dt$$

$$= f^{(n+1)}(x_2) \int_{x_0}^{x} \frac{(x-t)^n}{n!}\,dt = f^{(n+1)}(x_2) \frac{(x-x_0)^{n+1}}{(n+1)!},$$

and similarly

$$R_n(x) \geq f^{(n+1)}(x_1)\frac{(x-x_0)^{n+1}}{(n+1)!}.$$

Now for s between x_1 and x_2 look at the function

$$f^{(n+1)}(s)\frac{(x-x_0)^{n+1}}{(n+1)!}.$$

This is a continuous function of s, which for $s = x_1$ is smaller than $R_n(x)$, and for $s = x_2$ is bigger than $R_n(x)$. By the Intermediate Value Theorem there must be some point $s = c$ where

$$R_n(x) = f^{(n+1)}(c)\frac{(x-x_0)^{n+1}}{(n+1)!},$$

as desired. □

8.3.1 Calculating e

To illustrate the use of Taylor's Theorem, the decimal expansion for the number e will be considered. Using $x_0 = 0$, Taylor's Theorem gives

$$e^x = \sum_{k=0}^{n} \frac{x^k}{k!} + R_n(x).$$

An earlier crude estimate showed that $2 < e < 4$. Theorem 8.3.2 then implies

$$|R_n(x)| \leq \frac{4^{|x|}|x|^{n+1}}{(n+1)!}.$$

Notice that for any fixed value of x,

$$\lim_{n\to\infty} R_n(x) = 0,$$

so that the Taylor series for e^x based at $x_0 = 0$ converges to e^x,

$$e^x = \lim_{n\to\infty} \sum_{k=0}^{n} \frac{x^k}{k!}.$$

To compute the first few terms in the decimal representation of $e = e^1$, consider the case $n = 6$. Then

$$|R_n(1)| \leq \frac{4}{(7)!} = \frac{1}{1260}.$$

and

$$|e - (1 + 1 + \frac{1}{2!} + \frac{1}{3!} + \cdots + \frac{1}{6!})| < 10^{-3},$$

Thus with an error less than 10^{-3},

$$e \simeq 2 + \frac{1}{2} + \frac{1}{6} + \frac{1}{24} + \frac{1}{120} + \frac{1}{720} = 2.718\ldots.$$

The accuracy of this Taylor series approximation for e^x improves very rapidly as n increases.

8.3.2 Calculating π

The decimal expansion for the number π may also be obtained by using a power series. In this example algebraic manipulations are emphasized. The starting point is the calculus formula

$$\frac{d}{dx}\tan^{-1}(x) = \frac{1}{1+x^2}.$$

Since $\tan^{-1}(0) = 0$ this derivative formula may be integrated to give

$$\tan^{-1}(x) = \int_0^x \frac{1}{1+t^2}\, dt.$$

Use the geometric series identity

$$\frac{1}{1-x} = \sum_{k=0}^{m-1} x^k + \frac{x^m}{1-x}$$

to derive the identity

$$\frac{1}{1+t^2} = \frac{1}{1-(-t^2)} = \sum_{k=0}^{m-1}(-1)^k t^{2k} + \frac{(-1)^m t^{2m}}{1-(-t^2)}.$$

The sum coming from the geometric series is easily integrated, giving

$$\int_0^x \sum_{k=0}^{m-1}(-1)^k t^{2k}\, dt = \sum_{k=0}^{m-1}(-1)^k \frac{x^{2k+1}}{2k+1}.$$

To estimate the additional term, notice that for any real number t,

$$|\frac{(-1)^m t^{2m}}{1-(-t^2)}| = |\frac{t^{2m}}{1+t^2}| \le |t|^{2m}.$$

If $\tan^{-1}(x)$ is approximated by

$$\sum_{k=0}^{m-1}(-1)^k \frac{x^{2k+1}}{2k+1}$$

the error satisfies

$$|\tan^{-1}(x) - \sum_{k=0}^{m-1}(-1)^k \frac{x^{2k+1}}{2k+1}| \le |\int_0^x |\frac{(-1)^m t^{2m}}{1+t^2}|\, dt| \tag{8.2}$$

$$\le |\int_0^x |t^{2m}|\, dt| \le \frac{|x|^{2m+1}}{2m+1}.$$

The error estimate in (8.2) looks promising if $|x| < 1$. A convenient choice of x is determined by recalling a fact from trigonometry,

$$\tan(\pi/6) = \frac{1/2}{\sqrt{3}/2} = \frac{1}{\sqrt{3}},$$

or $\pi/6 = \tan^{-1}(1/\sqrt{3})$. With $x = 1/\sqrt{3}$ and, for instance, $m = 6$ we find that

$$|\pi - 6\sum_{k=0}^{5}(-1)^k \frac{1}{2k+1}(\frac{1}{\sqrt{3}})^{2k+1}| \le \frac{6}{13}\frac{1}{\sqrt{3}}\frac{1}{3^6} \le 4 \times 10^{-4},$$

and the computed value is

$$\pi \simeq \frac{6}{\sqrt{3}}[1 - \frac{1}{3\cdot 3} + \frac{1}{9\cdot 5} - \frac{1}{27\cdot 7} + \frac{1}{81\cdot 9} - \frac{1}{243\cdot 11}] \simeq 3.1413.$$

8.4 Additional results

8.4.1 Taylor series by algebraic manipulations

Taylor's Theorem gives an estimate of the form

$$|f(x) - \sum_{k=0}^{n} \frac{f^{(k)}(x_0)}{k!}(x-x_0)^k| \le C_1|x-x_0|^{n+1}.$$

As the ad hoc method for expanding $\tan^{-1}(x)$ illustrated, it is sometimes possible, as in (8.2), to come up with estimates of similar form

$$|f(x) - \sum_{k=0}^{n} a_k(x-x_0)^k| \le C_2|x-x_0|^{n+1}$$

without directly computing the derivatives $f^{(k)}(x_0)$. The next theorem says that the coefficients a_k must be the Taylor series coefficients.

A preparatory lemma notes that the magnitude of $p(x) = a_0 + a_1x + \cdots + a_nx^n$ can't be smaller than $|x|^{n+1}$ for all x in some interval containing 0 unless all the coefficients a_k are 0.

Lemma 8.4.1. *Suppose there is a polynomial*

$$p(x) = \sum_{k=0}^{n} b_k (x - x_0)^k$$

and a number $\delta > 0$ *such that*

$$|p(x)| \le C|x - x_0|^{n+1}$$

for $x_0 < x < x_0 + \delta$. *Then* $p(x) = 0$ *for all* x.

Proof. Suppose the polynomial $p(x)$ is not the zero function; then one or more of the coefficients b_k is not 0. Let m be the smallest index such that $b_m \ne 0$. Throwing away the terms whose coefficients are known to be 0, the inequality for $p(x)$ becomes

$$|\sum_{k=m}^{n} b_k (x - x_0)^k| \le C|x - x_0|^{n+1}.$$

When $x \ne x_0$ it is possible to divide by $|x - x_0|^m$ to get

$$|\sum_{k=m}^{n} b_k (x - x_0)^{k-m}| = |b_m + b_{m+1}(x - x_0) + \dots| \le C|x - x_0|^{n+1-m}.$$

By picking $|x-x_0|$ small enough, the expression $|b_m+b_{m+1}(x-x_0)+\dots|$ will be at least as big as $|b_m/2|$, while at the same time the expression $C|x-x_0|^{n+1-m}$ will be smaller than $|b_m/10|$. Since $b_m \ne 0$, it follows that $|b_m/2| < |b_m/10|$, or $1/2 < 1/10$. This contradiction shows that all the coefficients b_k are actually 0. □

Theorem 8.4.2. *Assume that the function* $f(x)$ *has* $n + 1$ *continuous derivatives on the open interval* (a, b), *and that* x_0 *is in this interval. Suppose in addition that there is a polynomial*

$$\sum_{k=0}^{n} a_k (x - x_0)^k$$

and numbers $\delta > 0$ *and* $C_0 > 0$ *such that*

$$|f(x) - \sum_{k=0}^{n} a_k (x - x_0)^k| \le C_0|x - x_0|^{n+1}, \quad x_0 < x < x_0 + \delta < b.$$

Then

$$a_k = \frac{f^{(k)}(x_0)}{k!}, \quad k = 0, \dots, n.$$

Proof. Theorem 8.3.2 tells us that for x in the interval $[x_0, x_0+\delta]$

$$|f(x) - \sum_{k=0}^{n} \frac{f^{(k)}(x_0)}{k!}(x-x_0)^k| \le C_1|x-x_0|^{n+1},$$

with

$$C_1 = \max_{x_0 \le x \le x_0+\delta} |f^{(n+1)}(x)|.$$

By the triangle inequality

$$|\sum_{k=0}^{n} a_k(x-x_0)^k - \sum_{k=0}^{n} \frac{f^{(k)}(x_0)}{k!}(x-x_0)^k|$$

$$= |[\sum_{k=0}^{n} a_k(x-x_0)^k - f(x)] + [f(x) - \sum_{k=0}^{n} \frac{f^{(k)}(x_0)}{k!}(x-x_0)^k]|$$

$$\le (C_0 + C_1)||x-x_0|^{n+1}.$$

That is, if $b_k = a_k - f^{(k)}(x_0)/k!$, the last inequality implies

$$|\sum_{k=0}^{n} b_k(x-x_0)^k| \le (C_0+C_1)|x-x_0|^{n+1}, \quad x_0 < x < x_0+\delta.$$

By Lemma 8.4.1 the coefficients b_k vanish, or

$$a_k = \frac{f^{(k)}(x_0)}{k!}, \quad k = 0, \dots, n.$$

□

This theorem may be applied to (8.2). The coefficients computed there using the geometric series are the Taylor coefficients for $\tan^{-1}(x)$ at $x_0 = 0$. In addition this analysis shows that

$$\tan^{-1}(x) = \lim_{m\to\infty} \sum_{k=0}^{m-1} (-1)^k \frac{x^{2k+1}}{2k+1}, \quad \text{if} \quad |x| \le 1.$$

8.4.2 The binomial series

Consider the Taylor series centered at $x_0 = 0$ for the function

$$f(x) = (1+x)^\alpha, \quad \alpha \in \mathbb{R}.$$

This function has derivatives of all orders at $x_0 = 0$, with

$$f^{(1)}(x) = \alpha(1+x)^{\alpha-1}, \quad f^{(2)}(x) = \alpha(\alpha-1)(1+x)^{\alpha-2},$$

and by induction

$$f^{(k)}(x) = \alpha(\alpha-1)\cdots(\alpha-[k-1])(1+x)^{\alpha-k}.$$

The Taylor polynomials centered at $x_0 = 0$ for this function are

$$P_n(x) = \sum_{k=0}^{n} \frac{f^{(k)}(0)}{k!}x^k = \sum_{k=0}^{n} \alpha(\alpha-1)\cdots(\alpha-[k-1])\frac{x^k}{k!}.$$

The basic question is, for which values of x does $R_n(x) \to 0$ as $n \to \infty$? This problem will be treated by two methods: first with Lagrange's form for the remainder, and second with the original integral form.

The Lagrange form for the remainder in this case is

$$R_n(x) = f^{(n+1)}(c)\frac{x^{n+1}}{(n+1)!} \tag{8.3}$$

$$= \frac{\alpha(\alpha-1)\cdots(\alpha-n)}{(n+1)!}(1+c)^{\alpha-n-1}x^{n+1},$$

where c lies between 0 and x. The main challenge in estimating the remainder is to understand the factor

$$\frac{\alpha(\alpha-1)\cdots(\alpha-n)}{(n+1)!} = \frac{\alpha}{1}\frac{(\alpha-1)}{2}\cdots\frac{(\alpha-n)}{(n+1)}.$$

Notice that

$$\lim_{n\to\infty}\frac{\alpha-n}{n+1} = \lim_{n\to\infty}\frac{\alpha}{n+1} - \frac{n}{n+1} = -1.$$

This means that for any $\epsilon > 0$ there is an M such that

$$|\frac{(\alpha-n)}{n+1}| < 1+\epsilon \quad \text{whenever} \quad n \ge M.$$

Going back to the more complex expression, if $n \ge M$ then

$$\frac{\alpha}{1}\frac{(\alpha-1)}{2}\cdots\frac{(\alpha-n)}{(n+1)}$$

$$= \Big[\frac{\alpha}{1}\frac{(\alpha-1)}{2}\cdots\frac{(\alpha-[M-1])}{M}\Big]\Big[\frac{(\alpha-M)}{(M+1)}\cdots\frac{(\alpha-n)}{(n+1)}\Big].$$

The term

$$\frac{\alpha}{1}\frac{(\alpha-1)}{2}\cdots\frac{(\alpha-[M-1])}{M}$$

is just some complicated constant, which we can call c_M. Thus

$$|\frac{\alpha}{1}\frac{(\alpha-1)}{2}\cdots\frac{(\alpha-n)}{(n+1)}|$$

$$\leq c_M(1+\epsilon)^{n-M} = [c_M(1+\epsilon)^{-M}](1+\epsilon)^n.$$

Again the term

$$c_M(1+\epsilon)^{-M}$$

is an awkward constant, which will be denoted C_M. The bottom line of this analysis is that for any $\epsilon > 0$ there is a constant C_M such that

$$\left|\frac{\alpha(\alpha-1)\cdots(\alpha-n)}{(n+1)!}\right| \leq C_M(1+\epsilon)^n.$$

The other factor in the remainder (8.3) was

$$(1+c)^{\alpha-n-1}x^{n+1} = (1+c)^{\alpha}(1+c)^{-n-1}x^{n+1}, \tag{8.4}$$

where c is between 0 and x. Notice that if $x \geq 0$ then $c \geq 0$ and

$$(1+c)^{-n-1} \leq 1.$$

The part $(1+c)^{\alpha}$ is another of those awkward constants.

Let's put everything together. If $x \geq 0$, then

$$|R_n(x)| = \left|\frac{\alpha(\alpha-1)\cdots(\alpha-n)}{(n+1)!}(1+c)^{\alpha-n-1}x^{n+1}\right|$$

$$\leq C(1+\epsilon)^n x^{n+1}.$$

Suppose that $0 \leq x < 1$. Pick $\epsilon > 0$ so that

$$0 \leq (1+\epsilon)x < 1.$$

Then $|R_n(x)| \to 0$ as $n \to \infty$ for $0 \leq x < 1$ since

$$(1+\epsilon)^n x^{n+1} = [(1+\epsilon)x]^n x \to 0.$$

The case $x < 0$ remains untreated. This can be partially rectified using (8.4), but a better result can be obtained by going back to the original integral form of the remainder. Since $x_0 = 0$,

$$R_n(x) = \int_0^x \frac{(x-t)^n}{n!} f^{(n+1)}(t)\, dt$$

$$= \int_0^x \frac{(x-t)^n}{n!} \alpha(\alpha-1)\cdots(\alpha-n)(1+t)^{\alpha-n-1}\, dt$$

$$= \frac{\alpha(\alpha-1)\cdots(\alpha-n)}{(n+1)!}\frac{n+1}{n}\int_0^x (x-t)^n(1+t)^{\alpha-n-1}\, dt.$$

Since $(n+1)/n \leq 2$ if $n \geq 1$, and the term $\alpha(\alpha-1)\cdots(\alpha-n)/(n+1)!$ is as before, the only new piece is

$$\int_0^x (x-t)^n(1+t)^{\alpha-n-1}\, dt = \int_0^x \left(\frac{x-t}{1+t}\right)^n (1+t)^{\alpha-1}\, dt.$$

It will be important to understand the function

$$|\frac{x-t}{1+t}|.$$

Let's focus on the case $-1 < x \le t \le 0$. Then

$$|\frac{x-t}{1+t}| = |x||\frac{1-t/x}{1+t}| = |x|\frac{1-t/x}{1-|t|}.$$

Since $|x| < 1$,

$$|t| \le t/x \le 1,$$

and

$$|\frac{x-t}{1+t}| \le |x|. \tag{8.5}$$

This last inequality also holds if $0 \le t \le x < 1$.

Using (8.5) in the remainder formula gives

$$|R_n(x)| = |\frac{\alpha(\alpha-1)\cdots(\alpha-n)}{(n+1)!}||\frac{n+1}{n}||\int_0^x \Big(\frac{x-t}{1+t}\Big)^n (1+t)^{\alpha-1}\, dt|$$

$$\le C(1+\epsilon)^n|x|^n|\int_0^x (1+t)^{\alpha-1}\, dt|.$$

Again picking ϵ so small that $|(1+\epsilon)x| < 1$, and noting that the last integral is independent of n, we have $\lim_{n\to\infty}|R_n| = 0$.

After all that work let's celebrate our success with a small theorem.

Theorem 8.4.3. *Suppose that R_n is the error made in approximating the function $(1+x)^\alpha$ by the partial sum of the binomial series*

$$P_n = \sum_{k=0}^n \alpha(\alpha-1)\cdots(\alpha-[k-1])\frac{x^k}{k!}.$$

Then as long as $-1 < x < 1$ we have

$$\lim_{n\to\infty}|R_n| = 0.$$

8.5 Problems

8.1. *Find the Taylor series for* e^x *based at* $x_0 = 1$.

8.2. *Find the Taylor series for* $\sin(x)$ *based at* $x_0 = \pi$.

8.3. *Find the Taylor series for* $\log(1 + x)$ *based at* $x_0 = 0$.

8.4. *Show by induction that if*

$$g(x) = \sum_{k=0}^{\infty} c_k (x - x_0)^k$$

then term by term differentiation gives

$$g^{(n)}(x) = \sum_{k=n}^{\infty} k(k-1)\cdots(k-(n-1))c_k(x-x_0)^{k-n} = \sum_{k=n}^{\infty} \frac{k!}{(k-n)!} c_k(x-x_0)^{k-n}.$$

Now evaluate at $x = x_0$ *to get a formula for* c_n.

8.5. *For those with some exposure to complex numbers, show by formal manipulation of power series that if* $i^2 = -1$, *then*

$$e^{ix} = \cos(x) + i\sin(x).$$

(Hint: find the Taylor series based at $x_0 = 0$.*)*

8.6. *a) Suppose you are trying to calculate*

$$y(x) = \frac{\sin(x) - x}{x} \tag{8.6}$$

on a computer which stores 12 digits for each number. If the calculation of $y(x)$ *is carried out as indicated, for which values of* x *(approximately) will the computer tell you that* $y(x) = 0$.

b) Find the Taylor polynomial $P_5(x)$ *of degree 5 for* $\sin(x)$ *based at* $x_0 = 0$. *How much can you improve the calculation of* (8.6) *if you use* $P_5(x)$ *and carry out the division symbolically?*

8.7. *Use Theorem 8.2.1 to show that if a function* $f(x)$ *has* $n+1$*-st derivative equal to* 0 *at every point of the interval* (a, b), *then* f *is a polynomial of degree at most* n *on* (a, b).

8.8. *Use Theorem 8.2.1 to show that if* $p(x)$ *and* $q(x)$ *are two polynomials of degree* n *with*

$$p^{(k)}(0) = q^{(k)}(0), \quad k = 0, \dots, n,$$

then $p(x) = q(x)$ *for all values of* x.

8.9. *Show that if $p(x)$ is a polynomial of degree at most n,*

$$p(x) = \sum_{k=0}^{n} c_k x^k,$$

then for any real number x_0 this polynomial may also be written in the form

$$p(x) = \sum_{k=0}^{n} a_k (x - x_0)^k.$$

Can you express a_k in terms of the coefficients $c_0, \dots, c_n$?

8.10. *Suppose that $p(x)$ is a polynomial of degree at most n, and $p^{(k)}(x_0) = 0$ for $k = 0, \dots, m-1$. Show that there is another polynomial $q(x)$ such that*

$$p(x) = (x - x_0)^m q(x).$$

8.11. *Using the Taylor series remainder, how many terms of the Taylor series for e^x centered at $x_0 = 0$ should you use to compute the number e with an error at most 10^{-12}.*

8.12. *Compare the number of terms of the Taylor series needed to compute $\sin(13)$ with an accuracy of 10^{-6} if the centers are $x_0 = 0$ and $x_0 = 4\pi$.*

8.13. *Use the result of Theorem 8.3.3 to give a second proof of Theorem 8.3.2.*

8.14. *From calculus we have*

$$\log(1 + x) = \int_0^x \frac{1}{1+t}\, dt.$$

(a) Use the partial sums of the geometric series to obtain a Taylor-like formula, with remainder, for $\log(1+x)$. For which values of x can you show that $|R_n| \to 0$ as $n \to \infty$. (Be sure to check $x = 1$. The polynomials are the Taylor polynomials.)

(b) Use the algebraic identity

$$\frac{1}{10 + x} = \frac{1}{10} \frac{1}{1 + x/10}$$

and the method of part (a) to obtain a Taylor-like series with remainder for $\log(10 + x)$. Again determine the values of x for which $|R_n| \to 0$ as $n \to \infty$.

8.15. *Find the Taylor-like series, with remainder, for*

$$f(x) = \log\left(\frac{1+x}{1-x}\right)$$

centered at $x_0 = 0$. (Hint: Use algebraic manipulations.) Assuming that this series is the Taylor series, what is $f^{(11)}(0)$?

8.16. *Consider approximating the function $(1+x)^\alpha$ by the partial sum of the binomial series*

$$\sum_{n=0}^{N} \alpha(\alpha-1)\cdots(\alpha-[n-1])\frac{x^n}{n!}.$$

Use (8.4) to show that Lagrange's form of the remainder does imply

$$\lim_{N\to\infty} |R_N| = 0 \quad \text{if} \quad -1/2 < x \le 0.$$

8.17. *Beginning with the Taylor series with remainder for e^x centered at $x_0 = 0$, then taking $x = -z^2$, find an expression for the Taylor series with remainder for e^{-z^2}. Use this expression to find*

$$\int_0^x e^{-z^2}\,dz, \quad 0 \le x \le 1,$$

with an error no greater than 10^{-3}.

8.18. *If $|x| < 1$ the function $\ln(1+x)$ may be written as the following series:*

$$\ln(1+x) = \sum_{k=1}^{\infty} (-1)^{k+1}\frac{x^k}{k}.$$

You may assume that the sequence of partial sums

$$s_n = \sum_{k=1}^{n} (-1)^{k+1}\frac{x^k}{k}$$

has a limit s. If $0 \le x < .1$, how big should you take n to make sure that

$$|s_n - s| < 10^{-10}?$$

What is the answer to the same question if $x = .5$? (Your reasoning is more important than the specific number.)

8.19. *Prove Theorem 8.3.1.*

8.20. *Use Taylor's Theorem to show that*

$$\lim_{x\to\infty} \frac{x^n}{e^x} = 0, \quad n = 0, 1, 2, \ldots.$$

8.21. *Show that*

$$f(x) = \begin{Bmatrix} 0, & x \le 0, \\ \exp(-1/x^2), & x > 0, \end{Bmatrix}$$

has derivatives of all orders for every $x \in \mathbb{R}$. Show in particular that $f^{(k)}(0) = 0$ for $k = 0, 1, 2, \ldots$.

Conclude that the function $g(x) = f(x) + f(-x)$ is strictly positive except at $x = 0$, but the coefficients of the Taylor series for g with center 0 are all zero.

9

Uniform convergence

9.1 Introduction

The representation of functions by Taylor series provides a powerful tool for reducing certain computations like the evaluation of integrals to simple algebraic operations. As an example, consider the recalcitrant problem of finding

$$\int_0^x \frac{\sin(t)}{t} \, dt.$$

The series formula

$$\sin(t) = t - t^3/3! + t^5/5! + \dots,$$

suggests

$$\int_0^x \frac{\sin(t)}{t} \, dt = \int_0^x \frac{t - t^3/3! + t^5/5! + \dots}{t} \, dt = \int_0^x 1 - t^2/3! + t^4/5! + \dots \, dt$$

$$= x - \frac{x^3}{3(3!)} + \frac{x^5}{5(5!)} + \dots.$$

When such a series is generated from a familiar function there is a good chance that the Taylor series remainders $R_n(x)$ can be estimated. Remainder estimates can justify the use of truncated series for practical computations. New problems arise when there is no known function $f(x)$ in the background.

As a concrete example, consider the functions which satisfy Airy's differential equation

$$\frac{d^2y}{dx^2} - xy = 0. \tag{9.1}$$

Consider a solution described by a power series (see Section 3.4),

$$y(x) = \sum_{k=0}^{\infty} a_k x^k.$$

Calculating derivatives formally, the equation (9.1) implies

$$\sum_{k=0}^{\infty} k(k-1)a_k x^{k-2} - \sum_{k=0}^{\infty} a_k x^{k+1} = 0. \tag{9.2}$$

Collecting terms for the first few powers of x produces

$$2a_2 + [3 \cdot 2a_3 - a_0]x + [4 \cdot 3a_4 - a_1]x^2 + [5 \cdot 4a_5 - a_2]x^3 + \cdots = 0.$$

If the series were a polynomial with only finitely many terms the coefficients for each x^k would be zero. Assuming the same holds for power series, the first few resulting equations are

$$2a_2 = 0, \quad 3 \cdot 2a_3 - a_0 = 0, \quad 4 \cdot 3a_4 - a_1 = 0, \quad 5 \cdot 4a_5 - a_2 = 0.$$

It appears that the coefficients $a_0 = y(0)$ and $a_1 = y'(0)$ are unconstrained, but the remaining coefficients will be determined. In fact (9.2) may be rewritten as

$$2a_2 + \sum_{j=1}^{\infty} [(j+2)(j+1)a_{j+2} - a_{j-1}]x^j = 0. \tag{9.3}$$

Setting the coefficients of each power of x to 0 leads to the recursion relations

$$(j+2)(j+1)a_{j+2} - a_{j-1} = 0, \quad j = 1, 2, 3, \dots.$$

This is equivalent to

$$a_{m+3} = \frac{a_m}{(m+3)(m+2)}, \quad m = 0, 1, 2, \dots. \tag{9.4}$$

Since $a_2 = 0$, once a_0 and a_1 are fixed, then all of the other coefficients are determined by the relations (9.4).

More explicit formulas for the a_k are available. The relation (9.4) together with $a_2 = 0$ shows that $a_{3k+2} = 0$ for $k = 0, 1, 2, \dots$. In addition,

$$a_3 = \frac{a_0}{3 \cdot 2}, \quad a_6 = \frac{a_3}{6 \cdot 5} = \frac{a_0}{6 \cdot 5 \cdot 3 \cdot 2},$$

and

$$a_4 = \frac{a_1}{4 \cdot 3}, \quad a_7 = \frac{a_4}{7 \cdot 6} = \frac{a_1}{7 \cdot 6 \cdot 4 \cdot 3}.$$

An induction argument can then be used to show that

$$a_{3k} = \frac{a_0}{2 \cdot 3 \cdot 5 \cdot 6 \cdots (3k-1) \cdot (3k)}, \tag{9.5}$$

$$a_{3k+1} = \frac{a_1}{3 \cdot 4 \cdot 6 \cdot 7 \cdots (3k) \cdot (3k+1)},$$

$$a_{3k+2} = 0.$$

This computation puts us in a strange situation. The idea of using a power series to look for solutions of (9.1) seems very successful. There is a unique power series solution for every choice of $y(0)$ and $y'(0)$. On the other hand, it is not clear if the function represented by the power series has the required number of derivatives, or is even continuous.

Let's expand on this point. Our earliest treatment of power series (see Section 3.4) considered convergence, but not whether the sum of the series produced a continuous or differentiable function. Taylor's Theorem provides more explicit information about the convergence of some power series; when Taylor series were computed for elementary functions $f(x)$, the series was expected to converge to $f(x)$. In contrast, the power series $y(x)$ coming from (9.2) has no explicit target function $f(x)$, so a different approach is required to determine its continuity and differentiability.

9.2 Uniform Convergence

Power series are one of several methods for representing functions $f(x)$ by a sequence or series of elementary functions. The elementary functions often have good properties, like ease of integration or differentiation. Hopefully, this ease of computation will extend to the limit function. The uniform convergence property is often a critical tool in such a process.

Suppose ι is an interval in the real line $\iota \subset \mathbb{R}$. If $\{f_n : \iota \to \mathbb{R}\}$ is a sequence of functions, there are several ways to describe the convergence of $\{f_n\}$ to a limit function f. Say that $\{f_n\}$ *converges pointwise* to f if $\lim_{n\to\infty} f_n(x) = f(x)$ for each $x \in \iota$.

Pointwise convergence is a rather weak notion of convergence, as the next example shows. Take $\iota = [0,1]$ and $f_n(x) = x^n$. Convergence is tested for each x independently. If $x = 1$, then $f_n(1) = 1$ and the sequence converges trivially to 1. For x strictly less than one,

$$\lim_{n\to\infty} f_n(x) = 0, \quad 0 \le x < 1.$$

This sequence of functions does converge pointwise, to the function

$$f(x) = \left\{\begin{matrix} 0, & 0 \le x < 1, \\ 1, & x = 1. \end{matrix}\right\}$$

Although the functions x^n have derivatives of all orders, and the sequence converges pointwise to a function $f(x)$, the limit function is not even continuous.

Uniform convergence is a much more satisfactory notion of convergence, as our analysis will soon show. For functions $f_n : \iota \to \mathbb{R}$, the sequence $\{f_n\}$ *converges uniformly* to $f : \iota \to \mathbb{R}$ if for every $\epsilon > 0$ and every $x \in \iota$ there is an N (depending on ϵ) such that

$$|f(x) - f_n(x)| < \epsilon, \quad \text{for all } n \ge N, \quad \text{and all } x \in \iota.$$

Speaking informally, $f_n(x)$ should be within the "ϵ -tube" around $f(x)$ for all $n \ge N$. (See the tube in Figure 9.1.)

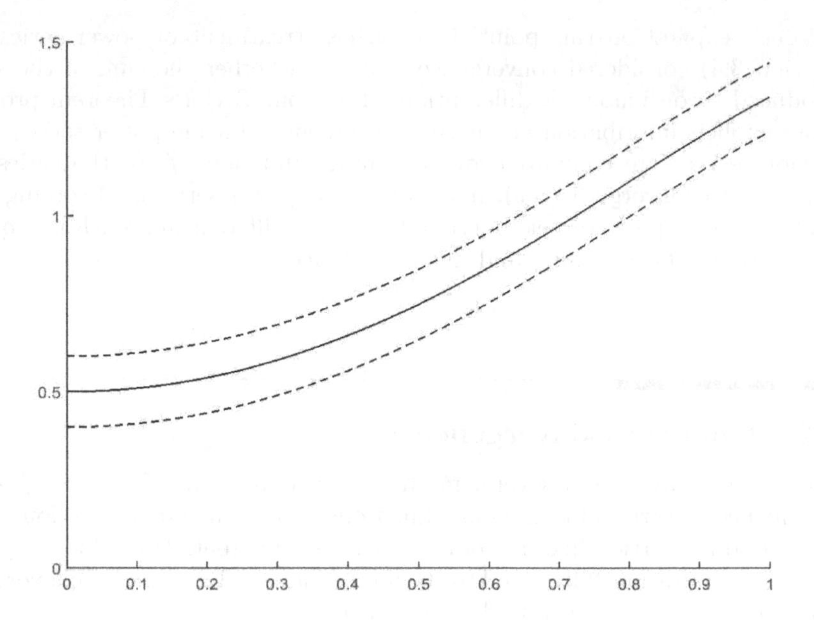

FIGURE 9.1
An ϵ- tube around f.

It is worth returning to the example $f_n(x) = x^n$ on $\iota = [0, 1]$ to clarify the distinction between uniform convergence and pointwise convergence. Suppose $0 < \beta < 1$. On the interval $[0, \beta]$ the sequence $\{f_n\}$ converges to $f(x) = 0$. The functions x^n are increasing on $[0, \beta]$, so

$$|f_n(x) - f(x)| = |x^n - 0| \le \beta^n.$$

For any $\epsilon > 0$ there is an N such that $\beta^n < \epsilon$ whenever $n \ge N$. This implies

$$|f_n(x) - f(x)| \le \beta^n < \epsilon, \quad n \ge N, \quad x \in [0, \beta].$$

On the interval $\iota = [0, 1]$ this uniform estimation breaks down. For any $\epsilon > 0$ and any N, there are values of x close to 1 such that $|x^N - 0| > \epsilon$ since $\lim_{x \to 1} x^N = 1$.

The convergence of an infinite series of functions $\sum_{k=0}^{\infty} f_k(x)$ is defined in the same way as convergence of a series of numbers. First define the sequence of partial sums

$$S_n(x) = \sum_{k=0}^{n-1} f_k(x).$$

Say that the series $\sum f_k$ converges pointwise to $f : \iota \to \mathbb{R}$ if the sequence $S_n(x) = \sum_{k=0}^{n-1} f_k(x)$ *converges pointwise* to $S(x)$. The series $\sum f_k$ *converges uniformly* to $S : \iota \to \mathbb{R}$ if the sequence $S_n(x) = \sum_{k=0}^{n-1} f_k(x)$ converges uniformly to $S(x)$ on ι.

The geometric series $\sum_{k=0}^{\infty} x^k$ is an illustrative example. Recall that

$$S_n(x) = \sum_{k=0}^{n-1} x^k = \left\{ \begin{matrix} \frac{1-x^n}{1-x}, & x \ne 1, \\ n, & x = 1. \end{matrix} \right\}.$$

For $|x| < 1$ the sequence of partial sums has a limit, $S_n(x) \to S(x) = 1/(1-x)$. Since the difference between the limit function and the partial sums is

$$|S_n(x) - S(x)| = |\frac{1-x^n}{1-x} - \frac{1}{1-x}| = |\frac{x^n}{1-x}|,$$

our previous discussion of the sequence x^n shows that $S_n(x)$ converges uniformly to $S(x)$ on any interval $-\beta < x < \beta$ if $0 \le \beta < 1$, while the convergence is pointwise, but not uniform, on the interval $-1 < x < 1$.

Another example is the series $S(x) = \sum_{k=1}^{\infty} \sin(kx)/k^2$. Since $|\sin(kx)| \le 1$, the summands satisfy

$$|\sin(kx)/k^2| \le 1/k^2,$$

implying that the series converges absolutely for all $x \in \mathbb{R}$. The estimation

$$|S_n(x) - S(x)| = |\sum_{k=n+1}^{\infty} \frac{\sin(kx)}{k^2}| \le |\sum_{k=n+1}^{\infty} \frac{1}{k^2}|,$$

suggests that the series converges uniformly for all $x \in \mathbb{R}$. The next result confirms this.

Theorem 9.2.1. *(Weierstrass M-test) Suppose $\{f_k : \iota \to \mathbb{R}\}$ is a sequence of functions satisfying $|f_k(x)| \le c_k$ for all $x \subset \iota$. If $\sum_k c_k$ converges, then*

$$\sum_{k=1}^{\infty} f_k(x), \tag{9.6}$$

converges uniformly on ι.

Proof. Since $|f_k(x)| \le c_k$ and $\sum_k c_k$ converges, the series (9.6) converges absolutely for each $x \in \iota$ by the comparison test (Theorem 3.2.2). Since absolute convergence implies convergence, the series (9.6) converges pointwise to a function

$$s(x) = \sum_{k=1}^{\infty} f_k(x).$$

Let

$$s_n(x) = \sum_{k=1}^{n} f_k(x), \quad \sigma_n = \sum_{k=1}^{n} c_k,$$

be the partial sums of the series, with $\sigma = \lim_{n\to\infty} \sigma_n$. For any $\epsilon > 0$ there is an N such that $m \ge N$ implies $|\sigma - \sigma_m| < \epsilon/2$. If $n > m \ge N$ then

$$|\sigma_n - \sigma_m| = \big|[\sigma_n - \sigma] + [\sigma - \sigma_m]\big| \le |\sigma_n - \sigma| + |\sigma - \sigma_m| < \epsilon.$$

This now implies that

$$|s_n(x) - s_m(x)| = \Big|\sum_{k=m+1}^{n} f_k(x)\Big| \le \sum_{k=m+1}^{n} c_k = \sigma_n - \sigma_m < \epsilon,$$

or, fixing m,

$$|s(x) - s_m(x)| \le \epsilon, \quad m \ge N.$$

Since N is independent of x, the sequence $s_m(x)$ converges uniformly to $s(x)$ as desired. □

In the example above, the function $S(x) = \sum_{k=1}^{\infty} \sin(kx)/k^2$ is not presented as a familiar function, so its continuity or differentiability is not immediately obvious. The next result provides a partial rectification.

Theorem 9.2.2. *Suppose $f_n : \iota \to \mathbb{R}$ is a sequence of continuous functions converging uniformly on ι to a function $f(x)$. Then $f : \iota \to \mathbb{R}$ is continuous.*

Proof. Suppose $\epsilon > 0$. Since the convergence is uniform, there is an N_1 such that

$$|f_n(x) - f(x)| < \epsilon/3, \quad n \ge N_1, \quad x \in \iota.$$

Let x_0 be any point of ι. Fix $M \ge N_1$. Since f_M is continuous, there is a $\delta > 0$ such that

$$|f_M(x) - f_M(x_0)| < \epsilon/3, \quad |x - x_0| < \delta.$$

Finally, using this δ we have

$$|f(x) - f(x_0)| = \Big|[f(x) - f_M(x)] + [f_M(x) - f_M(x_0)] + [f_M(x_0) - f(x_0)]\Big|$$

$$\le |f(x) - f_M(x)| + |f_M(x) - f_M(x_0)| + |f_M(x_0) - f(x_0)| < \epsilon.$$

Thus f is continuous at each $x_0 \in \iota$. □

As a corollary, if a series of continuous functions converges uniformly, then the limit function is continuous. This result is often used for power series

$$f(x) = \sum_k a_k (x - x_0)^k,$$

or series of trigonometric functions

$$g(x) = \sum_k b_k \cos(kx).$$

The next results show that uniform convergence and integration play well together.

Theorem 9.2.3. *Suppose $f_n : [a,b] \to \mathbb{R}$ are Riemann integrable functions converging uniformly to $f : [a,b] \to \mathbb{R}$. Then f is integrable and*

$$\lim_{n\to\infty} \int_a^b f_n(x)\, dx = \int_a^b f(x)\, dx.$$

Proof. Pick $\epsilon > 0$ and n so large that

$$|f(x) - f_n(x)| < \epsilon.$$

Then pick a partition $\mathcal{P}$ of $[a,b]$ such that the difference between the upper and lower sums for f_n is less than ϵ,

$$U(f_n, \mathcal{P}) - L(f_n, \mathcal{P}) < \epsilon.$$

Since f and f_n are everywhere close, the upper and lower sums are close:

$$U(f, \mathcal{P}) - L(f, \mathcal{P}) = U(f_n, \mathcal{P}) + [U(f, \mathcal{P}) - U(f_n, \mathcal{P})]$$

$$-L(f_n, \mathcal{P}) - [L(f, \mathcal{P}) - L(f_n, \mathcal{P})] \le \epsilon + 2[b-a]\epsilon.$$

The result follows by Lemma 6.2.2. □

Applying this result to a series of functions produces an immediate corollary.

Corollary 9.2.4. *Suppose $[a,b]$ is a bounded interval and $g_n : [a,b] \to \mathbb{R}$ are Riemann integrable functions. If $\sum_n g_n(x)$ converges uniformly to $g(x)$, then*

$$\sum_{n=1}^{\infty} \int_a^b g_n(x)\, dx = \int_a^b g(x)\, dx = \int_a^b [\sum_{n=1}^{\infty} g_n(x)]\, dx.$$

In general, uniform convergence of a sequence of functions is not sufficient to imply convergence of derivatives. As a simple example, consider the sequence

$$f_n(x) = \frac{\sin(n^2 x)}{n}, \quad -\infty < x < \infty.$$

This sequence obviously converges uniformly to 0. However

$$f_n'(x) = n\cos(n^2 x).$$

The sequence of derivatives diverges at $x = 0$. Another problem arises because uniform convergence of derivatives $g_n'(x)$ is not sufficient to imply convergence of the functions $g_n(x)$. The simple example $g_n(x) = n$ for $x \in \mathbb{R}$ illustrates the problem. Here is a positive result.

Theorem 9.2.5. *Suppose the functions $f_n : (a,b) \to \mathbb{R}$ are continuously differentiable, and $f_n' \to g$ uniformly on (a,b). Further assume that $f_n(c) \to A$ for some $c \in (a,b)$. Then f_n converges uniformly to a function $f : (a,b) \to \mathbb{R}$. The function f is continuously differentiable, and $f' = g$.*

Proof. The function g is continuous since it is the uniform limit of a sequence of continuous functions.

By the Fundamental Theorem of Calculus,

$$\int_c^x f_n'(t)\, dt = f_n(x) - f_n(c).$$

If we define $f(c) = A$, then Theorem 9.2.3 tells us that

$$f(x) = A + \lim_n \int_c^x f_n'(t))\, dt = A + \int_c^x g(t)\, dt.$$

Differentiation gives $f' = g$. Moreover,

$$|f(x) - f_n(x)| = |\int_c^x f'(t) - f_n'(t)\, dt|$$

$$\le \int_c^x |f'(t) - f_n'(t)|\, dt \le (b-a) \sup_{a<t<b} |f'(t) - f_n'(t)|,$$

so $\{f_n\}$ converges uniformly to f. □

9.3 Convergence of power series

Uniform convergence results may be applied to power series; the consequences are surprisingly strong. It will help to extract a bit more information in Theorem 9.3.3.

Theorem 9.3.1. *Suppose the series*

$$\sum_{n=0}^{\infty} c_n(x - x_0)^n$$

converges for $x = x_1 \ne x_0$. *Then the series converges uniformly on each interval*

$$x_0 - r \le x \le x_0 + r, \quad 0 \le r < |x_1 - x_0|.$$

Also, for each $0 \le r < |x_1 - x_0|$, *there is a number* M *such that*

$$|c_n(x - x_0)^n| \le M(\frac{r}{|x_1 - x_0|})^n, \quad |x - x_0| \le r < |x_1 - x_0|.$$

Proof. Convergence of the series at x_1 implies

$$\lim_{n\to\infty} c_n(x_1 - x_0)^n \to 0.$$

Thus there is an M such that

$$|c_n(x_1 - x_0)^n| \le M.$$

For $|x - x_0| < |x_1 - x_0|$ we have

$$|c_n(x - x_0)^n| = |c_n(x_1 - x_0)^n| \frac{|x - x_0|^n}{|x_1 - x_0|^n} \le M \frac{|x - x_0|^n}{|x_1 - x_0|^n},$$

and the series converges uniformly on $|x - x_0| \le r < |x_1 - x_0|$ by comparison with the geometric series.

□

Lemma 9.3.2. *Suppose the series*

$$f(x) = \sum_{n=0}^{\infty} c_n(x - x_0)^n$$

converges for $|x - x_0| < R$ for some $R > 0$. Then $f(x)$ and $f'(x)$ are continuous on $|x - x_0| < R$ and

$$f'(x) = \sum_{n=1}^{\infty} nc_n(x - x_0)^{n-1}, \quad |x - x_0| < R.$$

Proof. Choose x_1 with $|x_1 - x_0| < R$, pick $r < R$, and assume that $|x - x_0| \le r$. Theorem 9.2.5 is applied. Let

$$s_n(x) = \sum_{k=0}^{n-1} c_k(x - x_0)^k, \quad s_n'(x) = \sum_{k=1}^{n-1} kc_k(x - x_0)^{k-1}.$$

The sequence s_n converges for $|x - x_0| < R$, in particular for $x = x_1$. For the derivative series,

$$|kc_k(x - x_0)^{k-1}| \le k|c_k|r^{k-1} = \frac{k}{r} M \frac{r^n}{|x_1 - x_0|^n}.$$

Let

$$u_n = \frac{k}{r} M \frac{r^n}{|x_1 - x_0|^n}.$$

The series $\sum u_n$ converges by the ratio test, so the sequence $s_n'(x)$ converges uniformly by the M-test, and

$$f'(x) = \sum_{n=1}^{\infty} nc_n(x - x_0)^{n-1}, \quad |x - x_0| < R.$$

□

It is now straightforward to establish the next result.

Theorem 9.3.3. *Suppose the series*

$$f(x) = \sum_{n=0}^{\infty} c_n (x - x_0)^n$$

converges for $|x - x_0| < R$. *Then*

(i) $f(x)$ *has derivatives of all orders on* $|x - x_0| < R$, *and these may be computed by termwise differentiation of the series,*

(ii) if

$$F(x) = \int_{x_0}^{x} f(t)\, dt, \quad |x - x_0| < R,$$

then F *is given by termwise integration of the series for* f,

(iii) the constants c_n *are*

$$c_n = \frac{f^{(n)}(x_0)}{n!}.$$

Proof. For (i), repeatedly apply the previous lemma. Part (ii) just uses the result on integration for uniformly convergent sequences of functions. Finally, (iii) follows by evaluation of the derivatives at $x = x_0$.

□

9.4 The Weierstrass Approximation Theorem

Taylor's Theorem and the subsequent study of power series have provided us with a large collection of functions $f(x)$ which are well approximated by polynomials. But this approach assumes that $f(x)$ has derivatives of all orders, and in many cases the approximation by polynomials is only accurate in a small neighborhood of the center. In the late nineteenth century, Karl Weierstrass (1815–1897) took up the question of uniform approximation of a continuous function $f : [0,1] \to \mathbb{R}$ by polynomials. That is, given $f(x)$ and any $\epsilon > 0$, is there a polynomial $p(x)$ such that

$$|f(x) - p(x)| < \epsilon, \quad 0 \le x \le 1?$$

His positive result was published in 1885.

There are several distinct proofs of the Weierstrass Approximation Theorem. In addition to the method presented below [7, p. 146–148], there is an interesting alternative in [6, p. 305–306]. There are two important ideas used in the following proof.

Suppose first that a function $f : [0,1] \to \mathbb{R}$ and polynomials $p_n(x)$ are given. The Binomial Theorem Theorem 1.2.1 describes how to express $(t-x)^k$, $k = 0, 1, 2, \dots$ as a polynomial in x with coefficients depending on t. This observation extends to the functions $p_n(t-x)$. As a result, an integral of the form

$$q_n(x) = \int_0^1 f(t)p_n(t-x)\,dt$$

will produce from $f(t)$ a polynomial $q_n(x)$.

The second idea is that $p_n(x)$ should be positive, have integral 1, and be highly concentrated near $x = 0$. Then for each x the value of $q_n(x)$ will be a weighted average of the values of $f(t)$, with most of the weight concentrated near $t = x$. That is, $q_n(x)$ should be approximately $f(x)$ at each $x \in [0, 1]$.

Theorem 9.4.1. *Suppose $f : [0, 1] \to \mathbb{R}$ is continuous. For any $\epsilon > 0$ there is a polynomial $p_\epsilon(x)$ such that*

$$|f(x) - p_\epsilon(x)| < \epsilon, \quad 0 \le x \le 1. \tag{9.7}$$

Proof. The proof starts with two simplifying assumptions. First, assume that $f(0) = f(1) = 0$. If $g(x) : [0, 1] \to \mathbb{R}$ is any continuous function, then the function $f(x) = g(x) - g(0) - [g(1) - g(0)]x$ vanishes at $x = 0$ and $x = 1$. If $p_\epsilon(x)$ is a polynomial satisfying (9.7), then $q(x) = p_\epsilon(x) + g(0) + [g(1) - g(0)]x$ is also a polynomial, with

$$|g(x) - q(x)| = |f(x) - p_\epsilon(x)| < \epsilon, \quad 0 \le x \le 1.$$

Thus the assumption that $f(0) = f(1) = 0$ can be made without any loss of generality.

As a continuous function on $[0, 1]$, $f(x)$ is uniformly continuous. For the second assumption, suppose that $f(x)$ is extended to all $x \in \mathbb{R}$ by defining $f(x) = 0$ for $x \notin [0, 1]$. The extended function is uniformly continuous on $\mathbb{R}$.

The polynomials $r_n(x) = (1 - x^2)^n$ are positive if $-1 \le x \le 1$, with $r_n(0) = 1$. In addition, for each δ with $0 < \delta < 1$, $r_n(x)$ converges to zero uniformly if $\delta \le |x| \le 1$. By taking

$$c_n = \int_{-1}^1 (1 - x^2)^n\,dx, \quad p_n(x) = \frac{(1 - x^2)^n}{c_n}, \tag{9.8}$$

the desired property

$$\int_{-1}^1 p_n(x)\,dx = 1,$$

is obtained.

It is easy to check that

$$c_n = \int_{-1}^1 (1 - x^2)^n\,dx = 2\int_0^1 (1 - x^2)^n\,dx \ge 2\int_0^1 (1 - x)^n\,dx = \frac{2}{n+1}.$$

For $\delta > 0$ and $\delta \le |x| \le 1$ this implies

$$p_n(x) \le (n+1)(1-\delta^2)^n/2, \tag{9.9}$$

so the renormalized polynomials $p_n(x)$ still converge to zero uniformly in this region.

Now define the polynomials

$$q_n(x) = \int_0^1 f(t)p_n(t-x)\,dt.$$

The substitution $u = t - x$ yields

$$q_n(x) = \int_{-x}^{1-x} f(u+x)p_n(u)\,du,$$

and since $f(x) = 0$ for $x \notin [0,1]$ this is the same as

$$q_n(x) = \int_{-1}^{1} f(t+x)p_n(t)\,dt.$$

Since $f(x)$ is uniformly continuous for $x \in \mathbb{R}$, for any $\epsilon > 0$ there is a $\delta > 0$ such that $|f(x_2) - f(x_1)| < \epsilon/2$ if $|x_2 - x_1| < \delta$. Picking an upper bound C for $|f(x)|$, and using this δ, the difference between $f(x)$ and $q_n(x)$ is estimated:

$$|q_n(x) - f(x)| = |\int_{-1}^{1} f(t+x)p_n(t)\,dt - \int_{-1}^{1} f(x)p_n(t)\,dt|$$

$$\le \int_{-1}^{1} |f(t+x) - f(x)|p_n(t)\,dt|.$$

Breaking up the interval $[-1,1]$ into three parts the estimate becomes

$$|q_n(x) - f(x)| \le 2C\int_{-1}^{-\delta} p_n(t)\,dt + \frac{\epsilon}{2}\int_{-\delta}^{\delta} p_n(t)\,dt + 2C\int_{\delta}^{1} p_n(t)\,dt$$

$$\le 2C(n+1)(1-\delta^2)^n + \frac{\epsilon}{2}.$$

For n sufficiently large $|q_n(x) - f(x)| < \epsilon$, as desired.

□

Theorem 9.4.1 says that any continuous function $f : [0,1] \to \mathbb{R}$, no matter how complicated, may be approximated as well as we like by a polynomial. This is especially remarkable when one considers another result of Weierstrass: there are continuous functions which fail to have a derivative at every point. Sketching such a function is really a challenge!

Although the proof of Theorem 9.4.1 provides an algorithm for finding a polynomial approximation to a continuous function $f(x)$, it serves more as a hunting license than a specific procedure. A number of additional questions are suggested by this result. For example, given $f(x)$ and ϵ, how would you find the polynomial $p(x)$ of minimal degree such that

$$|f(x) - p(x)| \le \epsilon, \quad 0 \le x \le 1?$$

Or, is there a good method for constructing approximating polynomials from sample values of $f(x)$ at points $0 = x_0 < x_1 < \dots < x_K = 1$? What additional information about $f(x)$ would help construct such an approximation?

One also confronts possible generalizations. Is there a corresponding theorem if f is a function of several variables? Can the class of polynomials be replaced with another convenient collection such as exponential functions $e^{\alpha_n x}$? In 1937 Marshall Harvey Stone (1903–89) discovered a remarkable generalization of Theorem 9.4.1 that answers many of these questions. In the next section we will take up the question of approximating periodic functions by functions of the form

$$g(x) = a_0 + \sum_{n=1}^{N} [a_n \cos(nx) + b_n \sin(nx)].$$

9.5 Trigonometric approximation

9.5.1 Solving a heat equation

The results on uniform convergence of series can also be applied to the solution of certain partial differential equations. A fundamental example is the model for heat flow in a thin metal ring. This problem arises in physics and engineering when a thin homogeneous metal ring has an initial temperature distribution described by $f(x)$. In this model x is the angular variable measured in radians, $-\pi \le x \le \pi$. The partial differential equation for the temperature $u(t,x)$ as a function of time and angular position is

$$u_t = k u_{xx}, \quad u(0,x) = f(x), \quad -\pi \le x \le \pi, \quad t \ge 0. \tag{9.10}$$

When x is extended beyond the interval $[-\pi, \pi]$ the geometry of the ring imposes the periodicity condition $u(t, x + 2\pi) = u(t,x)$ on solutions.

Ignoring the role of the initial temperature for the moment, the functions

$$1, e^{-kn^2 t}\cos(nx), e^{-kn^2 t}\sin(nx), \quad n = 1, 2, 3, \dots,$$

are solutions of the equation satisfying the periodicity conditions. More complex solutions

$$u(t,x) = a_0 + \sum_{n=1}^{N} e^{-kn^2 t}[a_n \cos(nx) + b_n \sin(nx)].$$

may be constructed by taking linear combinations of the initial solutions. Postponing concerns about convergence, formal solutions of this problem can be written as infinite series

$$u(t,x) = a_0 + \sum_{n=1}^{\infty} e^{-kn^2 t}[a_n \cos(nx) + b_n \sin(nx)].$$

Returning to the initial condition $u(0,x) = f(x)$, suppose

$$f(x) = a_0 + \sum_{n=1}^{\infty} [a_n \cos(nx) + b_n \sin(nx)]. \tag{9.11}$$

With some mild conditions on the coefficients, such as $\sum |a_n| < \infty$ and $\sum |b_n| < \infty$, these functions will solve our problem in the following sense. By the Weierstrass M-test the function $u(t,x)$ is continuous. For $t > 0$ all termwise derivatives in t and x converge uniformly, so $u(t,x)$ has derivatives of all orders. The equation is satisfied, again for $t > 0$.

9.5.2 Approximation by trigonometric functions

The technique used to find solutions of (9.10) has wide applicability, but also raises a problem similar to the one Weierstrass treated for polynomials using Theorem 9.4.1. Which functions $f(x)$ with period 2π can be written in the form (9.11)? A function having the form of a finite sum

$$f(x) = a_0 + \sum_{n=1}^{N} [a_n \cos(nx) + b_n \sin(nx)]$$

is said to be a *trigonometric polynomial* of degree at most N. Posing this problem another way, if $f : (-\infty, \infty) \to \mathbb{R}$ is a continuous function satisfying $f(x + 2\pi) = f(x)$, is there a sequence $\{f_N(x)\}$ of trigonometric polynomials converging uniformly to $f(x)$? By posing the approximation problem as one of uniform convergence, the method of proof used for Theorem 9.4.1 can be applied. The argument begins with some facts about trigonometric polynomials.

Lemma 9.5.1. *For* $n = 1, 2, 3, \dots$, *the functions* $\cos(x)^n$ *and* $\sin(x)^n$ *are trigonometric polynomials.*

Proof. The proof is based on the trigonometric identities (see Problem 9.3)

$$2\sin(a)\cos(b) = \sin(a+b) + \sin(a-b), \tag{9.12}$$

$$2\cos(a)\cos(b) = \cos(a+b) + \cos(a-b),$$

$$2\sin(a)\sin(b) = \cos(a-b) - \cos(a+b).$$

The argument now continues by induction. If $n = 1$ there is nothing to do. For $n \geq 1$ suppose

$$\cos^n(x) = a_0 + \sum_{m=1}^{n} [a_m \cos(mx) + b_m \sin(mx)].$$

Then

$$\cos(x)^{n+1} = \cos(x)\Big(a_0 + \sum_{m=1}^{n} [a_m \cos(mx) + b_m \sin(mx)]\Big).$$

But for each m,

$$\cos(x)\cos(mx) = \frac{1}{2}[\cos([m+1]x) + \cos([1-m]x)],$$

and

$$\cos(x)\sin(mx) = \frac{1}{2}[\sin([m+1]x) + \sin([1-m]x)],$$

so $\cos(x)^{n+1}$ is a trigonometric polynomial whenever $\cos(x)^n$ is. The argument is similar for $\sin(x)^n$.

□

Theorem 9.5.2. *Suppose $f : (-\infty, \infty) \to \mathbb{R}$ is continuous and satisfies $f(x + 2\pi) = f(x)$ for all $x \in \mathbb{R}$. For any $\epsilon > 0$ there is a trigonometric polynomial $p_\epsilon(x)$ such that*

$$|f(x) - p_\epsilon(x)| < \epsilon, \quad -\infty < x < \infty. \tag{9.13}$$

Proof. The trigonometric polynomials $r_n(x) = (\frac{1+\cos(x)}{2})^n$ are positive, with a maximum of 1 at $x = 0$. In addition these functions are even $r_n(-x) = r_n(x)$, and strictly decreasing on $[0, \pi]$, with $r_n(\pi) = 0$. If

$$c_n = \int_{-\pi}^{\pi} (\frac{1+\cos(x)}{2})^n \, dx, \quad p_n(x) = \frac{1}{c_n}(\frac{1+\cos(x)}{2})^n, \tag{9.14}$$

then

$$\int_{-\pi}^{\pi} p_n(x) \, dx = 1.$$

If $g(x) = (1 + \cos(x))/2$, then $g'(x) = -\sin(x)/2 \geq -1/2$. It follows that $(1 + \cos(x))/2 \geq (1 - x/2)$, so

$$c_n = \int_{-\pi}^{\pi} (\frac{1+\cos(x)}{2})^n \, dx \geq 2\int_0^2 (1 - x/2)^n \, dx = \frac{4}{n+1}.$$

For $\delta > 0$ and $\delta \leq |x| \leq \pi$ this implies

$$p_n(x) \leq (n+1)(1 + \cos(\delta))^n/4, \tag{9.15}$$

so the renormalized polynomials $p_n(x)$ converge to zero uniformly in this region.

The identities (9.12) also show that

$$\cos(t-x) = \cos(t)\cos(x) + \sin(t)\sin(x),$$

implying that $p_n(t-x)$ is a trigonometric polynomial in x with coefficients which are trigonometric polynomials in t. Consequently, the integrals

$$q_n(x) = \int_{-\pi}^{\pi} f(t)p_n(t-x)\, dt.$$

are trigonometric polynomials. The substitution $u = t - x$ yields

$$q_n(x) = \int_{-\pi-x}^{\pi-x} f(u+x)p_n(u)\, du.$$

Since f and p_n are periodic with period 2π, and the last integral extends over an interval of length 2π, the function $q_n(x)$ may also be expressed as

$$q_n(x) = \int_{-\pi}^{\pi} f(t+x)p_n(t)\, dt.$$

It is easy to check that $f(x)$ is bounded and uniformly continuous for $x \in \mathbb{R}$, so for any $\epsilon > 0$ there is a $\delta > 0$ such that $|f(x_2) - f(x_1)| < \epsilon/2$ if $|x_2 - x_1| < \delta$. Picking an upper bound C for $|f(x)|$, and using this δ, the difference between $f(x)$ and $q_n(x)$ is estimated:

$$|q_n(x) - f(x)| = |\int_{-\pi}^{\pi} f(t+x)p_n(t)\, dt - \int_{-\pi}^{\pi} f(x)p_n(t)\, dt|$$

$$\le \int_{-\pi}^{\pi} |f(t+x) - f(x)|p_n(t)\, dt|.$$

Breaking up the interval $[-\pi, \pi]$ into three parts the estimate becomes

$$|q_n(x) - f(x)| \le 2C\int_{-\pi}^{-\delta} p_n(t)\, dt + \frac{\epsilon}{2}\int_{-\delta}^{\delta} p_n(t)\, dt + 2C\int_{\delta}^{\pi} p_n(t)\, dt$$

$$\le 2C\pi(n+1)(\frac{1+\cos(\delta)}{2})^n + \frac{\epsilon}{2}.$$

For n sufficiently large $|q_n(x) - f(x)| < \epsilon$, as desired.

□

The uniform approximation of a periodic function $f(x)$ by trigonometric polynomials can be improved if $f(x)$ has continuous derivatives.

Theorem 9.5.3. *Suppose $f : (-\infty, \infty) \to \mathbb{R}$ has a K continuous derivatives and satisfies $f(x + 2\pi) = f(x)$ for all $x \in \mathbb{R}$. Then for any $\epsilon > 0$ there is a trigonometric polynomial $p(x)$ such that*

$$|p^{(k)}(x) - f^{(k)}(x)| < \epsilon, \quad k = 0, \dots, K, \quad -\infty < x < \infty.$$

Proof. Suppose $f(x)$ has one continuous derivative. Given $\epsilon > 0$, Theorem 9.5.2 assures us that there is a trigonometric polynomial

$$p_1(x) = a_0 + \sum_{n=1}^{N} [a_n \cos(nx) + b_n \sin(nx)]$$

with

$$|p_1(x) - f'(x)| < \epsilon/2, \quad -\infty < x < \infty.$$

The main idea of the proof is to show that $|a_0|$ is small enough to discard, after which the trigonometric polynomial can be integrated to approximate $f(x)$.

Integrate $p_1(x)$ to obtain the function

$$p_0(x) = f(-\pi) + \int_{-\pi}^{x} p_1(t)\, dt$$

$$= f(-\pi) + a_0(x + \pi) + \int_{-\pi}^{x} \sum_{n=1}^{N} [a_n \cos(nt) + b_n \sin(nt)]\, dt.$$

Since

$$f(x) = f(-\pi) + \int_{-\pi}^{x} f'(t)\, dt,$$

the difference between $f(x)$ and $p_0(x)$ at $x = \pi$ is estimated by

$$|f(\pi) - p_0(\pi)| = \Big|\int_{-\pi}^{\pi} f'(t) - p_1(t)\, dt\Big| \le \int_{-\pi}^{\pi} |f'(t) - p_1(t)|\, dt \le 2\pi\epsilon/2. \quad (9.16)$$

Now $f(-\pi) = f(\pi)$ since $f(x)$ has period 2π. Also,

$$p_0(\pi) = f(-\pi) + 2\pi a_0 + \int_{-\pi}^{\pi} \sum_{n=1}^{N} [a_n \cos(nt) + b_n \sin(nt)]\, dt.$$

Since

$$\int_{-\pi}^{\pi} \cos(nt) dt = 0 = \int_{-\pi}^{\pi} \sin(nt)\, dt,$$

we find that $p_0(\pi) = f(-\pi) + 2\pi a_0$. The inequality (9.16) yields

$$|f(\pi) - p_0(\pi)| = |f(-\pi) - [f(-\pi) + 2\pi a_0]| \le 2\pi\epsilon/2,$$

or

$$|a_0| \le \epsilon/2.$$

Since $|a_0|$ is small, it can be dropped from $p_1(x)$ with minimal effect. Let $q_1(x) = p_1(x) - a_0$. Then

$$|q_1(x) - f'(x)| = |p_1(x) - a_0 - f'(x)|$$

$$\le |p_1(x) - f'(x)| + |a_0| < \epsilon, \quad -\infty < x < \infty.$$

Finally, if

$$p(x) = f(-\pi) + \int_{-\pi}^{x} q_1(t)\, dt$$

$$= f(-\pi) + \sum_{n=1}^{N} [\frac{a_n \sin(nx)}{n} - \frac{b_n \cos(nx)}{n}] + \sum_{n=1}^{N} \frac{b_n \cos(-n\pi)}{n},$$

then $p(x)$ is a trigonometric polynomial. For $-\pi \le x \le \pi$,

$$|f(x) - p(x)| = |[f(-\pi) + \int_{-\pi}^{x} f'(t)\, dt] - [f(-\pi) + \int_{-\pi}^{x} q_1(t)\, dt]|$$

$$\le \int_{-\pi}^{\pi} |f'(t) - q_1(t)|\, dt < 2\pi\epsilon.$$

Since both $f(x)$ and $p(x)$ have period 2π, this estimate actually holds for all x, concluding the argument in case f has one continuous derivative.

The remaining cases follow by induction. Suppose $f(x)$ has $K+1$ continuous derivatives. Then $f'(x)$ is periodic with K continuous derivatives, so by the induction hypothesis there is a trigonometric polynomial $p_1(x)$ such that

$$|p_1^{(k-1)}(x) - f^{(k)}(x)| < \epsilon, \quad k = 1, \dots, K+1.$$

The earlier argument says that the constant coefficient a_0 may be dropped from $p_1(x)$. If $q_1(x) = p_1(x) - a_0$, the higher derivatives of $q_1(x)$ agree with those of $p_1(x)$, so the higher derivative estimates remain the same.. Now $q_1(x)$ may be integrated as before, finishing the proof.

□

9.5.3 Fourier series

The use of trigonometric sums like (9.11) dates back to at least the 1750's. While Theorem 9.5.2 assures us that approximations by trigonometric series exist, it does not provide a convenient method for finding them. A more direct approach was championed by Joseph Fourier (1768–1830) as part of his study of heat flow in the early nineteenth century. In recognition of Fourier's contributions these series are called Fourier series.

To motivate the description of a Fourier series for a periodic function

$f(x)$, consider the problem of finding a formula that produces a trigonometric polynomial $p_N(X)$ of degree N which approximates $f(x)$. It is desirable for such a formula to exactly reproduce functions $f(x)$ which are trigonometric polynomials of degree N. A key role is played by the following remarkable integral identities.

Lemma 9.5.4. *For positive integers m, n,*

$$\frac{1}{\pi}\int_{-\pi}^{\pi} \sin(mx)\sin(nx)\ dx = \begin{Bmatrix} 0, & m \ne n, \\ 1, & m = n \end{Bmatrix}, \tag{9.17}$$

$$\frac{1}{\pi}\int_{-\pi}^{\pi} \cos(mx)\cos(nx)\ dx = \begin{Bmatrix} 0, & m \ne n, \\ 1, & m = n \end{Bmatrix},$$

$$\frac{1}{\pi}\int_{-\pi}^{\pi} \sin(mx)\cos(nx)\ dx = 0.$$

In addition

$$\frac{1}{2\pi}\int_{-\pi}^{\pi} \cos(nx)\ dx = \begin{Bmatrix} 0, & n \ne 0, \\ 1, & n = 0 \end{Bmatrix}.$$

Proof. The first case is typical. Using (9.12), when $m \ne n$

$$2\int_{-\pi}^{\pi} \sin(mx)\sin(nx)\ dx = \int_{-\pi}^{\pi} \cos([m-n]x) - \cos([m+n]x)\ dx$$

$$= \frac{\sin([m-n]x)}{m-n}\Big|_{-\pi}^{\pi} - \frac{\sin([m+n]x)}{m+n}\Big|_{-\pi}^{\pi} = 0,$$

while if $m = n$ the integral has the value 2π. □

Suppose now that $f(x)$ is a trigonometric polynomial of degree at most N,

$$f(x) = a_0 + \sum_{n=1}^{N} [a_n \cos(nx) + b_n \sin(nx)].$$

The integral formulas of Lemma 9.5.4 provide a method for picking out the coefficients a_n, b_n. The simple proof of the next lemma is left as an exercise.

Lemma 9.5.5. *If $f(x)$ is a trigonometric polynomial of degree at most M, then for all $N \ge M$*

$$f(x) = \frac{1}{2\pi}\int_{-\pi}^{\pi} f(x)\ dx \tag{9.18}$$

$$+\sum_{n=1}^{N}\Big[\frac{1}{\pi}\int_{-\pi}^{\pi} f(t)\cos(nt)\ dt\cos(nx) + \frac{1}{\pi}\int_{-\pi}^{\pi} f(t)\sin(nt)\ dt\sin(nx).\Big]$$

Notice that (9.18) can be used to produce approximating trigonometric polynomials for any integrable function $f : [-\pi, \pi] \to \mathbb{R}$. One can then go further, hoping that by letting $N \to \infty$ we will obtain a convergent series representation for $f(x)$. The first step is simply to associate the *Fourier series* to $f(x)$,

$$f(x) \sim \frac{1}{2\pi} \int_{-\pi}^{\pi} f(x)\, dx \tag{9.19}$$

$$+ \sum_{n=1}^{\infty} \Big[\frac{1}{\pi} \int_{-\pi}^{\pi} f(t) \cos(nt)\, dt \cos(nx) + \frac{1}{\pi} \int_{-\pi}^{\pi} f(t) \sin(nt)\, dt \sin(nx). \Big]$$

The resulting coefficients

$$a_0 = \frac{1}{2\pi} \int_{-\pi}^{\pi} f(x)\, dx,$$

$$a_n = \frac{1}{\pi} \int_{-\pi}^{\pi} f(t) \cos(nt)\, dt, \quad n = 1, 2, 3, \dots,$$

$$b_n = \frac{1}{\pi} \int_{-\pi}^{\pi} f(t) \sin(nt)\, dt, \quad n = 1, 2, 3, \dots,$$

are the *Fourier coefficients* of $f(x)$.

Determining when the association of (9.19) is actually an equality, even assuming that $f(x)$ is continuous, is surprisingly challenging. Suppose, for instance, that we want to check if the series in (9.19) is convergent at $x = 0$. Since $\sin(0) = 0$ the b_n terms are not relevant. In the absence of a convenient formula for $f(x)$, our inclination is to try a simple estimate for a_n; an obvious try is

$$|a_n| = |\frac{1}{\pi} \int_{-\pi}^{\pi} f(t) \cos(nt)\, dt| \le \frac{1}{\pi} \int_{-\pi}^{\pi} |f(t) \cos(nt)|\, dt \le \max_{-\pi \le t \le \pi} |f(t)|.$$

This estimate doesn't even show that $|a_n| \to 0$ as $n \to \infty$, so it is far too weak to establish convergence of the Fourier series when $x = 0$.

Peter Gustav Lejeune-Dirichlet (1805–59) produced the first careful, fairly general treatment of pointwise convergence of Fourier series in 1829. This event initiated a long series of investigations, including studies of pointwise convergence under more general conditions, and sufficient conditions for uniform convergence. Some reasonable expectations were dashed by the discovery of striking counterexamples. Some fundamental questions were not settled until 1966. Related problems are being investigated to this day.

Uniform convergence of the Fourier series to $f(x)$ can be established if some conditions on derivatives are assumed. An important observation is that differentiability of $f(x)$ implies decay of the Fourier coefficients.

Lemma 9.5.6. *Suppose $f : (-\infty, \infty) \to \mathbb{R}$ has two continuous derivatives and satisfies $f(x + 2\pi) = f(x)$ for all $x \in \mathbb{R}$. Then the Fourier series (9.19) converges uniformly to a continuous function $S(x)$.*

Proof. For $n \ge 1$ integration by parts gives

$$\pi a_n = \int_{-\pi}^{\pi} f(t)\cos(nt)\, dt = f(t)\frac{\sin(nt)}{n}\Big|_{-\pi}^{\pi} - \int_{-\pi}^{\pi} f'(t)\frac{\sin(nt)}{n}\, dt.$$

$$\pi b_n = \int_{-\pi}^{\pi} f(t)\sin(nt)\, dt = -f(t)\frac{\cos(nt)}{n}\Big|_{-\pi}^{\pi} + \int_{-\pi}^{\pi} f'(t)\frac{\cos(nt)}{n}\, dt.$$

The boundary terms disappear because of the periodicity of $f(t)$ and the trigonometric functions, leaving

$$\pi a_n = -\int_{-\pi}^{\pi} f'(t)\frac{\sin(nt)}{n}\, dt.$$

$$\pi b_n = \int_{-\pi}^{\pi} f'(t)\frac{\cos(nt)}{n}\, dt.$$

Another integration gives

$$\pi a_n = \int_{-\pi}^{\pi} f''(t)\frac{\cos(nt)}{n^2}\, dt. \tag{9.20}$$

$$\pi b_n = -\int_{-\pi}^{\pi} f''(t)\frac{\sin(nt)}{n^2}\, dt.$$

Since $f'(t)$ and $f''(t)$ are continuous on $[-\pi,\pi]$ the Fourier coefficients satisfy

$$|a_n| \le C/n^2, \quad |b_n| \le C/n^2$$

for some constant C. Since $|\sin(nx)| \le 1$, and $|\cos(nx)| \le 1$, while the numerical series $\sum 1/n^2$ is convergent, the Fourier series for $f(x)$ converges uniformly to some continuous function $S(x)$ by the Weierstrass M-test (Theorem 9.2.1). □

Let

$$S_N(f,x) = \frac{1}{2\pi}\int_{-\pi}^{\pi} f(x)\, dx$$

$$+\sum_{n=1}^{N}\Big[\frac{1}{\pi}\int_{-\pi}^{\pi} f(t)\cos(nt)\, dt\cos(nx) + \frac{1}{\pi}\int_{-\pi}^{\pi} f(t)\sin(nt)\, dt\sin(nx)\Big]$$

denote the N-th partial sum of the Fourier series for $f(x)$.

Lemma 9.5.7. *Suppose $f : (-\infty,\infty) \to \mathbb{R}$ has two continuous derivatives and satisfies $f(x+2\pi) = f(x)$ for all $x \in \mathbb{R}$. For every $\epsilon > 0$ there is a $\delta > 0$ such that the inequalities*

$$|f^{(j)}(x)| < \delta, \quad j = 0,1,2, \quad -\infty < x < \infty,$$

imply

$$|S_N(f,x)| < \epsilon, \quad N = 0,1,2,3,\dots, \quad -\infty < x < \infty,$$

Proof. The inequality $|a_0| \le \delta$ follows immediately from

$$a_0 = \frac{1}{2\pi}\int_{-\pi}^{\pi} f(x)\, dx.$$

For $n \ge 1$ the formulas (9.20) give

$$|a_n| \le 2\delta/n^2, \quad |b_n| \le 2\delta/n^2.$$

The partial sums of the Fourier series satisfy the simple estimate

$$|S_N(f,x)| \le |a_0| + \sum_{n=1}^{N}[|a_n| + |b_n|] \le \delta\Big[1 + \sum_{n=1}^{N}\frac{4}{n^2}\Big],$$

so it suffices to take

$$\delta\Big[1 + \sum_{n=1}^{\infty}\frac{4}{n^2}\Big] < \epsilon.$$

□

Theorem 9.5.8. *Suppose $f : (-\infty,\infty) \to \mathbb{R}$ has two continuous derivatives and satisfies $f(x+2\pi) = f(x)$ for all $x \in \mathbb{R}$. Then the Fourier series* (9.19) *converges uniformly to $f(x)$.*

Proof. Pick $\epsilon > 0$. By Theorem 9.5.3 there is a trigonometric polynomial $p(x)$ satisfying

$$|f^{(j)}(x) - p^{(j)}(x)| < \delta, \quad j = 0,1,2, \quad -\infty < x < \infty,$$

where δ is chosen as in Lemma 9.5.7 so that

$$|S_N(f-p,x)| < \epsilon/3, \quad N = 0,1,2,3,\dots, \quad -\infty < x < \infty.$$

Suppose the degree of $p(x)$ is at most M.

The formula for the Fourier coefficients shows that

$$S_N(p,x) - S_N(f,x) = S_N(p-f,x).$$

Using the uniform convergence of these Fourier series, there is an $N \ge M$ such that

$$|S(x) - S_n(x)| < \epsilon/3, \quad |S_n(p,x) - S_n(f,x)| < \epsilon/3, \quad n \ge N.$$

Also, for $n \ge M$ the Fourier series for $p(x)$ reproduces $p(x)$, so $|p(x) - S_n(p,x)| = 0$. Finally, by the triangle inequality

$$|f(x) - S(x)|$$

$$= |[f(x) - p(x)] + [p(x) - S_n(p,x)] + [S_n(p,x) - S_n(f,x)] + [S_n(f,x) - S(x)]|$$
$$\le |f(x) - p(x)| + |p(x) - S_n(p,x)| + |S_n(p,x) - S_n(f,x)| + |S_n(f,x) - S(x)|$$
$$< \epsilon/3 + 0 + \epsilon/3 + \epsilon/3 = \epsilon,$$

so $f(x) = S(x)$.

□

9.6 Problems

9.1. *Following the treatment of Airy's Equation* (9.1), *find a power series solution* $\sum_{k=0}^{\infty} a_k x^k$ *for the equation*

$$\frac{dy}{dx} = y.$$

Express the coefficients a_k *in terms of the first coefficient* a_0. *Now treat the slightly more general equation*

$$\frac{dy}{dx} = \alpha y,$$

where α *is a constant. For which* x *do the series converge?*

9.2. *Show that the series solutions* (9.5) *for Airy's equation converge for all values of* x.

9.3. *(a) Show that the initial value problem*

$$y'' + y = 0, \quad y(0) = a_0, \quad y'(0) = a_1,$$

has unique power series solutions. For which x *do the series converge?*

(b) Show that if $y_1(x), y_2(x)$ *are the solutions satisfying*

$$y_1(0) = 1, \quad y_1'(0) = 0,$$

$$y_2(0) = 0, \quad y_1'(0) = 1,$$

then $y_1(x) = \cos(x)$ *and* $y_2(x) = \sin(x)$.

(c) To establish the trigonometric identities

$$\cos(a + x) = \cos(a)\cos(x) - \sin(a)\sin(x),$$

$$\sin(a + x) = \sin(a)\cos(x) + \cos(a)\sin(x),$$

show that both sides of the equations are solutions of $y'' + y = 0$ *satisfying the same initial conditions.*

9.4. *Using the treatment of Airy's equation* (9.1) *as a model, find a power series solution* $\sum_{k=0}^{\infty} a_k x^k$ *for the Hermite equation*

$$\frac{d^2y}{dx^2} - 2x\frac{dy}{dx} + 2\alpha y = 0,$$

where α *is a constant. Express the coefficients* a_k *in terms of the first two coefficients* a_0, a_1. *What happens if* α *happens to be a positive integer?*

9.5. *Consider the sequence of functions $f_n : [0,1] \to \mathbb{R}$ given by*

$$f_n(x) = \begin{cases} n^2 x, & 0 \le x \le 1/n, \\ n^2(2/n - x), & 1/n \le x \le 2/n, \\ 0, & 2/n \le x \le 1. \end{cases}$$

Show that the sequence $\{f_n(x)\}$ converges pointwise to the function $f(x) = 0$. Compare the integrals of $f_n(x)$ and $f(x)$.

9.6. *Say that $s : [\alpha, \beta] \to \mathbb{R}$ is a step function if there is a partition $\alpha = x_0 < x_1 < x_2 < \dots < x_N = \beta$ of $[\alpha, \beta]$ such that $s(x)$ is constant on each subinterval $[x_n, x_{n+1})$, $n = 0, \dots, N-1$. Show that if $f : [\alpha, \beta] \to \mathbb{R}$ is continuous, then there is a sequence $\{s_k(x)\}$ of step functions converging uniformly to $f(x)$.*

9.7. *Say that $g : [\alpha, \beta] \to \mathbb{R}$ is a piecewise linear function if there is a partition $\alpha = x_0 < x_1 < x_2 < \dots < x_N = \beta$ of $[\alpha, \beta]$ such that $g(x)$ is a polynomial of degree at most one on each subinterval $[x_n, x_{n+1})$, $n = 0, \dots, N-1$. Show that if $f : [\alpha, \beta] \to \mathbb{R}$ is continuous, then there is a sequence $\{g_k(x)\}$ of continuous piecewise linear functions converging uniformly to $f(x)$.*

9.8. *(a) Show there is a sequence $\{f_n : \mathbb{R} \to \mathbb{R}\}$ of positive continuous functions such that $\int_{\mathbb{R}} f_n(x)\ dx < \infty$ for each n, $f_n(x)$ converges uniformly to 0, but the sequence $\{\int_{\mathbb{R}} f_n(x)\ dx\}$ diverges.*

(b) Assume in addition that there is a positive continuous function $g(x)$ such that $\int_{\mathbb{R}} g(x)\ dx < \infty$ and $f_n(x) \le g(x)$. Show that

$$\lim_{n \to \infty} \int_{\mathbb{R}} f_n(x)\ dx = 0.$$

9.9. *Suppose $f : \mathbb{R} \to \mathbb{R}$ is a nonconstant bounded function. Show there is no sequence of polynomials $p_n(x)$ converging uniformly to $f(x)$ on $\mathbb{R}$.*

9.10. *(a) Suppose $\alpha \le x_0 < x_1 < x_2 < \dots < x_N \le \beta$. By considering the polynomials*

$$p_k(x) = \prod_{n \ne k} (x - x_n) = [\prod_{n=0}^{N} (x - x_n)]/(x - x_k),$$

show that for any real numbers $c_0, \dots, c_N$ there is a unique polynomial $p(x)$ (the Lagrange interpolating polynomial) with degree at most N satisfying $p(x_n) = c_n$.

(b) Suppose $p_k(x)$ is a sequence of polynomials of degree at most N converging pointwise to a function $f(x)$. Show that $f(x)$ is a polynomial of degree at most N, and that the convergence is uniform.

(c) State and prove a similar result about interpolation by trigonometric polynomials on the interval $[-\pi, \pi]$.

9.11. *Suppose* $\{f_n : [0,1] \to \mathbb{R}\}$ *is a sequence of increasing functions converging pointwise to* $f(x)$.

(a) Show that $f(x)$ *is increasing.*

(b) Show that if $f(x)$ *is continuous, then* $\{f_n(x)\}$ *converges uniformly to* $f(x)$.

9.12. *Given that* $\tan^{-1}(0) = 0$ *and* $\frac{d}{dx}\tan^{-1}(x) = \frac{1}{1+x^2}$*, find the power series for* $\tan^{-1}(x)$ *centered at* $x_0 = 0$*. What is the radius of convergence of the series?*

9.13. *Suppose* $f : [a,b] \to \mathbb{R}$ *is continuous. Show that for any* $\epsilon > 0$ *there is a polynomial* $p_\epsilon(x)$ *such that*

$$|f(x) - p_\epsilon(x)| < \epsilon, \quad a \le x \le b.$$

(Hint: consider $t = (x-a)/(b-a)$*.)*

9.14. *Suppose* $f : [0,\infty) \to \mathbb{R}$ *is continuous, and* $\lim_{t\to\infty} f(t)$ *exists. Show that for any* $\epsilon > 0$ *there is a function*

$$p_N(t) = \sum_{n=0}^{N} a_n e^{-nt}$$

with

$$|f(t) - p_N(t)| < \epsilon, \quad 0 \le t < \infty.$$

(Hint: Consider $x = e^{-t}$*.)*

9.15. *(a) Suppose* $x_1 < x_2 < x_3 < x_4$*. Assume* $f_1(x)$ *is continuous on* $[x_1, x_2]$ *and* $f_2(x)$ *is continuous on* $[x_3, x_4]$*. Show that for every* $\epsilon > 0$ *there is a polynomial* $p(x)$ *such that*

$$|f_1(x) - p(x)| < \epsilon, \quad x_1 \le x \le x_2,$$

and

$$|f_2(x) - p(x)| < \epsilon, \quad x_3 \le x \le x_4.$$

(b) Can you extend this result to any finite collection of pairwise disjoint closed bounded intervals?

9.16. *Verify the formulas in Lemma 9.5.4.*

9.17. *Suppose* $f : [0,1] \to \mathbb{R}$ *is continuous, with*

$$\int_0^1 f(x)x^n \, dx = 0, \quad n = 0, 1, 2, \ldots.$$

Show that $f(x) = 0$ *for* $x \in [0,1]$.

9.18. *Prove Lemma 9.5.5.*

9.19. *(a) Suppose* $[\alpha, \beta] \subset [-\pi, \pi]$ *and*

$$g(x) = \begin{Bmatrix} 1, & x \in [\alpha, \beta], \\ 0, & x \notin [\alpha, \beta]. \end{Bmatrix}$$

By direct computation show that the Fourier coefficients for $g(x)$ *satisfy*

$$\lim_{n \to \infty} a_n = 0, \quad \lim_{n \to \infty} b_n = 0.$$

(b) Show that for any continuous function $f : [-\pi, \pi] \to \mathbb{R}$ *the Fourier coefficients for* $f(x)$ *satisfy*

$$\lim_{n \to \infty} a_n = 0, \quad \lim_{n \to \infty} b_n = 0.$$

9.20. *The partial differential equation*

$$u_{tt} = c^2 u_{xx}, \quad -\pi \le x \le \pi, \quad t \ge 0, \tag{9.21}$$

with initial conditions

$$u(0, x) = f(x), \quad u_t(0, x) = g(x)$$

and with the boundary conditions

$$u(t, 0) = u(t, \pi) = 0,$$

is a model for the vibrations of a string which is pinned at $x = 0, \pi$.

Show that the functions

$$\cos(nct)\sin(nx), \sin(nct)\sin(nx), \quad n = 1, 2, 3, \dots,$$

are solutions of the equation satisfying the boundary conditions.

Describe the corresponding infinite series solutions. What are $f(x), g(x)$ *in terms of these series.*

9.21. *(a) Show that the Fourier coefficients* a_n *vanish if* $f(-x) = -f(x)$.
(b) Show that the Fourier coefficients b_n *vanish if* $f(-x) = f(x)$.

A

Solutions to odd numbered problems

A.1 Chapter 1 Solutions

Problem 1.1

a) Either $a = 0$ or $a \neq 0$. If $a = 0$ we're done. If not, multiply by a^{-1} to get

$$a^{-1} \cdot (a \cdot b) = (a^{-1} \cdot a) \cdot b) = 1 \cdot b = b = a^{-1} \cdot 0 = 0.$$

b) There is an additive inverse for $a \cdot b$,

$$(a \cdot b) + (-(a \cdot b)) = 0$$

and by the distributive law

$$(a \cdot b) + (-a) \cdot b = (a + (-a)) \cdot b = 0 \cdot b = 0.$$

Thus $-(a \cdot b)$ and $(-a) \cdot b$ are additive inverses of $a \cdot b$.

Let's show that there is always a unique additive inverse in a field. Suppose

$$a + b = 0,$$

and

$$a + c = 0.$$

Adding b we get

$$a + b + c = 0 + b = b,$$

but

$$(a + b) + c = 0 + c = c,$$

so $b = c$, and the additive inverse is unique. This implies $-(a \cdot b) = (-a) \cdot b$.

c) By definition of the additive inverse,

$$(-a) + (-(-a)) = 0,$$

and

$$(-a) + a = 0.$$

Since $-(-a)$ and a are both additive inverses of $-a$,they are equal by the observation in part b).

Problem 1.3

Since

$$(-a) \cdot (-a) + (-a) \cdot a = (-a) \cdot ((-a) + a) = 0,$$

adding $a \cdot a$ gives

$$(-a) \cdot (-a) = a \cdot a.$$

Either $a > 0$ or $-a > 0$ by Proposition 1.1.7, so $a \cdot a > 0$ by O.6.

Problem 1.5

Notice that

$$4 = 1 + 1 + 1 + 1 = (1 + 1) \cdot (1 + 1) = 2 \cdot 2.$$

If $4 = 2 \cdot 2 = 0$, then

$$2^{-1} \cdot 2 \cdot 2 = 2 = 0.$$

The contrapositive of this last statement is if $2 \ne 0$ in a field $\mathbb{F}$, then $4 \ne 0$ in $\mathbb{F}$.

Problem 1.7

Suppose that p is prime and

$$\sqrt{p} = \frac{m}{n},$$

where m and n are positive integers with no common factors. Then

$$m^2 = pn^2,$$

and m^2 is divisible by p.

Suppose that the prime factorization (in increasing order) of m is

$$m = q_1 q_2 \dots q_k.$$

Then

$$m^2 = q_1 q_1 q_2 q_2 \dots q_k q_k.$$

Since the factorization into primes is unique, every prime factor of m^2 is a prime factor of m. In particular p is a prime factor of m, and p^2 is a prime factor of m^2. Since

$$n^2 = m^2/p,$$

n^2 has at least one factor p. Of course this implies that n also has a factor p. We now have a contradiction, since both m and n have the common prime factor p.

Problem 1.9

By the Archimedean Property O.7 (see Lemma 2.2.4) there is a positive integer n with $1/n < b - a$. Similarly there are integers L and M such that

$$L/n < a < b < M/n.$$

Let $(m-1)/n$ be the largest multiple of $1/n$ which is less than or equal to a,

$$(m-1)/n \le a, \quad m/n > a.$$

Since

$$0 < \frac{m}{n} - \frac{m-1}{n} = \frac{1}{n} < b - a,$$

we have

$$\frac{m}{n} = \frac{m-1}{n} + \frac{1}{n} \le a + \frac{1}{n} < a + (b-a) = b.$$

Thus

$$a \le \frac{m}{n} < b.$$

To get strict inequality, pick another number $k > 0$ such that

$$\frac{1}{k} < b - \frac{m}{n}.$$

Then

$$a < \frac{m}{n} + \frac{1}{k} < b,$$

and

$$q = \frac{m}{n} + \frac{1}{k} < b$$

is the desired rational number.

Problem 1.11

First check the formula when $n = 1$. In this case

$$\sum_{k=1}^{1} k = 1 = \frac{1 \cdot 2}{2}.$$

Thus the formula is correct when $n = 1$.

Now suppose the formula is true in the n-th case. Consider the $(n+1)$-st case, using the n-th case.

$$\sum_{k=1}^{n+1} k = (n+1) + \sum_{k=1}^{n} k = (n+1) + \frac{n(n+1)}{2} = (n+1)\frac{2+n}{2} = \frac{(n+1)(n+2)}{2}.$$

Thus the formula is true in case $n+1$ whenever it is true in case n.

Problem 1.13

The induction argument is presented in a slightly different way.

For the case $n = 1$,

$$\sum_{k=0}^{0} k2^{-k} = 0 = 2 - 2,$$

so the first case is correct.

Now suppose the formula holds in case $n = m$, and consider the formula in case $n = m + 1$.

$$\sum_{k=0}^{(m+1)-1} k2^{-k} = \sum_{k=0}^{m} k2^{-k} = \sum_{k=0}^{m-1} k2^{-k} + m2^{-m}$$

$$= 2 - (m+1)2^{1-m} + m2^{-m} = 2 - 2m2^{-m} - 2 \cdot 2^{-m} + m2^{-m}$$

$$= 2 - (2m + 2 - m)2^{-m} = 2 - (m+2)2^{-m} = 2[1 - (m+2)2^{-m-1}].$$

But this is the desired formula in case $n = m + 1$.

Problem 1.15

The argument is logically correct until the statement "Now bring back the ejected horse, toss out another one, repeat the argument, and all $K+1$ horses are white." Here is the problem. If $K = 1$ the only horse we can remove is the only one known to be white. The induction breaks down as we try to conclude that the first statement implies the second.

Problem 1.17

Consider the complete truth tables. The first is

A	$\neg A$	$A \vee \neg A$
T	F	T
F	T	T

.

For the second case, use the abbreviation C for the composite formula,

$$C: \quad (A \Rightarrow B) \Leftrightarrow (\neg B \Rightarrow \neg A).$$

Then

A	B	$A \Rightarrow B$	$\neg B \Rightarrow \neg A$	C
T	T	T	T	T
F	T	T	T	T.
T	F	F	F	T
F	F	T	T	T

Since the truth value of the formula is true for every choice of truth values for the formula's atoms, the two formulas are tautologies.

Problem 1.19

This is a counting problem. Functions of this type may be completely described by a truth table with four rows, for example

A	B	$f(A,B)$
T	T	T
F	T	T
T	F	T
F	F	F

.

The functions are the same if and only if the last column is the same. There are four positions in the last column, and each position may have the value T or F. This gives $2^4 = 16$ possible functions.

To construct any function using the given four, start with the functions $g_1(A,B) = A \wedge B$, $g_2(A,B) = \neg A \wedge B$, $g_3(A,B) = A \wedge \neg B$, and $g_4(A,B) = \neg A \wedge \neg B$. These functions are described in the following table.

A	B	$A \wedge B$	$\neg A \wedge B$	$A \wedge \neg B$	$\neg A \wedge \neg B$
T	T	T	F	F	F
F	T	F	T	F	F
T	F	F	F	T	F
F	F	F	F	F	T

Notice that the function g_k has the value T in row k; otherwise its value is F. The function with value T in row j and row k, but F otherwise, may be constructed as $g_j(A,B) \vee g_k(A,B)$. To get T in rows i,j,k, but F otherwise, use $(g_i(A,B) \vee g_j(A,B)) \vee g_k(A,B)$. Finally, a function with all row values F is $g_1(A,B) \wedge g_2(A,B)$, and $\neg[g_1(A,B) \wedge g_2(A,B)]$ is always T.

Problem 1.21

The argument may be represented as

$$[(A \Rightarrow \neg C) \wedge (\neg B \Rightarrow C)] \Rightarrow [\neg A \vee B].$$

Let D denote this statement We consider the truth table

A	B	C	$A \Rightarrow \neg C$	$\neg B \Rightarrow C$	$\neg A \vee B$	D
T	T	T	F	T	T	T
F	T	T	T	T	T	T
T	F	T	F	T	F	T
F	F	T	T	T	T	T
T	T	F	T	T	T	T
F	T	F	T	T	T	T
T	F	F	T	F	F	T
F	F	F	T	F	T	T

The statement D is a tautology, so the argument is sound.

Problem 1.23

a) The statement "William will marry either Jane or Mary if she loves him" could also be written "William will marry Jane if she loves him, and William will marry Mary if she loves him." Of course this leaves open the possibility that both Jane and Mary love William, and then William runs into legal problems.

Symbolically we have

$$\Big([(A \sqcup B) \wedge (C \sqcup D)] \wedge [(A \Rightarrow E) \wedge (C \Rightarrow F)] \wedge [\neg C \Rightarrow H]\Big)$$

$$\Rightarrow \Big([(\neg E \wedge \neg F) \vee (\neg G \wedge \neg H)] \Rightarrow [C \vee (B \wedge D)]\Big)$$

Denote this entire statement by Z. Since Z has the form

$$\Big(W\Big) \Rightarrow \Big(X\Big)$$

and the premises W of the left side are *True*, the whole of Z is *True* if and only if the right side X is *True*.

Formula X is

$$[(\neg E \wedge \neg F) \vee (\neg G \wedge \neg H)] \Rightarrow [C \vee (B \wedge D)]$$

If both $\neg E \wedge \neg F$ and $\neg G \wedge \neg H$ are *False*, then X is *True*.

Suppose $\neg E \wedge \neg F$ is *True*. We also know that

$$(A \Rightarrow E) \wedge (C \Rightarrow F)$$

is *True*, so A is *False* and C is *False*. Since $A \sqcup B$ is *True*, B is *True*, and since $C \sqcup D$ is *True*, D is *True*; thus X is *True*.

Suppose $\neg G \wedge \neg H$ is *True*. Then H is *False* and since we assumed $\neg C \Rightarrow H$, we must have that C is *True*. Since C is *True*, so is X.

We conclude that X is *True* if the premises W are true, so the reasoning is valid.

b) To avoid possible polygamy, one can use the following analysis: If A, then if $\neg C$ then E, and if C then $E \sqcup F$, and if C, then if $\neg A$ then F, and if A then $E \sqcup F$. Symbolically,

$$\Big(A \Rightarrow [(\neg C \Rightarrow E) \wedge (C \Rightarrow E \sqcup F)]\Big) \wedge \Big(C \Rightarrow [(\neg A \Rightarrow F) \wedge (A \Rightarrow E \sqcup F)]\Big)$$

.

Problem 1.25

For the *modus ponens* case, let D represent $A \wedge (A \Rightarrow B)$. The truth table for $D \Rightarrow B$ is

A	B	$A \Rightarrow B$	D	$D \Rightarrow B$
T	T	T	T	T
F	T	T	F	T
T	F	F	F	T
F	F	T	F	T

(A.1)

For the case $([A \vee B] \Rightarrow C) \wedge \neg C) \Rightarrow (\neg A \wedge \neg B)$, let D represent $[A \vee B] \Rightarrow C$, and let E represent $\neg A \wedge \neg B$. The truth table for $D \wedge \neg C \Rightarrow E$ is

A	B	C	$A \vee B$	D	E	$D \wedge \neg C$	$D \wedge \neg C \Rightarrow E$
T	T	T	T	T	F	F	T
F	T	T	T	T	F	F	T
T	F	T	T	T	F	F	T
F	F	T	F	T	T	F	T
T	T	F	T	F	F	F	T
F	T	F	T	F	F	F	T
T	F	F	T	F	F	F	T
F	F	F	F	T	T	T	T

(A.2)

Problem 1.27

Since there is no integer which is both positive and negative, the statement

$$\exists x(P(x) \wedge Q(x))$$

is always false. On the other hand the statements

$$\exists x P(x)$$

and

$$\exists x Q(x)$$

are both true if $x = 1$ in the first statement and $x = -1$ in the second. Since $T \Rightarrow F$ is false, the implication of the problem is not valid.

The values of x may be different in the statements $\exists x P(x)$ and $\exists x Q(x)$. An equivalent version would be

$$[\exists x P(x) \wedge \exists y Q(y)] \Rightarrow [\exists z(P(z) \wedge Q(z))].$$

(b) Consider the domain $\mathcal{D}$ equal to the set of integers. Suppose the predicate $P(x)$ is $x >= 0$ while the predicate $Q(x)$ is $x <= 0$. The statement $\forall x(P(x) \vee Q(x))$ is true since each integer is either nonnegative or nonpositive. On the other hand the statements $\forall x P(x)$ and $\forall x Q(x)$ are both false, since not all integers are nonnegative and not all are nonpositive. Thus the implication is false in this case.

Problem 1.29

Since $x - x = 0$, and 0 is even, $P(x, x)$ is true.

If $P(x, y)$ is false then $P(x, y) \Rightarrow P(y, x)$ is true, so we only have to check the case when $P(x, y)$ is true. If $P(x, y)$ is true, then $x - y = 2n$ is even, and so is $-(x - y) = y - x = 2(-n)$. In this case $P(x, y) \Rightarrow P(y, x)$ is true.

Finally, suppose that $P(x, y)$ is true and $P(y, z)$ is true. Then $x - y = 2n$ and $y - z = 2m$. This gives $x - z = (2n + y) - (y - 2m) = 2(n + m)$, and $P(x, z)$ is true. If either $P(x, y)$ or $P(y, z)$ is false the implication $[P(x, y) \wedge P(y, z)] \Rightarrow P(x, z)$ is true, completing the proof.

A.2 Chapter 2 Solutions

Problem 2.1

a) For all $a \in \mathbb{F}$ we have $|a| = |-a|$.

Suppose $0 \le a$ and $0 \le b$. The cases $|a+b| = |a| + |b| \le |a| + |b|$ and $|-(a+b)| \le |a| + |b|$ are immediate. From $0 \le b$ we get $-b \le 0 \le b$. Then $a - b \le a + b$, so $|a + (-b)| \le |a| + |b|$.

b) Start with $|a| = |a - b + b|$ and use the triangle inequality to get $|a| \le |a - b| + |b|$, or $|a| - |b| \le |a - b|$.

Problem 2.3

Suppose $\epsilon > 0$. Since $\lim_{k\to\infty} a_k = L$, there is an N_1 such that $|a_k - L| < \epsilon$ for all $k \ge N_1$. Take $N = \max(N_1, M)$. Then for $k \ge N$ the terms of the sequences are the same, so

$$|b_k - L| = |a_k - L| < \epsilon.$$

Problem 2.5

a) The sequence

$$x_k = (-1)^k$$

is bounded, since $|x_k| \le 1$, but this sequence has no limit.

To carefully justify the last statement you can take the following approach. If there were a limit L, then for any $\epsilon > 0$ there would be an N such that

$$|x_k - L| < \epsilon, \quad k \ge N.$$

This would mean, in particular, that for $k \ge N$

$$|x_{k+1} - x_k| = |(x_{k+1} - L) + (L - x_k)| \le |x_{k+1} - L| + |L - x_k| < 2\epsilon, \quad k \ge N.$$

But in this example we actually find that for all k

$$|x_{k+1} - x_k| = 2,$$

so as soon as $\epsilon < 1$ the sequence does not have the required property.

b) The sequence $x_k = k$ is unbounded, so cannot have a limit.

Problem 2.7

The answers are: *a*) 1, *b*) 8, *c*) π , *d*) 1.

Problem 2.9

This is one of those problems which is best worked backwards, so you may want to start at eqrefA.1 below, understand how the problem is decomposed, and then check the proof.

Pick $\epsilon > 0$, and let

$$\epsilon_1 = \epsilon_2 = \epsilon_3 = \epsilon/3.$$

Since $\lim_{n\to\infty} b_n = b$, there is an N_1 such that

$$|b_n - b| < \epsilon_1, \quad n \ge N_1.$$

Since $b_1 + \cdots + b_{N_1}$ is a fixed number, there is an N_2 such that

$$(b_1 + \cdots + b_{N_1})/n < \epsilon_2, \quad n \ge N_2.$$

Finally, there is an N_3 such that

$$bN_1/n < \epsilon_3, \quad n \ge N_3.$$

We now take $N = \max(N_1, N_2, N_3)$ and suppose $n \ge N$. Then

$$\frac{\sum_{k=1}^n b_k}{n} = \frac{b_1 + \cdots + b_n}{n} = \frac{b_1 + \cdots + b_{N_1}}{n} + \frac{b_{N_1+1} + \cdots + b_n}{n}, \tag{A.3}$$

$$= \frac{b_1 + \cdots + b_{N_1}}{n} + \frac{(n - N_1)b}{n} + \frac{(b_{N_1+1} - b) + \cdots + (b_n - b)}{n},$$

so

$$|\frac{\sum_{k=1}^n b_k}{n} - b| < |\frac{b_1 + \cdots + b_{N_1}}{n}| + |\frac{(N_1)b}{n}| + |\frac{(b_{N_1+1} - b) + \cdots + (b_n - b)}{n},$$

$$< \epsilon_1 + \epsilon_2 + \epsilon_3 = \epsilon.$$

Problem 2.11

a) By Lemma 2.2.6, if L is the least upper bound for a set S, but $L \notin S$, then for any $\epsilon > 0$ there is a point $x \in S$ with $x < L$ and such that $L - x < \epsilon$. To generate the desired sequence of points x_k, let $\epsilon_1 = 1$, and pick $x_1 \in S$ such that $L - x_1 < \epsilon_1$. Since $L \notin S$ we have $x_1 < L$.

Now for $k \ge 2$ define

$$\epsilon_k = \min(1/k, L - x_{k-1}).$$

Pick $x_k \in S$ such that $L - x_k < \epsilon_k$. Since $L \notin S$ we have $x_k < L$. Since

$$L - x_k < L - x_{k-1},$$

we have $x_k > x_{k-1}$. Thus the sequence $\{x_k\}$ is strictly increasing. Finally, since $L - x_k < 1/k$, we also have

$$\lim_{k\to\infty} x_k = L.$$

b) Suppose that S consists of a single point L. Then L is the least upper bound for S, but S has no strictly increasing sequences. In fact any finite set S will be a counterexample.

Problem 2.13

Suppose $U \subset \mathbb{R}$ which is nonempty, and $x \le M$ for all $x \in U$. We will construct two sequences $\{a_n\}$ and $\{b_n\}$. The points a_n will form an increasing sequence from U. The points b_n will be a decreasing sequence of upper bounds for U.

Start with any number $a_0 \in U$, and let $b_0 = M$. Now for $n \ge 0$ suppose $a_n \in U$ and b_n is an upper bound for U. Let $c_n = (a_n + b_n)/2$. If c_n is an upper bound for U define $b_{n+1} = c_n$ and $a_{n+1} = a_n$. If c_n is not an upper bound for U, pick $b_{n+1} = b_n$ and $a_{n+1} \in U$ with $a_{n+1} > c_n$. In either case $a_{n+1} \ge a_n$, $b_{n+1} \le b_n$, and the length of the interval $[a_n, b_n]$ satisfies $|b_n - a_n| \le 2^{-n}|b_0 - a_0|$.

By the Bounded Monotone Sequence property the sequence $\{b_n\}$ has a limit L_2. Since $|b_n - L_2| \le 2^{-n}|b_0 - a_0|$, no element of U can be bigger than L_2, so L_2 is an upper bound for U. Since $|L_2 - a_n| \le 2^{-n}|b_0 - a_0|$, L_2 is the least upper bound for U.

Problem 2.15

The sequence whose terms are

$$x_{2n} = 0, \quad x_{2n+1} = n$$

will do.

Problem 2.17

Write $1/x = 1 + \delta$ with $\delta > 0$. By the Binomial Theorem, $1/x^n = 1 + n\delta + p$ with $p > 0$. By Proposition 1.1.14, for any $\epsilon > 0$ there is a positive integer N such that

$$1/x^n \ge 1 + N\delta \ge 1/\epsilon, \quad n \ge N,$$

or $0 < x^n < \epsilon$.

Problem 2.19

Suppose first that the sequence $\{x_k\}$ is unbounded. Let $y_1 = x_1$. Now for each positive integer $n > 1$ there is a term $y_n = x_{k(n)}$ such that $|y_n| \ge |y_{n-1}|$. The sequence $\{y_n\}$ either has an infinite subsequence which is positive, hence increasing, or negative and decreasing.

Suppose now that the sequence $\{x_k\}$ is bounded. This sequence has a subsequence $\{y_n\}$ which converges to a number y_0. The sequence $\{y_n\}$ either has an infinite subsequence which is $\ge y_0$ or an infinite subsequence which is $\le y_0$. Let's suppose that the first case holds, and let the subsequence be z_m. Let $w_1 = z_1$, and for each positive integer $j > 1$ pick a term $w_j = z_{m(j)}$ such that $w_{j-1} \ge w_j \ge y_0$. This is possible because if $w_{j-1} = y_0$ then we can make all subsequent choices equal to y_0, and if $w_{j-1} > y_0$ then there is a $w_{j-1} > w_j \geq y_0$. The sequence $\{w_j\}$ is the desired monotone (decreasing) subsequence.

Problem 2.21

Let's show that an unbounded set cannot be compact. Suppose that $U \subset \mathbb{R}$ is unbounded. Pick a sequence x_n from U satisfying $|x_n| \geq n$. This sequence has no convergent subsequence since any subsequence $y_k = x_{n(k)}$ also satisfies $|y_k| \geq k$, so the subsequence is unbounded, while any convergent sequence must be bounded. Thus if U is not bounded, then U is not compact.

Problem 2.23

We're given the existence of a sequence $\{y_j\}$ with $y_j \neq z$ and $\lim_{j\to\infty} y_j = z$. Take $x_1 = y_1$. For each $k \geq 2$ pick a point y_J such that the index J is larger than any previously selected index j, and such that

$$|y_J - z| < \min\{|x_i - z|, i = 1, \dots, k-1\}.$$

Now let $x_k = y_J$. The sequence $\{x_k\}$ is a subsequence of $\{y_j\}$, so $\lim_{k\to\infty} x_k = z$. The method of selection also shows that $x_k \neq x_i$ for $i < k$, so the points x_k are distinct.

Problem 2.25

The sequence $\{1/(n+1)\}$ has limit 0, but $0 \notin B$, so B is not closed.

Since $0 \leq x \leq 1$ for all $x \subset [0,1]$, the same holds for limits of sequences from B. So B has all its limit points and is closed.

Problem 2.27

Suppose that $\{x_n\}$ is any sequence whose terms come from K. Since K is the union of a finite collection of intervals, there must be at least one interval, say $[a_1, b_1]$, which contains x_n for an infinite set of indices n. Let $\{y_k\}$ be the subsequence of $\{x_n\}$ consisting of those terms in $[a_1, b_1]$. Since $[a_1, b_1]$ is compact, the sequence $\{y_k\}$ has a convergent subsequence $\{z_m\}$, which is both a subsequence of $\{x_n\}$ and is also convergent in K.

Problem 2.29

Suppose that $\{x_m\}$ is a sequence of points from K. Then for each n the terms x_m are in K_n. Focus first on K_1. The sequence $\{x_m\}$ has a subsequence $\{y_j\}$ whose terms are in K_1 and which converges to some $y_0 \in K_1$. Since $y_j \in K$, it is also the case that $y_j \in K_n$. The sequence $\{y_j\}$ must have a subsequence which converges in the set K_n, since K_n is compact, but since $\{y_j\}$ has limit y_0, all subsequences converge to y_0. Thus y_0 is also in K_n for all n. We finally conclude that the subsequence $\{y_j\}$ of $\{x_m\}$ is in K, and converges to $y_0 \in K$.

Problem 2.31

Theorem 2.4.3 shows that a Cauchy sequence converges to some real number L. Since K is compact, it is closed, so $L \in K$.

Problem 2.33

If $x_0 = 2$ and $x_{n+1} = 2 + 1/x_n$, then $x_n \ge 2$ for all n. By Theorem 2.5.5 the sequence x_n has a limit $x \ge 2$. Thus

$$x = \lim_{n\to\infty} x_{n+1} = \lim_{n\to\infty} 2 + \frac{1}{x_n} = 2 + \frac{1}{x}.$$

That is,

$$x^2 - 2x - 1 = 0, \quad x = 1 + \sqrt{2}.$$

Problem 2.35

A periodic simple continued fraction is convergent. If the limit is x, then

$$x = [a_0, a_1, \dots, a_{K-1}, a_0, a_1, \dots] = [a_0, a_1, \dots, a_{K-1}, x].$$

Now

$$a_{K-1} + \frac{1}{x} = \frac{a_{K-1}x + 1}{x},$$

$$a_{K-2} + \frac{x}{a_{K-1}x + 1} = \frac{a_{K-2}(a_{K-1}x + 1) + x}{a_{K-1}x + 1}.$$

Continuing in this fashion,

$$x = \frac{r_1(x)}{r_2(x)},$$

where $r_1(x)$ and $r_2(x)$ are polynomials of degree one, and x satisfies

$$r_2(x)x - r_1(x) = 0.$$

A.3 Chapter 3 Solutions

Problem 3.1

Since

$$0 \le 2^{-k}\frac{1}{1+k^2} \le 2^{-k},$$

the series converges by comparison with the convergent geometric series $\sum_k (1/2)^k$.

Problem 3.3

We consider the general case using the ratio test.

$$\frac{c_{k+1}}{c_k} = \frac{(k+1)^m 2^{-k-1}}{k^m 2^{-k}} = \frac{(k+1)^m 2^{-1}}{k^m} = 2^{-1}(1 + 1/k)^m,$$

so

$$\lim_{k\to\infty} \frac{c_{k+1}}{c_k} = 2^{-1} < 1.$$

By the ratio test the series converges.

Problem 3.5

a) Try the ratio test.

$$\frac{c_{k+1}}{c_k} = \frac{2^{k+1}}{(k+1)!} \frac{k!}{2^k} = \frac{2}{k+1}.$$

In this case $\lim_{k\to\infty} c_{k+1}/c_k = 0$, so the series converges.

b) Notice that the terms

$$\frac{k^k}{k!} = \frac{k}{k} \frac{k}{k-1} \cdots \frac{k}{1}$$

are all at least 1. Since the terms do not have limit 0, the series can't converge.

It is also possible to use the ratio test, with the help of a limit formula you may have seen in calculus.

$$\frac{c_{k+1}}{c_k} = \frac{(k+1)^{k+1}}{(k+1)!} \frac{k!}{k^k} = \frac{(k+1)(k+1)^k}{k+1} \frac{1}{k^k} = (1+1/k)^k \to e.$$

In this case $\lim_{k\to\infty} c_{k+1}/c_k > 1$, so the series diverges.

Problem 3.7

Define the partial sums

$$s_n = \sum_{k=1}^{n} c_k, \quad \sigma_n = \sum_{k=1}^{n} a_k.$$

Since the series $\sum c_k$ converges, the sequence of partial sums $\{s_n\}$ is bounded, $|s_n| \le M$.

For $n \ge N$ we have

$$\sigma_n = \sum_{k=1}^{n} a_k = \sum_{k=1}^{N-1} a_k + \sum_{k=N}^{n} a_k \le \sum_{k=1}^{N-1} a_k + r \sum_{k=N}^{n} c_k \le \sum_{k=1}^{N-1} a_k + rM.$$

Since the sequence of partial sums $\{\sigma_n\}$ is increasing and bounded, it converges by the BMS Theorem.

Problem 3.9

The harmonic series

$$\sum_{k=1}^{\infty} \frac{1}{k}$$

satisfies $\lim_{k\to\infty} c_k = 0$, but the series diverges.

Problem 3.11

(a) Rewrite the series as

$$\sum_{k=1}^{\infty} \frac{k+1}{k^3+6} = \sum_{k=1}^{\infty} \frac{k}{k^3} \frac{1+1/k}{1+6/k^3} = \sum_{k=1}^{\infty} \frac{1}{k^2} \frac{1+1/k}{1+6/k^3}.$$

The terms

$$c_k = \frac{1+1/k}{1+6/k^3}$$

have limit 1, so for k sufficiently large

$$\frac{k+1}{k^3+6} = \frac{c_k}{k^2} \le \frac{2}{k^2}.$$

The series

$$\sum_{k=1}^{\infty} \frac{2}{k^2}$$

converges, as shown in the example following the comparison test. Thus the original series converges by the comparison test.

(b)

Theorem: Suppose

$$p(k) = a_M k^M + \cdots + a_1 k + a_0, \quad a_M \ne 0$$

is a polynomial of degree M and

$$q(k) = b_N k^N + \cdots + b_1 k + b_0, \quad b_N \ne 0$$

is a polynomial of degree N, with $q(k) \ne 0$ for $k = 1, 2, 3, \dots$.

The series

$$\sum_{k=1}^{\infty} \frac{p(k)}{q(k)}$$

converges if $N \ge M + 2$, and it diverges if $N \le M + 1$.

Proof: Factor out the highest degree terms from the numerator and denominator to get

$$\sum_{k=1}^{\infty} \frac{p(k)}{q(k)} = \sum_{k=1}^{\infty} \frac{a_M k^M}{b_N k^N} \frac{1 + \cdots + a_1 k^{1-M} + a_0 k^{-M}}{1 + \cdots + b_1 k^{1-N} + b_0 k^{-N}}$$

$$= \frac{a_M}{b_N} \sum_{k=1}^{\infty} k^{M-N} c_k,$$

where $\lim_{k\to\infty} c_k = 1$. There is a K such that $\frac{1}{2} \le c_k \le 2$ when $k \ge K$.

The convergence of the series is not affected by the sum of the first K

terms, since K is a fixed finite number. Also, convergence is not affected if the terms are multiplied by a nonzero constant.

If $N \ge M + 2$ then for $k \ge K$ we have

$$\frac{b_N}{a_M}\frac{p(k)}{q(k)} \le 2k^{M-N} \le 2k^{-2}.$$

The series converges by comparison with the series $\sum 1/k^2$.

Similarly, if $N \le M + 1$ then for $k \ge K$ we have

$$\frac{b_N}{a_M}\frac{p(k)}{q(k)} \ge \frac{1}{2}k^{M-N} \ge \frac{1}{2}k^{-1},$$

and the series diverges by the comparison with the series $\sum 1/k$.

Problem 3.13

The absolute convergence of $\sum c_k$ means that the partial sums

$$\sigma_n = \sum_{k=1}^{n} |c_k|$$

are a Cauchy sequence. Thus for any $\epsilon > 0$ there is an N such that

$$\sum_{k=m}^{n-1} |c_k| < \epsilon, \quad m, n \ge N.$$

Since

$$|s_n - s_m| = |(s_n - s_{n-1}) + (s_{n-1} - s_{n-2}) + \cdots + (s_{m+1} - s_m)| \le \sum_{k=m}^{n-1} |c_k|,$$

the partial sums of the series $\sum c_k$ are also a Cauchy sequence, so have a limit.

Problem 3.15

a) We argue by contradiction. Suppose that there are only finitely many $c_k < 0$. Then there is some N such that $c_k \ge 0$ for $k \ge N$. Since the series $\sum_{k=0}^{\infty} c_k$ converges, so does $\sum_{k=N}^{\infty} c_k$. This last series, being positive, converges absolutely, and adding $\sum_{k=0}^{N} |c_k|$ does not change the absolute convergence. We conclude that if a convergent series has only finitely many negative terms, then it converges absolutely. However, we have started by assuming that the convergence is conditional, so we can't have only finitely many negative terms. The other case is similar.

(b) Again argue by contradiction, assuming that $\sum c_k$ converges conditionally, but that $\sum b_k$ converges. By part (a) we can assume that the sequences $\{a_k\}$ and $\{b_k\}$ each have infinitely many terms.

Suppose $\sum_k b_k$ converges to B and $\sum_k c_k$ converges to C. Let

$$s_n = \sum_{k=0}^{n-1} c_k, \quad \sigma_m = \sum_{k=0}^{m-1} b_k,$$

be the partial sums of the convergent series. Examining the partial sums of the remaining series, we have

$$t_N = \sum_{k=0}^{N-1} a_k = \sum_{k=0}^{n-1} c_k - \sum_{k=0}^{m-1} b_k,$$

where $n \ge N$ and $n \ge m$.

Pick $\epsilon > 0$ and let $\epsilon_1 = \epsilon/2$. Find n_1 and m_1 such that

$$|s_n - C| < \epsilon_1, \quad n \ge n_1$$

and

$$|\sigma_m - B| < \epsilon_1, \quad m \ge m_1.$$

Now pick N_1 so that $N \ge N_1$ implies both $n \ge n_1$ and $m \ge m_1$. Then for $N \ge N_1$ we have

$$|t_N - (C + B)| = |(s_n - C) + (\sigma_m - B)| \le |s_n - C| + |\sigma_m - B| \le \epsilon.$$

We conclude that $\sum a_k$ converges as well, and so $\sum c_k$ converges absolutely.

Problem 3.17

First, test for absolute convergence using the ratio test.

$$\frac{c_{k+1}}{c_k} = \frac{|x-1|^{k+1}}{k+1} \frac{k}{|x-1|^k} = |x-1| \frac{k+1}{k}.$$

The limit is less than 1 if and only if $|x - 1| < 1$. In particular the series converges if $0 < x < 2$.

The cases $x = 0$ and $x = 2$ remain. When $x = 2$ the series is the divergent harmonic series $\sum 1/k$, while if $x = 0$ the series is the convergent alternating harmonic series $\sum (-1)^k 1/k$.

Problem 3.19

Suppose $|x| < 1$. Use the ratio test for absolute convergence.

$$\frac{c_{k+1}}{c_k} = \frac{|p(k+1)||x|^{k+1}}{|p(k)||x|^k} = \frac{|p(k+1)|}{|p(k)|}|x|.$$

If $p(k) = c_n k^n + \cdots + c_1 k + c_0$, with $c_n \ne 0$, then

$$\frac{p(k+1)}{p(k)} = \frac{(k+1)^n [c_n + \cdots + c_0/(k+1)^n]}{k^n [c_n + \cdots + c_0/k^n]} \to 1,$$

and so

$$\lim_{k\to\infty} \frac{|p(k+1)|}{|p(k)|}|x| = |x|.$$

This gives convergence if $|x| < 1$.

Here is a general observation worth recording. Suppose that $|c_k| > 0$ and

$$\lim_{k\to\infty} \frac{|c_{k+1}|}{|c_k|} > 1.$$

Then there is a number $L > 1$ and a number N such that

$$\frac{|c_{k+1}|}{|c_k|} > L.$$

By induction it follows that

$$|c_k| \ge |c_N| L^{k-N} \to \infty.$$

In particular the series $\sum c_k$ diverges, since its terms do not have limit 0.

Applying this to the problem at hand, we see that $\sum_{k=1}^{\infty} p(k)x^k$ diverges if $|x| > 1$.

Problem 3.21

If $L > 1$, pick a number L_1 satisfying $1 < L_1 < L$. Since

$$\lim_{k\to\infty} \frac{c_{k+1}}{c_k} = L,$$

there is an integer N such that

$$L_1 < \frac{c_{k+1}}{c_k}, \quad k \ge N.$$

This implies that for $k \ge N$ we have

$$c_k = c_N \frac{c_{N+1}}{c_N} \frac{c_{N+2}}{c_{N+1}} \cdots \frac{c_k}{c_{k-1}} \ge c_N L_1^{k-N}.$$

Since $L_1 > 1$,

$$\lim_{k\to\infty} |c_k| = \infty,$$

and so the series $\sum_k c_k$ diverges.

A.4 Chapter 4 Solutions

Problem 4.1

Let s_n denote the n-th partial sum of the series (4.4),

$$s_1 = 1, \quad s_2 = 1 + (-\frac{1}{2} + \frac{1}{3}), \quad s_3 = 1 + (-\frac{1}{2} + \frac{1}{3}) + (-\frac{1}{4} + \frac{1}{5} - \frac{1}{6}), \dots .$$

Each partial sum s_n matches a partial sum σ_m (with a larger index) of the alternating harmonic series,

$$s_2 = 1 - \frac{1}{2} + \frac{1}{3} = \sigma_3, \quad s_3 = 1 - \frac{1}{2} + \frac{1}{3} - \frac{1}{4} + \frac{1}{5} - \frac{1}{6} = \sigma_6, \dots .$$

That is, the sequence $\{s_n\}$ is a subsequence of $\{\sigma_m\}$. Since $\lim_{m\to\infty} \sigma_m$ exists, so does $\lim_{n\to\infty} s_n$.

Next, let s_n denote the n-th partial sum of the series (4.5). Now modify (4.5) by grouping adjacent terms of the same sign,

$$(1 + \frac{1}{2}) - (\frac{1}{3} + \frac{1}{4}) + (\frac{1}{5} + \frac{1}{6}) - (\frac{1}{7} + \frac{1}{8}) + \dots = \frac{3}{2} - \frac{7}{12} + \frac{11}{30} - \frac{15}{56} + \dots . \quad (A.4)$$

The series (A.4), with partial sums σ_n, converges by the Alternating Series Test. Notice that

$$\sigma_n = s_{2n}, \quad \sigma_{2n+1} = s_{2n} \pm \frac{1}{2n+1}.$$

Since

$$\lim_{n\to\infty} \sigma_n = \lim_{n\to\infty} s_{2n} = \lim_{n\to\infty} s_{2n+1},$$

the series (4.5) is convergent.

Problem 4.3

The idea will be to group the adjacent pairs of positive terms and study the modified series using the Alternating Series Test. Let c_n denote the n-th term of (4.3). The negative terms of this series are

$$c_{3n} = -\frac{1}{2n}, \quad n = 1, 2, 3, \dots .$$

Let b_n denote the n-th term of the modified series. Then $b_{2n} = -\frac{1}{2n} = c_{3n}$, while

$$b_{2n-1} = \frac{1}{4n-3} + \frac{1}{4n-1}.$$

Now

$$\frac{1}{4n-3} + \frac{1}{4n-1} \ge \frac{1}{4n} + \frac{1}{4n} = \frac{1}{2n} \ge \frac{1}{4n+1} + \frac{1}{4n+3},$$

so $b_{2n-1} \ge b_{2n} \ge b_{2n+1}$ and $b_n \to 0$. Thus the modified series converges by the Alternating Series Test.

The partial sums $s_n = \sum_{k=1}^{n} b_k$ of the modified series form a subsequence of the partial sums $\sigma_m = \sum_{k=1}^{m} c_k$ of the original series. In particular,

$$s_{2j} = \sigma_{3j}, \quad j = 1, 2, 3, \dots ,$$

and

$$\sigma_{3j+1} = s_{2j} + c_{3j+1}, \quad \sigma_{3j+1} = s_{2j} + c_{3j+1} + c_{3j+2}.$$

The sequence $\{s_n\}$ has a limit S, and since $\lim_{k\to\infty} c_k = 0$ the sequence $\{\sigma_m\}$ also converges to S.

The alternating harmonic series converges to a number N_1. The values of the second and third partial sums gives

$$\frac{1}{2} < N_1 < \frac{5}{6}.$$

The sum S of the series (4.3) satisfies $\frac{5}{6} < S$, so the rearrangement results in a different sum.

Problem 4.5

Break the series into N parts based on the value of $n = k \mod N$. That is, consider the sums S_n, where

$$S_1 = \sin(\pi/N)c_1 + \sin(\pi(N+1)/N)c_{N+1} + \sin(\pi(2N+1)/N)c_{2N+1} + \dots,$$

$$S_2 = \sin(\pi 2/N)c_2 + \sin(\pi(N+2)/N)c_{N+2} + \sin(\pi(2N+2)/N)c_{2N+2} + \dots,$$

$$\dots,$$

$$S_N = \sin(\pi N/N)c_N + \sin(\pi(2N)/N)c_{2N} + \sin(\pi(3N)/N)c_{3N} + \dots,$$

Since $\sin(x+\pi) = -\sin(x)$,

$$S_1 = \sin(\pi/N)c_1 - \sin(\pi/N)c_{N+1} + \sin(\pi/N)c_{2N+1} + \dots,$$

$$S_2 = \sin(\pi 2/N)c_2 - \sin(\pi 2/N)c_{N+2} + \sin(\pi 2/N)c_{2N+2} + \dots,$$

$$\dots,$$

$$S_N = 0.$$

After factoring out the common term $\sin(n\pi/N)$, the remaining series converges by the Alternating Series Test. Finally, apply Theorem 4.1.4.

Problem 4.7

Starting with the assumption $T(1) \le b$, the recurrence formula

$$T(2^m) \le aT(2^{m-1}) + b2^m, \quad m \ge 1,$$

with $m = 1$ gives

$$T(2) \le aT(1) + 2b \le ab + 2b = 2^1 b[1 + \frac{a}{2}],$$

which is the desired inequality for the first case $m = 1$. Assuming the result holds for $k < m$,

$$T(2^m) \le aT(2^{m-1}) + 2^m b \le ab2^{m-1} \sum_{k=0}^{m-1} (a/2)^k + 2^m b$$

$$= 2^m b[1 + \frac{a}{2} \sum_{k=0}^{m-1} (a/2)^k] = 2^m b \sum_{k=0}^{m} (a/2)^k.$$

Problem 4.9

$$a) \quad f(n+1) - f(n) = (n+1)^3 - n^3 = 3n^2 + 3n + 1$$

$$b) \quad g(n+1) - g(n) = \frac{1}{(n+2)^2} - \frac{1}{(n+1)^2} = \frac{(n+1)^2 - (n+2)^2}{(n+1)^2(n+2)^2} = \frac{-(2n+3)}{(n+1)^2(n+2)^2}$$

Problem 4.11

$$\Delta^+ \frac{f(n)}{g(n)} = \frac{f(n+1)}{g(n+1)} - \frac{f(n)}{g(n)} = \frac{f(n+1)g(n) - f(n)g(n+1)}{g(n)g(n+1)}$$

$$= \frac{f(n+1)g(n) - f(n)g(n) + f(n)g(n) - f(n)g(n+1)}{g(n)g(n+1)} = \frac{f^+(n)g(n) - f(n)g^+(n)}{g(n)g(n+1)}$$

Problem 4.13

If

$$f(n) = \frac{1}{n+1} + \frac{1}{n+2},$$

then

$$f^+(n) = \frac{1}{n+2} + \frac{1}{n+3} - \frac{1}{n+1} - \frac{1}{n+2} = \frac{1}{n+3} - \frac{1}{n+1} = \frac{-2}{(n+1)(n+3)}.$$

Let

$$g(n) = \frac{1}{n+1} + \frac{1}{n+2} + \cdots + \frac{1}{n+m}.$$

Then

$$g^+(n) = [\frac{1}{n+2} + \frac{1}{n+3} + \cdots + \frac{1}{n+1+m}] - [\frac{1}{n+1} + \frac{1}{n+2} + \cdots + \frac{1}{n+m}]$$

$$= \frac{1}{n+1+m} - \frac{1}{n+1} = -\frac{m}{(n+1)(n+m+1)}.$$

It follows from Theorem 4.2.3 that

$$\sum_{k=0}^{n-1} \frac{m}{(k+1)(k+m+1)} = g(0) - g(n)$$

$$= [\frac{1}{1} + \frac{1}{2} + \cdots + \frac{1}{m}] - [\frac{1}{n+1} + \frac{1}{n+2} + \cdots + \frac{1}{n+m}].$$

Problem 4.15

Take

$$f(n) = n^2, \quad g(n) = \frac{x^n}{x-1}.$$

Then Theorem 4.2.4 gives

$$\sum_{k=0}^{n-1} k^2 x^k = \sum_{k=0}^{n-1} f(k) g^+(k) = f(n)g(n) - f(0)g(0) - \sum_{k=0}^{n-1} f^+(k) g(k+1)$$

$$= n^2 \frac{x^n}{x-1} - \sum_{k=0}^{n-1} [2k+1] \frac{x^{k+1}}{x-1}.$$

Use (4.9) to conclude that

$$\sum_{k=0}^{n-1} k^2 x^k = n^2 \frac{x^n}{x-1} - \frac{2x}{x-1}[\frac{nx^n}{x-1} + \frac{x(1-x^n)}{(1-x)^2}] - \frac{x}{x-1}\frac{1-x^n}{1-x}$$

Problem 4.17

Express the function

$$p_4(n) = \sum_{k=0}^{n} k^4$$

as a polynomial in n. The recursion formula of Theorem 4.3.4 gives

$$5p_4(n) = (n+1)^5 - \Big[p_0(n) + 5p_1(n) + 10p_2(n) + 10p_3(n)\Big]$$

$$= (n+1)^5 - \Big[(n+1) + 5\frac{n(n+1)}{2} + 10\frac{n(n+1)(2n+1)}{6} + 10\frac{n^2(n+1)^2}{4}\Big].$$

Problem 4.19

a) If

$$q_k(n) = n(n+1)\cdots(n+k-1),$$

then

$$q_k(n+1) = (n+1)(n+2)\cdots(n+(k-1))(n+k)$$

and

$$nq_k(n+1) = n(n+1)(n+2)\cdots(n+(k-1))(n+k) = q_k(n)(n+k).$$

b) Using part a),

$$nq_k^+(n) = n[q_k(n+1) - q_k(n)] = (n+k)q_k(n) - nq_k(n) = kq_k(n).$$

c) Using Theorem 4.2.4 and part b),

$$\sum_{j=0}^{n-1} q_k(j) = \frac{1}{k}\sum_{j=0}^{n-1} jq_k^+(j) = \frac{1}{k}nq_k(n) - \frac{1}{k}\sum_{j=0}^{n-1} q_k(j+1) = \frac{n}{k}q_k(n) - \frac{1}{k}\sum_{j=1}^{n} q_k(j)$$

$$= \frac{n}{k} q_k(n) - \frac{1}{k} q_k(n) + \frac{1}{k} q_k(0) - \frac{1}{k} \sum_{j=0}^{n-1} q_k(j).$$

Since $q_k(0) = 0$, solving for the left hand side gives

$$\sum_{j=0}^{n-1} q_k(j) = \frac{n-1}{k+1} q_k(n).$$

A.5 Chapter 5 Solutions

Problem 5.1

As real valued functions, rational functions

$$r(x) = \frac{p(x)}{q(x)}$$

are not defined at points where $p(x) \neq 0$ but $q(x) = 0$. (If $p(x) = 0$ there may be cancellations of common factors which allow definition of the function.) For instance the function

$$r(x) = \frac{1}{x - x_0}$$

is not defined at x_0. The problem with finding a common domain for all rational functions simultaneously is that no point $x \in \mathbb{R}$ will be in the domain of all rational functions.

For a single rational function $r(x)$, the function is first written as

$$r(x) = \frac{p(x)}{q(x)}$$

where p and q have no common roots. The domain is the set of x where $q(x) \neq 0$. If we have a fixed finite collection of rational functions, we can take the intersection of their domains as the common domain. This set will include all but a finite set of real numbers.

Problem 5.3

The statement

$$\lim_{x \to \infty} f(x) = \infty$$

means that for any $M > 0$ there is an $N \in \mathbb{R}$ such that $f(x) > M$ whenever $x \geq N$.

Let $\epsilon > 0$. Pick $M > 0$ such that

$$\frac{1}{M} < \epsilon.$$

Now let $x \ge N$. Then

$$|\frac{1}{f(x)} - 0| < \frac{1}{M} < \epsilon,$$

which means

$$\lim_{x \to \infty} 1/f(x) = 0.$$

The definition of

$$\lim_{x \to \infty} f(x) = \infty$$

shows that the set of x where $f(x) = 0$ is bounded above.

Problem 5.5

Let $\epsilon = |M|/2$. There is a $\delta > 0$ such that for $0 < |x - x_0| < \delta$ we have

$$|f(x) - M| < |M|/2,$$

or

$$-|M|/2 < f(x) - M < |M|/2,$$

or

$$M - |M|/2 < f(x) < M + |M|/2.$$

If $M > 0$ then we have

$$f(x) > M/2,$$

while if $M < 0$ then $M = -|M|$ and

$$f(x) < -|M|/2.$$

In either case $f(x) \ne 0$ when $0 < |x - x_0| < \delta$.

b) State and prove an analogous result if $\lim_{x \to \infty} f(x) = M$ *and* $M \ne 0$.

The statement is there is an N such that $f(x) \ne 0$ for $x \ge N$. The proof is essentially identical.

Problem 5.7

a) Take $\epsilon = M/2$. There is a $\delta > 0$ such that for $0 < |x - x_0| < \delta$ we have

$$|f(x) - M| < M/2,$$

or

$$-M/2 < f(x) - M < M/2,$$

or

$$M/2 < f(x) < 3M/2 < 2M.$$

b) Use part (a). This shows there is a $\delta > 0$ such that

$$1/2 \le \frac{\sin(x)}{x} \le 2$$

for $0 < |x| < \delta$. If $x > 0$ we may multiply by x to get

$$x/2 \le \sin(x) \le 2x.$$

If $x < 0$ the multiplication by x reverses the inequalities, so

$$2x \le \sin(x) \le x/2.$$

Problem 5.9

a) Pick any $\epsilon > 0$. Take $\delta = \epsilon$. If

$$0 \le |x - x_0| = |x - 0| = |x| < \delta = \epsilon,$$

then

$$|f(x) - f(x_0)| = \big||x| - 0\big| = |x| < \epsilon.$$

As desired we have shown that

$$\lim_{x \to 0} |x| = |0|.$$

b) In part a) we've shown that $|x|$ is continuous at $x_0 = 0$. If $x_0 \ne 0$, for any $\epsilon > 0$ let

$$\delta = \min(|x_0|, \epsilon).$$

Now suppose

$$0 \le |x - x_0| < \delta.$$

If $x_0 > 0$, then

$$|f(x) - f(x_0)| = |x - x_0| < \epsilon.$$

If $x_0 < 0$, then

$$|f(x) - f(x_0)| = |(-x) - (-x_0)| = |x - x_0| < \epsilon.$$

Thus $f(x) = |x|$ is continuous at any real number x_0.

Problem 5.11

First consider $a < x_0 < b$. Since f is continuous on $[a, b]$ we know that for any $\epsilon > 0$ there is a $\delta_1 > 0$ such that $0 \le |x - x_0| < \delta_1$ and $a \le x \le b$ implies

$$|f(x) - f(x_0)| < \epsilon.$$

We may simply take

$$\delta = \min(\delta_1, |x_0 - a|, |x_0 - b|).$$

That is, use the same δ as long as it is small enough so that only points in (a, b) are considered.

The same idea works for $b < x_0 < c$, and a similar plan applies for $x_0 = a$ or $x_0 = c$.

Suppose that $x_0 = b$. Since f is continuous on $[a, b]$ we know that for any $\epsilon > 0$ there is a $\delta_1 > 0$ such that $0 \le |x - x_0| < \delta_1$ and $a \le x \le b$ implies

$$|f(x) - f(b)| < \epsilon.$$

Since f is continuous on $[b, c]$ we know that for any $\epsilon > 0$ there is a $\delta_2 > 0$ such that $0 \le |x - x_0| < \delta_2$ and $b \le x \le c$ implies

$$|f(x) - f(b)| < \epsilon.$$

Pick

$$\delta = \min(\delta_1, \delta_2)$$

to get the desired inequality.

It is not enough to assume that f is continuous on the intervals (a, b) and (b, c). As a counterexample, consider

$$f(x) = \begin{Bmatrix} 0, & x < 0, \\ 1, & x \ge 0 \end{Bmatrix}.$$

This function is continuous on $(-1, 0)$ and $(0, 1)$, but not on $[-1, 1]$.

Problem 5.13

A polynomial $p(x)$ is continuous at every point $x \in \mathbb{R}$ by Theorem 5.2.1 and the observation that $g(x) = x$ is continuous at every $x \in \mathbb{R}$. The function $|x|$ is also continuous at every $x \in \mathbb{R}$; see Problem 5.9. By Theorem 5.2.5 the composition of continuous functions is continuous, so $f(x) = |p(x)|$ is continuous for all $x \in \mathbb{R}$.

Problem 5.15

First recall that the function x^2 is continuous on $\mathbb{R}$. Pick any x_0, and let $\{q_k\}$ be a sequence of rationals with

$$\lim_{k \to \infty} q_k = x_0.$$

Since g is continuous at x_0 we have

$$g(x_0) = \lim_{k \to \infty} g(q_k) = \lim_{k \to \infty} (q_k)^2 = x_0^2.$$

Problem 5.17

Consider the function $g(x) = f(x) - x$. Then $g(x)$ is continuous on $[0, 1]$ and

$$g(0) = f(0) < 0, \quad g(1) = f(1) - 1 > 0.$$

By the intermediate value theorem there is some point $x_0 \in [0, 1]$ where $g(x_0) = f(x_0) - x_0 = 0$, or

$$f(x_0) = x_0.$$

Problem 5.19

Every iteration halves the length of the interval. To get $|b-r| \le 10^{-10}$ we want

$$[b-a]2^{-n} = 2^{1-n} < 10^{-10}$$

or

$$n > 10\log_2(10) + 1.$$

Problem 5.21

If $C = 0$ then f is a constant, and so is uniformly continuous. Thus we may suppose that $C > 0$. Pick any $\epsilon > 0$, and take $\delta = \epsilon/C$. Then for any x_1 and x_2 such that $|x_1 - x_2| < \delta$ we have

$$|f(x_1) - f(x_2)| \le C|x_1 - x_2| < C\delta = \epsilon.$$

Problem 5.23

In order to have a derivative, the limit

$$f'(0) = \lim_{x \to 0} \frac{f(x) - f(0)}{x - 0}$$

has to exist. If $f(x) = |x|$ this limit is

$$\lim_{x \to 0} \frac{|x|}{x}.$$

Notice however, that

$$\lim_{x \to 0+} \frac{|x|}{x} = 1,$$

while

$$\lim_{x \to 0-} \frac{|x|}{x} = -1.$$

Since these values are different, the limit does not exist.

Problem 5.25

a) This is where the h version of the definition is a bit more convenient.

$$\frac{d}{dx}x^2 = \lim_{h \to 0} \frac{(x+h)^2 - x^2}{h} = \lim_{h \to 0} \frac{2xh + h^2}{h} = 2x.$$

Similarly,

$$\frac{d}{dx}x^3 = \lim_{h \to 0} \frac{(x+h)^3 - x^3}{h} = \lim_{h \to 0} \frac{3x^2h + 3xh^2 + h^3}{h} = 3x^2.$$

b) The method of part a) can be extended. Alternatively, induction and

Theorem 5.3.3 will do the job. The case $n = 1$ is easy. If the formula holds for x^n, that is

$$\frac{d}{dx}x^n = nx^{n-1},$$

then the product rule establishes the formula for x^{n+1},

$$\frac{d}{dx}x^{n+1} = \frac{d}{dx}xx^n = x^n + xnx^{n-1} = (n+1)x^n.$$

Problem 5.27

a) Since f is differentiable at $x_0 = 0$, it is also continuous at $x_0 = 0$. By Theorem 7.3.2 we may evaluate the limit along the sequence x_n, getting

$$f(0) = \lim_{n\to\infty} f(x_n) = \lim_{n\to\infty} 0 = 0.$$

We may apply the same idea to evaluate the limit in the derivative, since we know the limit exists. Thus

$$f'(0) = \lim_{n\to\infty} \frac{f(x_n) - f(0)}{x_n - 0} = \lim_{n\to\infty} 0 = 0.$$

To verify the logic, simply run through the first half of the proof of Theorem 7.3.2, replacing $f(x)$ by the difference quotient

$$\frac{f(x) - f(x_0)}{x - x_0}$$

and replacing $f(x_0)$ with $f'(x_0)$.

b) The function is not differentiable. To see this, look at the difference quotient,

$$\frac{f(x) - f(0)}{x - 0} = \sin(1/x).$$

We want to see if this function has a limit as $x \to 0$. First consider the sequence $x_n = 1/n\pi$. Here we find $g(x_n) = \sin(n\pi) = 0$. Next let $t_n = 1/(2n\pi + \pi/2)$. For these values we get

$$g(t_n) = \sin(2n\pi + \pi/2) = \sin(\pi/2) = 1.$$

Since both sequences have limit 0, but

$$\lim_{n\to\infty} g(x_n) \neq \lim_{n\to\infty} g(t_n),$$

there is not limit for the difference quotient,

$$\frac{f(x) - f(0)}{x - 0} = \sin(1/x).$$

Problem 5.29

Let x_1 and x_2 be any two distinct points in (a, b); without loss of generality assume that $x_1 < x_2$. Since f is differentiable on (a, b) it is continuous on $[x_1, x_2]$. By the Mean Value theorem there is a point $\xi \in (x_1, x_2)$ such that

$$f'(\xi) = \frac{f(x_2) - f(x_1)}{x_2 - x_1}.$$

By assumption $f'(\xi) = 0$, so

$$f(x_2) = f(x_1).$$

Thus $f(x)$ is the constant $f(x_1)$.

Problem 5.31

By a previous problem, a polynomial with real coefficients and odd degree has at least one real root. Suppose f has two (or more) real roots, x_1 and x_2. Since polynomials $f(x)$ are differentiable for all real x, Rolle's theorem may be applied. By Rolle's theorem there must be a point $\xi \in (x_1, x_2)$ with $f'(\xi) = 0$. Calculating thc derivative we find that

$$f'(x) = 5x^4 + 3x^2 + 1,$$

which is always at least as big as 1. Since $f'(x)$ has no zeroes, f cannot have two distinct zeroes.

Problem 5.33

This is another combination of the Mean Value Theorem and an argument by contradiction. We claim that

$$-x - 1 \le f(x) \le x + 1, \quad x > 0.$$

Suppose that for some $x_1 > 0$ the inequality

$$f(x_1) > x_1 + 1$$

holds. The Mean Value Theorem then gives

$$f'(\xi) = \frac{f(x_1) - f(0)}{x_1 - 0} > \frac{x_1 + 1 - 1}{x_1 - 0} = 1,$$

contradicting the bound on the derivative of f. The argument is similar for the inequality

$$f(x_1) < -x_1 - 1.$$

Of course the function $f(x) = x + 1$ achieves the bound.

Problem 5.35

The argument is by induction. If $n = 1$ and $f'(x) = 0$, then $f(x)$ is a constant by the Mean Value Theorem. Suppose the result is true if $f^{(N)}(x) =$

0, and assume now that $f^{(N+1)}(x) = 0$. Notice that there is a constant c with $f^{(N)}(x) = c$. Using the fact that

$$\frac{d^N}{dx^N} x^N = N!,$$

consider the function

$$g(x) = f(x) - c\frac{x^N}{N!}.$$

Since

$$g^{(N)}(x) = c - c = 0,$$

the function $g(x)$ is a polynomial of degree at most $N-1$, and $f(x)$ is a polynomial of degree at most N.

Problem 5.37

For any positive integer n the function x^n has a positive derivative for $x \in (0, \infty)$. The inverse function is $x^{1/n}$. By Theorem 5.3.13 the function $x^{1/n}$ is differentiable, with

$$\frac{d}{dx} x^{1/n} = \frac{1}{n(x^{1/n})^{n-1}} = \frac{1}{n} x^{1/n-1}.$$

Problem 5.39

a) Take $[c, d] = [1, 2]$ and $f(x) = 0$ for all $x \in [1, 2]$. The function $f(x)$ has no fixed point in the interval.

b) Take $f(x) = x/2$. The only real number x with $f(x) = x$ is $x = 0$, which is not in the interval $(0, 1)$.

Problem 5.41

a) The function $f(x)$ is strictly decreasing on $(0, \infty)$, with $\lim_{x \to 0^+} f(x) = \infty$ and $\lim_{x \to \infty} f(x) = a_n$. Consequently, the function $f(x) - x$ is strictly decreasing on $(0, \infty)$, with $\lim_{x \to 0^+} f(x) = \infty$ and $\lim_{x \to \infty} f(x) = -\infty$. By the Intermediate Value Theorem there is some $x_1 > 0$ with $f(x_1) - x_1 = 0$, and since $f(x) - x$ is strictly decreasing, this fixed point is unique.

b) Since $a_k > 0$ for $k = 0, \dots, n$, if $x \ge a_n$ then $f(x) = a_n + \sum_{k=0}^{n-1} a_k x^{k-n} > a_n$. In addition the functions x^{k-n} are decreasing, so $f(x) \le f(a_n)$ if $x \ge a_n$. In particular $f : [a_n, f(a_n)] \to [a_n, f(a_n)]$.

c) Notice that

$$f'(x) = \sum_{k=0}^{n-1} (k-n) a_k x^{k-n-1},$$

so for $x \in [a_n, f(a_n)]$

$$|f'(x)| \le |\sum_{k=0}^{n-1} (k-n) \frac{a_k}{a_n^{n-k+1}}|.$$

If the sum on the right is less than 1, the result follows from Theorem 5.3.16.

Problem 5.43

a) Since the constants c_k are positive, a simple calculation gives

$$f''(x) = \sum_{k=1}^{n} c_k a_k^2 \exp(a_k x) > 0.$$

If $\lim_{x \to \pm\infty} f(x) = \infty$ there must be at least one minimizer, but since f is strictly convex there is at most one minimizer.

b) The function $f(x) = e^x$ is a strictly convex function with no global minimizer.

Problem 5.45

Suppose $f(x_i) = 0$ for $i = 0, 1, 2$ with $x_0 < x_1 < x_2$. Write

$$x_1 = t_1 x_2 + (1 - t_1) x_0, \quad t_1 = \frac{x_1 - x_0}{x_2 - x_0}.$$

If f were strictly convex, then since $0 < t_1 < 1$ the inequality

$$f(x_1) = f(t_1 x_2 + (1 - t_1) x_0) < t_1 f(x_2) + (1 - t_1) f(x_0)$$

should hold. But it doesn't.

Problem 5.47

(a) If f is continuous at x_k then

$$\lim_{x \to x_k^+} f(x) = f(x_k) = \lim_{x \to x_k^-} f(x),$$

and f has no jump at x_k.

Since f is increasing,

$$f(x) \le f(x_k), \quad x \le x_k,$$

and

$$f(x) \ge f(x_k), \quad x \ge x_k,$$

so

$$\lim_{x \to x_k^+} f(x) \le f(x_k) \le \lim_{x \to x_k^-} f(x).$$

If f has no jump at x_k then

$$\lim_{x \to x_k^+} f(x) = \lim_{x \to x_k^-} f(x),$$

so the limits equal $f(x_k)$, and f is continuous at x_k.

(b) For any $\epsilon > 0$ there is a positive integer N such that $1/N < \epsilon$. Suppose there are N points $0 < x_1 < \cdots < x_N < 1$ with $s(x_k) > \epsilon$. Pick points $y_0, \dots, y_N$ with $0 < y_0 < x_1 < y_1 < \cdots < x_N < y_N < 1$. Since f is increasing,

$$1 \geq f(y_N) - f(y_0) = [f(y_N) - f(y_{N-1}] + \cdots + [f(y_2) - f(y_1)] + [f(y_1) - f(y_0)].$$

Moreover $f(y_n) - f(y_{n-1}) \geq s(x_n)$, so $1 \geq N\epsilon > 1$. This contradiction shows the N points x_k can't exist.

A.6 Chapter 6 Solutions

Problem 6.1

Since the function $f(x) = x^3$ is increasing for $x \geq 0$, the minimum and maximum values of the function f on any subinterval $[x_k, x_{k+1}]$ are at x_k and x_{k+1} respectively. In this new case

$$f(x_k) = x_k^3 = \frac{k^3 b^3}{n^3}$$

and

$$A_o = \sum_{k=1}^{n} \frac{b}{n} \frac{k^3 b^3}{n^3} = \frac{b^4}{n^4} \sum_{k=1}^{n} k^3 = \frac{b^4}{n^4} \frac{n^2(n+1)^2}{4} = \frac{b^4}{4}[1 + \frac{2}{n} + \frac{1}{n^2}].$$

Similarly

$$A_i = \sum_{k=0}^{n-1} \frac{b}{n} \frac{k^3 b^3}{n^3} = \frac{b^4}{n^4} \frac{(n-1)^2 n^2}{4} = \frac{b^4}{4}[1 - \frac{2}{n} + \frac{1}{n^2}].$$

Again $A_i < A < A_o$ for every positive integer n, so the area under the graph of x^3 is bigger than any number smaller than $b^4/4$, and smaller than any number bigger than $b^4/4$, or $A = b^4/4$.

Problem 6.3

a) For each k with $1 \leq k \leq n-1$ the term

$$f(x_k)[x_k - x_{k-1}] = f(k\frac{b-a}{n})\frac{b-a}{n}$$

from the right endpoint sum exactly matches

$$f(x_k)[x_{k+1} - x_k] = f(k\frac{b-a}{n})\frac{b-a}{n}$$

from the left endpoint sum. These terms cancel when $L_n - R_n$ is computed, leaving

$$L_n - R_n = f(x_0)\frac{b-a}{n} - f(x_n)\frac{b-a}{n} = \frac{b-a}{n}[f(a) - f(b)].$$

b) The inequalities

$$R_n \le \int_a^b f(x)\,dx \le L_n,$$

give

$$|\int_a^b f(x)\,dx - L_n| = L_n - \int_a^b f(x)\,dx \le L_n - R_n = \frac{b-a}{n}[f(a) - f(b)],$$

and

$$|\int_a^b f(x)\,dx - R_n| \le L_n - R_n = \frac{b-a}{n}[f(a) - f(b)].$$

c) Since $T_1 = L_n - \int_a^b f(x)\,dx \ge 0$, $T_2 = \int_a^b f(x)\,dx - R_n \ge 0$, and

$$T_1 + T_2 = [L_n - \int_a^b f(x)\,dx] + [\int_a^b f(x)\,dx - R_n] = L_n - R_n = \frac{b-a}{n}[f(a) - f(b)],$$

it is not possible for both T_1 and T_2 to be smaller than half the size of $L_n - R_n$.

d) One should pick n such that

$$\frac{b-a}{n}[f(a) - f(b)] \le 10^{-6},$$

or

$$\frac{9}{n}[1 - \frac{1}{10}] \le 10^{-6}.$$

That is, $n \ge 81 \cdot 10^5$.

Problem 6.5

Suppose to the contrary, that $f(x_1) > 0$ for some $x_1 \in [a, b]$. Since f is continuous there is an interval $[c, d]$ containing x_1 where

$$f(x) \ge f(x_1)/2, \quad x \in [c, d].$$

If $\mathcal{P}$ is any partition containing the points c, d, then since each $m_k \ge 0$ it follows that the lower sum for this partition satisfies

$$\mathcal{L}(f, \mathcal{P}) \ge \frac{f(x_1)}{2}[d - c].$$

Since all partitions have a refinement which contains the points c, d, and taking refinements raises lower sums, it follows that

$$\int_a^b f(x)\,dx > \frac{f(x_1)}{2}[d - c] > 0.$$

This contradiction is what we needed.

The function

$$f(x) = \begin{cases} 0, & x \neq 1/2 \\ 1, & x = 1/2 \end{cases}$$

is integrable, nonnegative, and has integral 0 over any interval $[a, b]$, so the conclusion is false in this case.

Problem 6.7

First, let

$$A = \int_a^b t^2 \, dt = \frac{b^3}{3} - \frac{a^3}{3}$$

be the area under the graph of t^2. Notice that the function

$$F(x) = \int_a^x t^2 \, dt = \frac{x^3}{3} - \frac{a^3}{3}$$

is a polynomial, which is continuous, and that

$$F(a) = 0, \quad F(b) = A.$$

By the intermediate value theorem there are points x_k such that

$$F(x_k) = kA/n, \quad k = 1, \dots, n-1.$$

If $x_0 = a$, $x_n = b$, then

$$\int_{x_k}^{x_{k+1}} t^2 \, dt = F(x_{k+1}) - F(x_k) = A/n.$$

Thus we get the desired property

$$\int_{x_j}^{x_{j+1}} t^2 \, dt = \int_{x_k}^{x_{k+1}} t^2 \, dt.$$

Problem 6.9

The idea is quite similar to that of Problem 6.5.

Suppose to the contrary, that $f(x) > 0$ for some $x_1 \in [a, b]$. Since f is continuous, there is an interval $[c, d]$ containing x_1 where

$$f(x) \geq f(x_1)/2, \quad x \in [c, d].$$

If $\mathcal{P}$ is any partition of $[c, d]$, then the lower sum for this partition satisfies

$$\mathcal{L}(f, \mathcal{P}) \geq \frac{f(x_1)}{2}[d - c].$$

It follows that

$$\int_c^d f(x) \, dx > \frac{f(x_1)}{2}[d - c] > 0.$$

This contradiction is what we needed.

Problem 6.11

Apply Theorem 6.2.6 repeatedly with $f_1 = f_2 = f$.

Problem 6.13

Since f is continuous on the interval $[a, b]$, there are points $x_1, x_2 \in [a, b]$ with

$$f(x_1) \le f(x) \le f(x_2), \quad a \le x \le b.$$

Since $w(x) \ge 0$ we have

$$f(x_1)w(x) \le f(x)w(x) \le f(x_2)w(x), \quad a \le x \le b,$$

so

$$\begin{aligned} f(x_1) \int_a^b w(x)\, dx &= \int_a^b f(x_1)w(x)\, dx \\ &\le \int_a^b f(x)w(x)\, dx \\ &\le \int_a^b f(x_2)w(x)\, dx = f(x_2) \int_a^b w(x)\, dx. \end{aligned}$$

Now for t in the interval with endpoints x_1 and x_2 define

$$g(t) = f(t) \int_a^b w(x)\, dx.$$

This function is continuous, and by the Intermediate Value Theorem there is some ξ such that

$$g(\xi) = f(\xi) \int_a^b w(x)\, dx = \int_a^b f(x)w(x)\, dx.$$

Problem 6.15

There is a $C > 0$ such that $|f(x)| \le C$ for all $x \in [a, b]$. For any $\epsilon > 0$ there is a $\delta > 0$ such that $a+\delta < b-\delta$ and $4\delta C < \epsilon/2$. By Theorem 6.2.4 the function $f(x)$ is integrable on $[a + \delta, b - \delta]$, so there is a partition $\mathcal{P}_1 = \{x_0, \dots, x_n\}$ of $[a + \delta, b - \delta]$, with

$$\mathcal{U}(f, \mathcal{P}_1) - \mathcal{L}(f, \mathcal{P}_1) < \epsilon/2.$$

Extend $\mathcal{P}_1$ to a partition $\mathcal{P} = \{\xi_0, \dots, \xi_{n+2}\}$ of $[a, b]$ with $\xi_0 = a$, $\xi_{n+2} = b$, and $\xi_k = x_{k-1}$ for $k = 1, \dots, n + 1$. Since

$$\mathcal{U}(f, \mathcal{P}) - \mathcal{L}(f, \mathcal{P}) < \epsilon/2 + 2 \cdot 2 \cdot \delta \cdot C < \epsilon,$$

the function f is integrable on $[a, b]$.

Problem 6.17

Start with the function $h : [0,1] \to \mathbb{R}$ given by $h(x) = 0$ if x is rational, and $h(x) = 1$ if x is irrational. Now take

$$f(x) = \left\{ \begin{matrix} h(x), & 0 \le x < 1/2, \\ 0, & 1/2 \le x \le 1. \end{matrix} \right\}, \quad g(x) = \left\{ \begin{matrix} h(x), & 1/2 < x \le 1, \\ 0, & 0 \le x \le 1/2. \end{matrix} \right\}.$$

The functions f and g are not integrable, $f(x)g(x) = 0$ for all $x \in [0,1]$.

Problem 6.19

Since f is integrable there is a constant C such that $|f(x)| \le C$ for all $x \in [a,b]$. Suppose $x_0 \in [a,b]$ and $x > x_0$. By Theorem 6.2.6,

$$F(x) = \int_a^{x_0} f(t)\ dt + \int_{x_0}^{x} f(t)\ dt = F(x_0) + \int_{x_0}^{x} f(t)\ dt.$$

The set $\mathcal{P} = \{x_0, x\}$ serves as a partition for $[x_0, x]$, and

$$\mathcal{L}(f, \mathcal{P}) \ge -C|x - x_0|, \quad \mathcal{U}(f, \mathcal{P}) \le C|x - x_0|.$$

Thus

$$|F(x) - F(x_0)| \le C|x - x_0|,$$

and $\lim_{x \to x_0^+} F(x) = F(x_0)$. The argument is similar if $x < x_0$, so $F(x)$ is continuous at x_0.

Problem 6.21 Integration by substitution .

By the Fundamental Theorem of Calculus

$$\int_{g(a)}^{g(b)} f(x)\ dx = F(g(b)) - F(g(a)).$$

On the other hand, the chain rule gives

$$\frac{d}{dt} F(g(t)) = F'(g(t))g'(t)) = f(g(t))g'(t),$$

so

$$\int_a^b f(g(t))g'(t)\ dt = F(g(b)) - F(g(a)).$$

Problem 6.23

Suppose $f'(x)$ is continuous and nonnegative for $x \in [a,b]$. For a partition $\mathcal{P} = \{x_0, \dots, x_n\}$, let $\Delta x_k = x_{k+1} - x_k$.

a) This is just

$$\int_a^b f(x)\ dx = \sum_{k=0}^{n-1} \int_{x_k}^{x_{k+1}} f(x)\ dx$$

together with

$$\sum_{k=0}^{n-1} f(x_k)\Delta x_k = \sum_{k=0}^{n-1} \int_{x_k}^{x_{k+1}} f(x_k)\ dx.$$

b) Since f is increasing on the interval $[x_k, x_{k+1}]$, Theorem 5.3.12 gives

$$f(x) - f(x_k) \ge C_k(x - x_k) \ge 0, \quad x_k \le x \le x_{k+1}.$$

Integration gives

$$\int_{x_k}^{x_{k+1}} f(x) - f(x_k)\ dx \ge \int_{x_k}^{x_{k+1}} C_k(x - x_k)\ dx = C_k \frac{(x - x_k)^2}{2}\Big|_{x_k}^{x_{k+1}} = C_k(\Delta x_k)^2/2.$$

c) If $f' \le 0$ we can simply consider $g(x) = -f(x)$, and let $0 < C \le |f'|$. Use the above analysis to get

$$-\int_a^b f(x)\ dx + \sum_{k=0}^{n-1} f(x_k)\Delta x_k \ge C\frac{(b-a)^2}{2n},$$

or

$$\sum_{k=0}^{n-1} f(x_k)\Delta x_k - \int_a^b f(x)\ dx \ge C\frac{(b-a)^2}{2n}.$$

In this case the left endpoint Riemann sums overestimate the integral.
For the example given,

$$f' = -2xe^{-x^2} \le -e^{-1}, \quad 1/2 \le x \le 1.$$

Thus the Riemann sum estimate is off by at least

$$e^{-1}\frac{(1 - 1/2)^2}{2n} = e^{-1}\frac{1}{8n}.$$

Problem 6.25

a) Recall from Calculus that

$$\int_0^1 \pi\sin(\pi x)\ dx = -\cos(\pi x)\Big|_0^1 = 2.$$

The left endpoint Riemann sum for this integral with n equally spaced sample points $x_k = k/n$ is

$$\sum_{k=0}^{n-1} \frac{\pi}{n}\sin(k\pi/n),$$

establishing the claim.
b) Approximate the integral

$$\int_0^1 \frac{1}{1+x^2}\ dx = \tan^{-1}(1) = \pi/4.$$

by a left endpoint Riemann sum

$$\sum_{k=0}^{n-1} \frac{1}{n} \frac{1}{1+(k/n)^2} = \sum_{k=0}^{n-1} \frac{n}{n^2+k^2}.$$

with n equally spaced sample points. Since the Riemann sums converge to the integral,

$$\lim_{n\to\infty} \sum_{k=0}^{n-1} \frac{n}{n^2+k^2} = \pi/4.$$

Problem 6.27

Since $f(0) = 0$,

$$f(x) = \int_0^x f'(t)\, dt.$$

Next,

$$|f(x)| \le \int_0^x |f'(t)| \cdot 1\, dt.$$

By the Cauchy–Schwarz inequality Theorem 6.5.4

$$|f(x)| \le \Big[\int_0^x |f'(t)|^2\, dt \int_0^x 1^2\, dt\Big]^{1/2} \le \Big[\int_0^x |f'(t)|^2\, dt\Big]^{1/2} \sqrt{x},$$

so the result holds with

$$C = \Big[\int_0^\infty |f'(t)|^2\, dt\Big]^{1/2}.$$

A.7 Chapter 7 Solutions

Problem 7.1

The equation of the line joining $(1.5, 0.1761)$ and $(1.6, 0.2041)$ is

$$y = 0.28(x - 1.5) + 0.1761.$$

Thus

$$\log_{10}(1.53) \simeq 0.28 \cdot 0.03 + 0.1761 = 0.1845$$

The actual value to four digits is 0.1847.

Problem 7.3

For $t > 0$ the function $1/t$ is positive and decreasing, so for $n = 2, \dots, N$

$$\int_{n-1}^{n} \frac{1}{t}\, dt \ge \frac{1}{n}.$$

Thus

$$\int_1^N \frac{1}{t}\,dt = \sum_{n=2}^{N} \int_{n-1}^{n} \frac{1}{t}\,dt > \sum_{n=2}^{N} \frac{1}{n}.$$

Since the harmonic series is divergent, $\lim_{N\to\infty} \log(N) = \infty$. The function $\log(x)$ is increasing, so $\lim_{x\to\infty} \log(x) = \infty$

Problem 7.5

Starting with

$$\log(x) = \int_1^x \frac{1}{t}\,dt,$$

follow the proof of Theorem 7.2.2, replacing e with x. The main step is to show

$$\log(x^{1/n}) = \frac{1}{n}\log(x), \quad n = 1, 2, 3, \dots.$$

Problem 7.7

Examine

$$\log(p_N) = \log(\prod_{n=1}^{N}(1 - 6^{-n})) = \sum_{n=1}^{N} \log(1 - 6^{-n}).$$

By Theorem 7.3.2 the sequence $\{p_N\}$ has a positive limit if $\sum_{n=1}^{\infty} 6^{-n}$ converges. This is a convergent geometric series, so the chance of winning is

$$P = \lim_{N\to\infty} \prod_{n=1}^{N}(1 - 6^{-n}) \simeq \exp(-\sum_{n=1}^{\infty} 6^{-n}) = e^{-1/5} \simeq 4/5.$$

The expected loss suggests that playing is inadvisable.

Problem 7.9

a) The integrand is decreasing so the integral is smaller than the corresponding left endpoint Riemann sum, and larger than the corresponding right endpoint Riemann sum.

b) On the interval $[k, k+1]$ the function $1/x$ is decreasing, so that

$$\int_k^{k+1} \frac{1}{x}\,dx < \frac{1}{k} \cdot 1,$$

which is the first claim. Each of the differences

$$\frac{1}{k} - [\log(k+1) - \log(k)]$$

is positive, so the sum

$$f(n) = \sum_{k=1}^{n}(\frac{1}{k} - [\log(k+1) - \log(k)]) = (\sum_{k=1}^{n} \frac{1}{k}) - \log(n+1)$$

increases with n. (Notice the cancellation of logarithmic terms in the sum.)

c) From (7.22) we see that

$$f(n) = (\sum_{k=1}^{n} \frac{1}{k}) - \log(n+1) \le 1 + \log(n) - \log(n+1) = 1 + \log(\frac{n}{n+1}).$$

Since $n/(n+1) < 1$ for $n \ge 1$ its logarithm is < 0, and so

$$f(n) < 1.$$

Problem 7.11

Stirling's formula gives the following estimates. If n is even then

$$\binom{n}{n/2} = \frac{n!}{(n/2)!(n/2)!} \simeq \frac{\sqrt{2\pi} n^{n+1/2} e^{-n}}{[\sqrt{2\pi}(n/2)^{n/2+1/2} e^{-n/2}]^2}$$

$$= \frac{n^{n+1/2}}{\sqrt{2\pi}(n/2)^{n+1}} = \frac{2^{n+1} n^{-1/2}}{\sqrt{2\pi}}.$$

For n odd we reduce to the even case as follows.

$$\binom{n}{(n-1)/2} = \frac{n!}{[(n-1)/2]![(n+1)/2]!}$$

$$= \frac{n(n-1)!}{[(n-1)/2]![(n+1)/2][(n-1)/2]!} = \frac{n}{(n+1)/2} \binom{n-1}{(n-1)/2}$$

$$\simeq \frac{n}{(n+1)} \frac{2^{n+1}(n-1)^{-1/2}}{\sqrt{2\pi}} \simeq \frac{2^{n+1} n^{-1/2}}{\sqrt{2\pi}}$$

as before.

Problem 7.13

Suppose that the sequence $p_n = \prod_{k=1}^{n}(1 + a_k)$ has a limit p, with $0 < p < \infty$, and $(1 + a_k) > 0$ for $k \ge N$. There is no loss of generality if we assume $(1 + a_k) > 0$ for all k. Since $\log(x)$ is continuous at p,

$$\log(p) = \log(\lim_{n\to\infty} \prod_{k=1}^{n}(1+a_k)) = \lim_{n\to\infty} \log(\prod_{k=1}^{n}(1+a_k)) = \lim_{n\to\infty} \sum_{k=1}^{n} \log(1+a_k).$$

Thus the series $\sum_{k=1}^{\infty} \log(1 + a_k)$ converges to $\log(p)$.

Suppose the terms a_k are all positive (the negative case is similar). Since $\lim_{k\to\infty} \log(1 + a_k) = 0$, the inequality $0 < a_k < 1/2$ holds for k sufficiently large. For such k apply Lemma 7.3.1 to conclude that $a_k/2 \le \log(1+a_k) \le 2a_k$, so the series $\sum_{k=1}^{\infty} a_k$ converges by the comparison test.

A.8 Chapter 8 Solutions

Problem 8.1

Recall the general Taylor series formula for $f(x)$ based at x_0:

$$\sum_{k=0}^{\infty} \frac{f^{(k)}(x_0)}{k!}(x - x_0)^k.$$

In this case $f^{(k)}(x) = e^x$, so the Taylor series based at $x_0 = 1$ is

$$\sum_{k=0}^{\infty} \frac{e}{k!}(x-1)^k.$$

Problem 8.3

In this case

$$f(x) = \log(1+x), \quad f'(x) = \frac{1}{1+x} = [1+x]^{-1},$$

$$f^{(2)}(x) = -[1+x]^{-2}, \quad f^{(3)}(x) = 2[1+x]^{-3},$$

and by induction

$$f^{(k)}(x) = (-1)^{k-1}(k-1)![1+x]^{-k}, \quad k \ge 1.$$

We also have $f^{(0)}(0) = \log(1) = 0$. Plugging in to the general form we find that the Taylor series based at $x_0 = 0$ is

$$\sum_{k=1}^{\infty} (-1)^{k-1} \frac{(k-1)!}{k!} x^k = \sum_{k=1}^{\infty} (-1)^{k-1} \frac{1}{k} x^k.$$

Problem 8.5

Split the Taylor series into two pieces, containing respectively the even and odd values for k. Formal computation gives

$$e^{ix} = \sum_{m=0}^{\infty} \frac{(ix)^m}{m!} = \sum_{k=0}^{\infty} \frac{(ix)^{2k}}{(2k)!} + \sum_{k=0}^{\infty} \frac{(ix)^{2k+1}}{(2k+1)!}$$

$$= \sum_{k=0}^{\infty} \frac{(-1)^k x^{2k}}{(2k)!} + i \sum_{k=0}^{\infty} \frac{(-1)^k x^{2k+1}}{(2k+1)!} = \cos(x) + i\sin(x).$$

Problem 8.7

Taylor's Theorem gives us

$$f(x) = \sum_{k=0}^{n} \frac{f^{(k)}(x_0)}{k!}(x - x_0)^k + \int_{x_0}^{x} f^{(n+1)}(t)\frac{(x-t)^n}{n!}\, dt.$$

If the $n+1$-st derivative is equal to 0 at every point of the interval (a, b), then the remainder vanishes for all values of $x \in (a, b)$ and

$$f(x) = \sum_{k=0}^{n} \frac{f^{(k)}(x_0)}{k!}(x - x_0)^k,$$

which is a polynomial of degree at most n.

Problem 8.9

Writing the polynomial $p(x)$ as a Taylor polynomial based at x_0 we get

$$p(x) = \sum_{j=0}^{n} c_j x^j = \sum_{k=0}^{n} \frac{p^{(k)}(x_0)}{k!}(x - x_0)^k = \sum_{k=0}^{n} a_k (x - x_0)^k.$$

Thus

$$a_k = \frac{p^{(k)}(x_0)}{k!} = \sum_{j=0}^{n} j(j-1)\cdots(j-k+1)c_j x_0^{j-k}$$

$$= \sum_{j=k}^{n} j(j-1)\cdots(j-k+1)c_j x_0^{j-k}.$$

Problem 8.11

The remainder satisfies

$$|R_n(x)| \le M\frac{|x - x_0|^{n+1}}{(n+1)!}, \qquad M = \max_{x_0 \le \xi \le x} |f^{(n+1)}(\xi)|.$$

In this case $f(x) = e^x$, $x_0 = 0$, and $x = 1$. Thus $M \le 4$ and we want

$$\frac{4}{(n+1)!} \le 10^{-12},$$

which my calculator says is satisfied if $n \ge 15$.

Problem 8.13

Theorem 8.3.3 says that there is some ξ between x_0 and x such that

$$R_n(x) = f^{(n+1)}(\xi)\frac{(x - x_0)^{n+1}}{(n+1)!}.$$

Taking absolute values we get

$$|R_n(x)| = |f^{(n+1)}(\xi)|\frac{|x - x_0|^{n+1}}{(n+1)!}.$$

If $M = \max|f^{(n+1)}(\xi)|$ for all ξ between x_0 and x, then

$$|R_n(x)| \le M\frac{|x-x_0|^{n+1}}{(n+1)!}$$

as desired.

Problem 8.15

Recall that

$$\log(\frac{1+x}{1-x}) = \log(1+x) - \log(1-x).$$

The Taylor series for $\log(1+x)$ is

$$\log(1+x) = \sum_{k=0}^{n-1}(-1)^k\frac{x^{k+1}}{k+1} + R_n^+(x),$$

where

$$R_n^+(x) = \int_0^x \frac{(-t)^n}{1+t}.$$

Similarly,

$$\log(1-x) = -\sum_{k=0}^{n-1}\frac{x^{k+1}}{k+1} + R_n^-(x),$$

where

$$R_n^-(x) = \int_0^{-x} \frac{(-t)^n}{1+t}.$$

Subtracting the series we get cancellation of every other term.

$$\log(\frac{1+x}{1-x}) = \log(1+x) - \log(1-x)$$

$$= \sum_{k=0}^{n-1}(-1)^k\frac{x^{k+1}}{k+1} + R_n^+(x) + \sum_{k=0}^{n-1}\frac{x^{k+1}}{k+1} - R_n^-(x)$$

$$= 2\sum_{j=0}^{\lfloor n/2\rfloor}\frac{x^{2j+1}}{2j+1} + R_n^+(x) - R_n^-(x).$$

For the Taylor series the eleventh derivative appears in the coefficient of x^{11}. Thus we have (for $j = 5$)

$$\frac{f^{(11)}(0)}{11!}x^{11} = 2\frac{x^{11}}{11},$$

or

$$f^{(11)}(0) = 2\frac{11!}{11} = 2(10!).$$

Problem 8.17

Using Lagrange's form for the remainder we get

$$e^x = \sum_{k=0}^{n} \frac{x^k}{k!} + R_n(x),$$

where for some c between 0 and x

$$R_n(x) = e^c \frac{x^{n+1}}{(n+1)!}.$$

Writing nonpositive values of x as $-z^2$ we get

$$e^{-z^2} = \sum_{k=0}^{n} (-1)^k \frac{z^{2k}}{k!} + R_n(-z^2),$$

where

$$R_n(-z^2) = e^c \frac{(-1)^{n+1} z^{2n+2}}{(n+1)!}, \quad -z^2 \le c \le 0.$$

To evaluate the integral we have

$$\int_0^x e^{-z^2}\, dz = \sum_{k=0}^{n} \int_0^x (-1)^k \frac{z^{2k}}{k!}\, dz + \int_0^x R_n(-z^2)\, dz, \quad 0 \le x \le 1.$$

For $0 \le z \le 1$ we have $c \le 0$ in the remainder formula for $R_n(-z^2)$, so

$$|R_n(-z^2)| \le \frac{1}{(n+1)!}.$$

Thus

$$|\int_0^x R_n(-z^2)\, dz| \le \int_0^x \frac{1}{(n+1)!}\, dz \le \frac{1}{(n+1)!}, \quad 0 \le x \le 1.$$

To make the error smaller than 10^{-3}, pick $(n+1)! > 10^3$. It suffices to pick $n = 6$. Within the desired accuracy, and for $0 \le x \le 1$

$$\int_0^x e^{-z^2}\, dz \simeq \sum_{k=0}^{6} \int_0^x (-1)^k \frac{z^{2k}}{k!}\, dz = \sum_{k=0}^{6} (-1)^k \frac{x^{2k+1}}{(2k+1)(k!)}.$$

Problem 8.19

Taylor's Theorem says that

$$f(x) = \sum_{k=0}^{n} f^{(k)}(x_0) \frac{(x-x_0)^k}{k!} + R_n(x).$$

The sum on the right is the partial sum s_{n+1} for the Taylor series. Thus we have

$$f(x) = s_{n+1}(x) + R_n(x).$$

Suppose the Taylor series converges to $f(x)$. By the definition of convergence of the infinite series this means that

$$f(x) = \lim_{n\to\infty} s_{n+1}(x), \quad \text{or} \quad \lim_{n\to\infty} |f(x) - s_{n+1}(x)| = 0.$$

But this gives

$$\lim_{n\to\infty} |R_n(x)| = \lim_{n\to\infty} |f(x) - s_{n+1}(x)| = 0.$$

Conversely, suppose that $\lim_{n\to\infty} |R_n(x)| = 0$. Again,

$$\lim_{n\to\infty} |R_n(x)| = 0 = \lim_{n\to\infty} |f(x) - s_{n+1}(x)|,$$

and so $f(x)$ is the limit of the sequence of partial sums.

Thus the sequence of partial sums converges to $f(x)$ if and only if the remainders $R_n(x)$ have limit 0.

Problem 8.21

For $x > 0$ the function $f(x)$ has derivatives of all orders, with

$$f'(x) = 2x^{-3}\exp(-1/x^2), \quad f''(x) = -6x^{-4}\exp(-1/x^2) + 4x^{-6}\exp(-1/x^2).$$

By induction $f^{(k)}(x) = p(1/x)\exp(-1/x^2)$, where $p(x)$ is a polynomial. By Problem 8.20, for any positive integer m, a term $Cx^{-m}\exp(-1/x^2)$ satisfies

$$\lim_{x\to 0+} Cx^{-m}\exp(-1/x^2) = \lim_{x\to 0+} \frac{C}{x^m \exp(1/x^2)} = \lim_{t\to\infty} \frac{Ct^m}{\exp(t^2)} = 0.$$

Checking the derivatives at $x = 0$ we find for $h > 0$

$$\lim_{h\to 0+} \frac{f^{(k)}(0+h) - f(0)}{h} = \lim_{h\to 0+} \frac{p(1/h)\exp(-1/h^2)}{h} = 0.$$

The limit computation for $h < 0$ is easier. For every positive integer k we conclude that f has a k-th derivative at $x = 0$ with $f^{(k)}(0) = 0$.

A.9 Chapter 9 Solutions

Problem 9.1

Treating the more general problem of the second part, consider functions $y(x)$ and $y'(x)$, with

$$y(x) = \sum_{k=0}^{\infty} a_k x^k, \quad y'(x) = \sum_{k=1}^{\infty} k a_k x^{k-1} = \sum_{k=0}^{\infty} (k+1) a_{k+1} x^k.$$

Substituting the series displaying x^k into the equation $\frac{dy}{dx} - \alpha y = 0$ gives

$$\sum_{k=0}^{\infty} [(k+1)a_{k+1} - \alpha a_k] x^k$$

$$= (a_1 - \alpha a_0) + (2a_2 - \alpha a_1)x + (3a_3 - \alpha a_2)x^2 + \cdots = 0.$$

Setting the coefficient of each power of x to zero leads to

$$a_1 = \alpha a_0, \quad a_2 = \frac{\alpha a_1}{2} = \alpha^2 \frac{a_0}{2}, \quad a_3 = \frac{\alpha a_2}{3} = \alpha^3 \frac{a_0}{3 \cdot 2}.$$

An induction proof shows that

$$a_k = \alpha^k \frac{a_0}{k!}.$$

The series for $y(x) = a_0 e^{\alpha x}$ converges for all x by the ratio test.

Problem 9.3

(a) If

$$y(x) = \sum_{k=0}^{\infty} a_k x^k, \quad y''(x) = \sum_{k=2}^{\infty} k(k-1) a_k x^{k-2} = \sum_{k=0}^{\infty} (k+2)(k+1) a_{k+2} x^k,$$

then the equation $y'' + y = 0$ becomes

$$\sum_{k=0}^{\infty} [(k+2)(k+1)a_{k+2} + a_k] x^k = 0.$$

Setting the coefficients to zero leads to the recurrence relation

$$a_{k+2} = -\frac{a_k}{(k+2)(k+1)}.$$

This relation uniquely determines the remaining coefficients from a_0, a_1.

If the index k is even, then

$$a_2 = -\frac{a_0}{2 \cdot 1}, \quad a_4 = -\frac{a_2}{4 \cdot 3} = \frac{a_0}{4!}, \quad a_{2k} = (-1)^k \frac{a_0}{(2k)!}.$$

If the index k is odd, then

$$a_3 = -\frac{a_1}{3 \cdot 2}, \quad a_5 = -\frac{a_3}{5 \cdot 4} = \frac{a_1}{5!}, \quad a_{2k+1} = (-1)^k \frac{a_1}{(2k+1)!}.$$

For each k,

$$|a_k| \le \frac{\max(|a_0|, |a_1|)}{k!},$$

so the series converge for all $x \in \mathbb{R}$.

(b) The functions $y_1(x) = \cos(x)$ and $y_2(x) = \sin(x)$ satisfy the differential equation and the initial conditions. By Taylor's Theorem these functions have the indicated convergent power series.

(c) Consider the functions $f(x) = \cos(a + x)$ and $g(x) = \cos(a)\cos(x) - \sin(a)\sin(x)$. Both satisfy $y'' + y = 0$, with $f(0) = \cos(a) = g(0)$ and $f'(0) = -\sin(a) = g'(0)$. By Taylor's Theorem both $f(x)$ and $g(x)$ can be represented as power series convergent for all x. Since the power series solutions satisfying the same initial conditions are unique, $f(x) = g(x)$. The argument is similar for the second identity.

Problem 9.5

By definition $f_n(0) = 0$ for all n. For $x > 0$, $f_n(x) = 0$ for all $x > 2/n$ or $n > 2/x$. Thus $\{f_n(x)\}$ converges pointwise to the function $f(x) = 0$. A computation gives

$$\int_0^1 f(x)\,dx = 0, \quad \int_0^1 f_n(x)\,dx = n.$$

Problem 9.7

If $f : [\alpha, \beta] \to \mathbb{R}$ is continuous, then $f(x)$ is uniformly continuous, so for any $\epsilon > 0$ there is a $\delta > 0$ such that $|t_2 - t_1| < \delta$ implies $|f(t_2) - f(t_1)| < \epsilon$. Choose the partition $\alpha = x_0 < x_1 < x_2 < \cdots < x_N = \beta$ so that $|x_{n+1} - x_n| < \delta$ for $n = 0, \dots, N-1$. If $g_\epsilon(x)$ is linear between $(x_n, f(x_n))$ and $(x_{n+1}, f(x_{n+1}))$, that is

$$g_\epsilon(x) = f(x_n) + \frac{f(x_{n+1}) - f(x_n)}{x_{n+1} - x_n}(x - x_n), \quad x_n \le x \le x_{n+1},$$

then $g_\epsilon(x)$ is continuous and piecewise linear. Finally, since $g_\epsilon(x)$ is linear from x_n to x_{n+1},

$$|f(x) - g_\epsilon(x)| = |[f(x) - f(x_n)] + [f(x_n) - g_\epsilon(x)]|$$

$$\le |f(x) - f(x_n)| + |f(x_n) - g_\epsilon(x)| < \epsilon + |f(x_n) - f(x_{n+1})| < 2\epsilon, \quad x_n \le x \le x_{n+1}.$$

If $\epsilon = 1/k$ for $k = 1, 2, 3, \dots$, then the sequence $\{g_k(x)\}$ converges uniformly to $f(x)$.

Problem 9.9

If a sequence of polynomials $p_n(x)$ converges uniformly to a bounded $f(x)$ on $\mathbb{R}$, then in particular there is some index N such that

$$|f(x) - p_n(x)| \le 1, \quad n \ge N, \quad -\infty < x < \infty.$$

Assume that $|f(x)| \le M$. For $n \ge N$,

$$|p_n(x)| = |[p_n(x) - f(x)] + f(x)| \le |p_n(x) - f(x)| + |f(x)| \le M + 1.$$

Every polynomial $p(x)$ of positive degree is unbounded, so $p_n(x)$ must have degree zero for $n \ge N$. But then $f(x)$ is constant.

Problem 9.11

(a) Suppose $x_1 < x_2$ and $\epsilon > 0$. Since $f_n(x_1)$ converges to $f(x_1)$ and $f_n(x_2)$ converges to $f(x_2)$, there is an N such that

$$|f_n(x_1) - f(x_1)| < \epsilon/2, \quad |f_n(x_2) - f(x_2)| < \epsilon/2,$$

for all $n \ge N$. Thus

$$f(x_2) - f(x_1) = [f_n(x_2) - f_n(x_1)] + [f(x_2) - f_n(x_2)] - [f(x_1) - f - n(x_1)] \ge -\epsilon.$$

Since ϵ is arbitrary, $f(x_2) - f(x_1) \ge 0$.

(b) The function $f(x)$ is uniformly continuous, so for any $\epsilon > 0$ there is a $\delta > 0$ such that $|t_2 - t_1| < \delta$ implies $|f(t_2) - f(t_1)| < \epsilon$. Choose a partition $0 = x_0 < x_1 < x_2 < \cdots < x_M = 1$ so that $|x_{m+1} - x_m| < \delta$ for $m = 0, \dots, M-1$.

Since the functions $f_n(x)$ converge pointwise to $f(x)$, there is an N such that

$$|f(x_m) - f_n(x_m)| < \epsilon, \quad m = 0, \dots, M,$$

for all $n \ge N$. Suppose $x_m \le x \le x_{m+1}$. Since f_n is increasing, if $n \ge N$ then

$$f_n(x) - f_n(x_m) \le f_n(x_{m+1}) - f_n(x_m)$$

$$= [f_n(x_{m+1}) - f(x_{m+1})] + [f(x_{m+1}) - f(x_m)] + [f(x_m) - f_n(x_m)] \le 3\epsilon.$$

Finally,

$$|f(x) - f_n(x)| \le |f(x) - f(x_m)| + |f(x_m) - f_n(x_m)| + |f_n(x_m) - f_n(x)| \le 5\epsilon,$$

so $\{f_n(x)\}$ converges uniformly to $f(x)$.

Problem 9.13

Suppose $f : [a, b] \to \mathbb{R}$ is continuous. The function $t = (x-a)/(b-a)$ maps the interval $a \le x \le b$ one-to-one and onto $0 \le t \le 1$. Solving for x, define a new function $g(t) = f((b-a)t + a) = f(x)$. Since g is the composition of continuous functions, it is continuous. By Theorem 9.4.1, for any $\epsilon > 0$ there is a polynomial $p(t)$ such that

$$|g(t) - p(t)| < \epsilon, \quad 0 \le t \le 1.$$

Replace $p(t)$ with $q(x) = p((x-a)/(b-a))$. The function $q(x)$ is a polynomial in x, which also satisfies

$$|f(x) - q(x)| < \epsilon, \quad a \le x \le b.$$

Problem 9.15

Define a function

$$f(x) = \begin{cases} f_1(x), & x_1 \le x \le x_2, \\ f_1(x_2) + [f_2(x_3) - f_1(x_2)][x - x_2]/[x_3 - x_2], & x_2 \le x \le x_3, \\ f_2(x), & x_3 \le x \le x_4. \end{cases}$$

which joins $(x_2, f_1(x_2))$ to $(x_3, f_2(x_3))$ with a straight line. The function $f(x)$ is continuous on $[x_1, x_4]$, so there is a polynomial $p(x)$ such that

$$|f(x) - p(x)| < \epsilon, \quad x_1 \le x \le x_4.$$

The same idea works to extend this result to any finite collection of pairwise disjoint closed bounded intervals.

Problem 9.17

If

$$\int_0^1 f(x)x^n \, dx = 0, \quad n = 0, 1, 2, \dots,$$

then

$$\int_0^1 f(x)p(x) \, dx = 0$$

for any polynomial $p(x)$. Given $\epsilon > 0$, find a polynomial $p(x)$ such that $|f(x) - p(x)| < \epsilon$ for $0 \le x \le 1$. Then

$$\int_0^1 f(x)p(x) \, dx = 0 = \int_0^1 f(x)[f(x) + p(x) - f(x)] \, dx$$

$$= \int_0^1 f^2(x) \, dx + \int_0^1 f(x)[p(x) - f(x)] \, dx.$$

If $M = \max_{0 \le x \le 1} |f(x)|$, then

$$|\int_0^1 f(x)[p(x) - f(x)] \, dx| \le M\epsilon,$$

so for any $\epsilon > 0$

$$\int_0^1 f^2(x) \, dx \le M\epsilon.$$

That is ,

$$\int_0^1 f^2(x) \, dx = 0,$$

and since $f^2(x)$ is continuous, $f(x) = 0$ for all $x \in [0, 1]$.

Problem 9.19

(a) For the given function $g(t)$ and $n \geq 1$,

$$a_n = \frac{1}{\pi}\int_\alpha^\beta 1 \cdot \cos(nt)\ dt = \frac{\sin(n\beta)}{n} - \frac{\sin(n\alpha)}{n},$$

$$b_n = \frac{1}{\pi}\int_\alpha^\beta 1 \cdot \sin(nt)\ dt = -\frac{\cos(n\beta)}{n} + \frac{\cos(n\alpha)}{n}.$$

Since $|\sin(x)| \leq 1$ and $|\cos(x)| \leq 1$,

$$\lim_{n\to\infty} a_n = 0, \qquad \lim_{n\to\infty} b_n = 0.$$

(b) By Problem 9.6 any continuous function $f : [-\pi, \pi] \to \mathbb{R}$ may be uniformly approximated by a piecewise constant function. That is, for any $\epsilon > 0$ there is a partition $-\pi = x_0 < x_1 < \cdots < x_N = \pi$ and a function $g(x)$ which is constant on each subinterval $[x_n, x_{n+1}]$ such that

$$|f(x) - g(x)| < \epsilon, \quad -\pi \leq x \leq \pi.$$

For $n \geq 1$,

$$a_n = \frac{1}{\pi}\int_{-\pi}^{\pi} f(t)\cos(nt)\ dt = \frac{1}{\pi}\int_{-\pi}^{\pi} g(t)\cos(nt)\ dt + \frac{1}{\pi}\int_{-\pi}^{\pi} [f(t) - g(t)]\cos(nt)\ dt.$$

First, observe that

$$|\frac{1}{\pi}\int_{-\pi}^{\pi} [f(t) - g(t)]\cos(nt)\ dt| \leq |\frac{1}{\pi}\int_{-\pi}^{\pi} |f(t) - g(t)||\cos(nt)|\ dt \leq 2\epsilon.$$

For the other integral,

$$\frac{1}{\pi}\int_{-\pi}^{\pi} g(t)\cos(nt)\ dt = \sum_{n=0}^{N_1} \frac{g(x_n)}{\pi}\int_{x_n}^{x_{n+1}} 1 \cdot \cos(nt)\ dt,$$

and each summand has limit zero as $n \to \infty$ by part (a). The computations for b_n are similar.

Problem 9.21

(a) Let $g(t) = f(t)\cos(nt)$. Since $\cos(nt)$ is even while $f(t)$ is odd, $g(-t) = -g(t)$. For any odd function $g(t)$, the substitution $u = -t$ gives

$$\int_{-\pi}^{\pi} g(t)\ dt = \int_{-\pi}^{0} g(t)\ dt + \int_0^{\pi} g(t)\ dt = -\int_0^{\pi} g(t)\ dt + \int_0^{\pi} g(t)\ dt = 0.$$

(b) If $f(t)$ is even then $g(t) = f(t)\sin(nt)$ is odd.

Bibliography

[1] D. Bressoud, *A Radical Approach to Real Analysis.* The Mathematical Association of America, 1994

[2] L. Euler, *Introduction to Analysis of the Infinite.* Springer-Verlag, New York, 1988

[3] L. Euler, *Foundations of Differential Calculus.* Springer-Verlag, New York, 2000

[4] G. Hardy and E. Wright, *An Introduction to the Theory of Numbers.* Oxford University Press, 1984

[5] M. Kline, *Mathematical Thought from Ancient to Modern Times.* Oxford University Press, New York, 1972

[6] J. Marsden and M. Hoffman, *Elementary Classical Analysis.* W.H. Freeman, 1993

[7] W. Rudin, *Principles of Mathematical Analysis.* McGraw-Hill, New York, 1964

[8] P. Schaefer, Sum-Preserving Rearrangements of Infinite Series, *The American Mathematical Monthly,* 88 no. 1, 1981, pp. 33-40.

[9] D. Smith, *A Source Book in Mathematics,* Dover, New York, 1959

Index